Studies in
PLANT PHYSIOLOGY SERIES No. 2

PHOTOSYNTHESIS:

Physiological, Biochemical and Molecular Aspects

Studies in Plant Physiology Series No. 2

PHOTOSYNTHESIS:

Physiological, Biochemical and Molecular Aspects

S. S. Purohit
R. Ranjan

2002

Agrobios (India)

Dr. S. S. Purohit
Laboratory of Plant Physiology
Ex-Head, Department of Botany
Dungar College
Bikaner (India)

Dr. R. Ranjan
Department of Botany
T. P. Verma College
Narkatia Gunj 845 455 (Bihar)

Agrobios (India)

ISBN : [illegible]5
Rs. 695[illegible]00

9 788177 540475
Rs. 2500.00 / US$ .00
Published by:
AGROBIOS (INDIA)

Published by: Dr. Updesh Purohit for Agrobios (India), Jodhpur
Lasertypeset at: Yashee Computers, Jodhpur
Cover Design by: Reena
Printed at: Anmol Prints, Jodhpur

Preface

This book provides a simplied description of the photosynthesis at the molecular, biochemical, organelle, cell and organ levels of organization in plants, which contribute to the complex processes. It considers how photosynthesis and biochemistry determines the physiological characteristics of plants and production of dry matter. The text links the fundamentals of light capture by pigment molecules to the generation of high-energy organic molecules and their consumption in carbon dioxide reduction. The mechanisms are related to the control of energy and material fluxes.

The text is intended for post-graduate study and for non-specialists in other disciplines who wish to understand photosynthetic mechanism and control. References are mainly from the secondary or review literature, from which detailed arguments and the primary literature may be obtained. Several recent textbooks have considered more specialized aspects of photosynthesis and there are many excellent reviews; this book is intended to point the reads to them. We hope that this text will be enable readers to appreciate the hierarchy of organization that enables plants to produce dry matter within the confines of the environment. With current exciting possibilities for genetically engineering plants for specific purposes to give greater dry matter production or improved growth in particular environments for example, it is essential that the role of molecular events in the functioning of the whole system in relation to environment shall be well understood.

In production of the manuscript Dr. Updesh Purohit gave help with infinite patience over a long period. The unstinting support of Dr. Updesh Purohit is beyond thanks.

Authors

Contents

CHAPTER 1

The Growth of our Knowledge of Photosynthesis

It was only through the exploitation of light, for the development of matter, that life on earth was able to gather momentum and completely change the face of the planet. The early Greeks believed that plants obtained their organic material directly from the soil. The *soil eater* concept was perhaps further fostered by the observation that soil additives (fertilizers) increased crop production. The notion remained firmly centered until the 17th century when the simple but the elegant experiments performed by Jean Baptiste van Helmont revealed that plants were not soil eaters. The key knowledge Helmont lacked was, the atmosphere contains gases and that one of these is of prime importance in photosynthesis. A more complete understanding of the process of photosynthesis gradually evolved and showed that in sunlight, plants were absorbing carbon dioxide, utilizing the carbon (C) for the nourishment, and expelling oxygen (O_2). Theodore de Saussure (1767-1845) further confirmed these notions and was able to show that the bulk of the dry weight of a plant comes from the carbon taken as CO_2. He further recognized that water plays an important and stable role in the process.

Photosynthesis is the word used to describe a process fundamental to plant nutrition. This word came into being relatively recently. Absent in the 1880 edition of Pfeffer's great *Plant physiology,* it appeared in the 1897 edition as *photosynthetische assimilation*. It was translated by Ewart as *photosynthetic assimilation* (Pfeffer, 1900-1906). Between these two editions, Barnes in America (1893) had proposed the word *Photosyntax* and had rejected *photosynthesis* on etymological grounds; one of his colleagues, Macdougal, preferred photosynthesis and he used this form in his textbook on practical plant physiology. The true origin of the word does not seem to be stated in any of the numerous textbooks that we have. Judging from the literature there seems to have been a polemic about who had actually invented it.

This is essentially the reverse of the process of respiration. The green plants are called autotrophic or self-feeding because they require to take in no organic food for the increase of their substance. The process of photosynthesis enables plants to convert solar energy into chemical energy. The chemical energy is stored in the form of organic compounds (food), which generally constitute more that 90 per cent of dry weight of plants. This production of organic compounds provides the basis for all the varied food chains on the earth. Thus the daily calories intake by human beings is based on transformed solar energy and is acquired by eating either plant food or by eating animal food.

The plants cover on our planet and the algae in the oceans, every year produce ten times as much stored chemical energy as is consumed by mankind. The total energy of the presently known oil reserves represents only 100 years of photosynthesis. Coal, another of our geological energy sources, is also entirely of plant origin. During evolution, photosynthesis has changed from a non-oxygen evolving process, probably similar to that now used by a small number of bacterial strains, to the oxygen evolving process used by cyanobacteria, algae and higher plants. This change has gradually made oxygen a part of our atmosphere and has given organisms the possibility to efficiently combust organic compounds by respiration rather than fermentation. A complete renewal of the oxygen content of our atmosphere corresponds to 2000 years of photosynthetic oxygen production.

The radiation energy of visible light is used in the photosynthesis of green plants to synthesize organic compounds from the simple inorganic molecules- carbon dioxide and water. Oxygen is photochemically cleaved from water. The radiation energy is stored in carbohydrate in the form of Gibbs Free Energy (GD). The endergonic reaction can be summarized as follows:

$$6CO_2 + 12\ H_2O \xrightarrow{hv} C_6H_{12}O_6 + 6\ H_2O + 6O_2$$

$$GD = 2868\ KJ/mol\ C_6H_{12}O_6$$

Chemically this process is the reduction of carbon dioxide since the inverse reaction is the oxidation of carbohydrate. CO_2 is reduced by the reductant in the form of NADH and for thermodynamic reasons ATP is also required.

$$CO_2 + 2NADPH + 3ATP^{4-} + H_2O \longrightarrow [CH_2O] + 2\ NADP^+ + 3ADP + 3HPO_4^{2-} + H^+$$

For the purpose of this text and simplification, we have used NADP and NADPH instead of $NADP^+$ and $NADPH^+$ H^+. In green plants and algae, NADPH and ATP for carbohydrate synthesis are produced exclusively in specific organelles, the chloroplasts. Each chloroplast is enclosed by double membrane which separates its interior from the cellular cytoplasm. The inner phase is called *stroma* and contains all necessary enzymes for CO_2 fixation and carbohydrate synthesis. Within the stroma, membranous structure *thylakoids* are present in stacks (grana) where NADPH and ATP are produced. The production is light-dependent.

EARLY GREEK CONCEPTS OF EARTH (SOIL), WATER AND PLANT RELATIONS

In prehistoric times, in places where agriculture was developed, it would be clear that water and soil were both essential requirements for the living plants. It would have also become obvious that plants could be burnt, for, when the use of fire had been established, wood was used as fuel.

In Greek era people were knowing about the study of plant nutrition. The essay in this is inevitably distorted because people were so conditioned with scientific jargon and later conceptions. Greeks could grasp a relation between the *plant* and *water*. Nowadays, we could feel that there was some dim connection between a prehistoric conception of fire and perhaps what we now call energy. But energy in the precise scientific sense is a very recent conception; Partington gives 1850 as the date, virtually the last item to be placed correctly in the nineteenth-century formulation.

At any time when people simply think about things, what is variously held to be right and true shows wide discrepancies. So it was that quite independent and mutually conflicting views about the nature of things have come down to us from the Greeks. These influences, sometimes quite amazingly, are still to be recognized in current thought. In science, it is assumed that the *thought is tempered by results of experiments* and by *the experimental methods*. It would be fine to have a thought that would explain everything. Thales of Miletus (c. 640-546 B.C.) gave as the thought *all things are water* (*πάvτα vδωρ εστί*).

Water = Plant

The above thought took over the ultimate reality.

A different type of theory, which was associated with Democritus, c.360 B.C., was that everything was composed of characteristic atoms. These atoms would confer the characteristic to each plant. Here the atoms composing an olive tree would generally be supposed to be different from those composing the grape vine, that is Atoms (a) = Plant (a), and Atoms (b) = Plant (b).

According to Greek conception of plants took their nourishment from the earth. Each plant took the atoms appropriate to its kind. Thus the plants, unlike animals, produced no excreta. Plants, by contrast, the intelligent agent would have to be in the earth or soil. Thus a plant would have seemed to resemble a fetus drawing its nourishment from the Mother-Earth, where the appropriate atoms were elaborated in the soil. F.O. Bower (1835-1948) in his book *Botany of the Living Plant* described a growing plant as showing continuous embryology.

Concept of Aristotle

Aristotle gave the theory of the four elements. The four elements were defined by Aristotle in much the same way, as originally were the chemical elements of the present day. The differences in what we now understand as a mixture from that of a single compound was appreciated. Mercury metal was called silver-water; the different kinds of matter were described in terms of intermediate or combined qualities. It is in this question of matter and quality of attributes where the difficulties for scientific progress came in. Nowadays we should tend to think that the four elements would symbolize a manifestation of energy as fire, and the three states of matter, solid, liquid and gas. In evaporation something changed into air and, in liquification of a solid, part of the earth had to be changed into water.

Quite apart from this theoretical aspect very much practical experience on both agriculture and in chemical matters existed in early times. This can be gathered from the writings of Pliny the elder especially as regards soils and fertilizers. A famous experiment of a candle burning in a closed space over water was carried out by Philon of Byzantium (third century B.C.). After the candle became extinguished a contraction was observed. But experiments and material practice were usually subsequent to theory and often diverged from it.

THE BEGINNINGS OF MODERN EXPERIMENT

van Helmont studied relation between plant and nutrition. He not only invented the word *gas*, c. 1630, but also contributed much to the beginnings of chemistry, as distinct from alchemy, and again, indeed, in view of his conception of a *ferment*, to the beginnings of biochemistry. van Helmont decided to test the theory:

Water = Plant

His experiment completely satisfied the theory when judged by the knowledge at the time. The permanent value of the experiment lies in the fact that it showed how little of the actual matter composing the plant can be derived from the soil.

I took an Earthen Vessel, in which I put 200 pounds of Earth that had been dried in a Furnace, which I moistened with Rain-water, and I implanted therein the Trunk or Stem of a Willow Tree, Weighing five pounds; and at length, five years being finished, the Tree sprung from thence, did weigh 169 pounds, and about three ounces: But I moistened the Earthen Vessel with Rain-water, or distilled water (always when there was need) and it was large and implanted into the Earth, and least the dust that flew about should be co-mingled with the Earth, I covered the lip or mouth of the Vessel, with an Iron-Plate covered with Tin, and easily passable with many holes. I computed not the weight of the leaves that fell off in the four Autumnes. At length, I again dried the Earth of the Vessel, and there were found the same 200 pounds, wanting about two ounces. Therefore 164 pounds of Wood, Barks, and Roots, arose out of water only. (The quotation is taken from Partington, 1957. – A short History of Chemistry, pp 51-52).

Now the contemporary interpretation of this experiment, it could really involve the conflicting notions of conservation of the soil (earth) element and the transformation of the water element. Both the conservation and transmutation ideas had come from the Greek philosophy. Then again there was the idea of each plant requiring taking its specific kind of atoms from the soil. The different kinds of plants, especially those used in agriculture and in medicine, have been recognized, depicted and described from very early times. In fact, the purely descriptive side of botany has dominated in the study of plants on more than one occasion in the history of this science.

Deem Mariotte (c. 1620-84) also supported a Greek view of the earth nutrition of plants. He argued that - *I can take a definite amount of soil say 7-8 lb. In this soil I can grow a plant, which will weight say ¼ lb.-but it matters little out of 3,000-4,000 different kinds of plant which one I choose to grow. Thus if each plant has to take atoms or elements for its characteristic kind from the soil (and a small amount of rain water with its salts) there certainly could not be enough kinds of atoms to produce the material of 3,000 different plants.*

The Discovery of Photosynthesis

The experiments and studies of Robert Boyle (1627-91) and the foundation of the Royal Society in England (1645-62) are remarkable in this field. Eugene Rabinowitch in volume I of his classical book on photosynthesis (1945-58) has given a splendid historical account of the discovery of photosynthesis starting from the work of Stephen Hales (1677-1761) up to the time of the famous plant physiologist Julius von Sachs (1832-97). Rabinowitch gave the credit to Stephen Hales who showed that air must be an important constituent of all organic matter. He quoted from *Vegetable staticks* (1727) where Hales wrote *Plants very probably draw through their leaves some part of their nourishment from the air, may not light also by freely entering surfaces of leaves and flowers contribute much to ennobling the principles of vegetables.*

The importance of light in for plant growth and ripening fruit has been indicated by the scientist of this age. The necessity for light must have been obvious in very early days but perhaps only apparent to people working in horticultural practice.

The experimental work of van Helmont can not be ignored in the light of modern chemistry and biochemistry. He was able to distinguish the properties of different kinds of gas and different methods of producing them. He burnt charcoal and called the product *gas carbonum,* which also he called *gas silvestre* in circumstances where it comes from fermenting wine. This was the *fixed air* later described by Joseph Black (1728-99), which he obtained from the action of acid on limestone. The word air continued to be used to designate the different gases for a considerable time after van Helmont.

It was never possible to contain a gas in an open vessel and in some experiments a closed vessel would burst. The development of the study of the nutrition of plants, and indeed the development of chemistry itself largely depended on the experimental procedures, which involved gases. The postulate of conservation of material substance seems to have been the basis of both the atomic and transmutational theories in antiquity. This perhaps accounted for the tendency to regard all the four elements as substances. It could long have been observed that flame was burning smoke; this was stated by Aristotle and by van Helmont.

The word *caloric* dating from 1987 in France and the German 'warmstoff' may still be found in (not so old) dictionaries. The postulation of a material flame principle led to the 'phlogiston' theory, which is associated with J.J. Becher, 1635-1682 and with G.E. Stahl, 1660-1734. On this theory, when the candle, being very rich in phlogiston, was burnt in a closed space the air became 'phlogistigated' and then it was incapable of supporting combustion. This strange scientific jargon actually persisted among a few scientists until the beginning of the nineteenth century.

The Real Beginning

Joseph Priestley (1733-1804) carried out skilful experiments with gases-air. Priestley found that the 'phlogistigated' air, which no longer could support combustion and the respiration of a mouse, could be restored by a green plant. The green plant could 'dephlogistigate' the 'phlogistigated air'. The gas which Priestley called 'dephlogistigated air' we should now, following Lavoisier, call oxygen. In this discovery Priestley first showed the complementary relationship existing between vegetation and animals. Priestley was especially amazed by the vigorous combustion of a candle in his dephlogistigated air when he prepared it from oxide of mercury. He realized that the common air was less pure. Partington (1957) quotes him: *the air which Nature has provided for us is as good as we deserve.* Later, Priestley showed, by using his 'nitrous air' (nitric oxide), how to determine the amount of pure dephlogistigated air in the atmosphere. He found one-fifth of a volume present; this was an accurate result.

The discovery by Priestley seemed to have initiated at least three independent workers in the subject. The first was Jan Ingenhousz (1730-99) who while acknowledging Priestley continued to follow up his work with success. Rabinowitch considerd that Ingenhousz established that light was necessary and that he had shown that *something phlogistigated* was produced in the plant. Also it appeared that Jean Senebier (1742-1809) had discovered the requirement for fixed air, which originally had been denied by Ingenhousz. Later Nicolas Theodore de Saussure (1767-1845) obtained the first evidence that water was involved. We can be amazed at the progress made during this period of phlogiston and with the lack of a precise scientific terminology. In his *History of Botany* (1890), Julius von Sachs (1832-97) gives the credit for the discovery of photosynthesis to Ingenhousz. This is no doubt connected with the fact that in 1796 a book entitled *Food of plants and renovation of the soil* was published where Ingenhousz had adopted the new chemical theory of Lavoisier. Dephlogistigated air was called oxygen and fixed air was called carbonic acid.

The quantitative experiments of Antoine Laurent Lavoisier (1743-94) showed that the phlogiston theory was untenable. When a substance was burnt in air it combined with the oxygen; so that a metal, when heated in air, formed a calx with increase in weight while the volume of air showed a contraction.

A medical man Julius Robert Mayer (1814-78) defined process of photosynthesis. He introduced the conception of conservation of energy (as it would apply to the nutrition of the plant).

The early experiments with plant nutrition were carried out at a time when the theory of nature of chemical changes was in its infancy. By the time J. von Sachs was writing his *Plant physiology.* He wrote that many of the processes involving chemical and physical changes could be described by precise scientific terms. He found that the first carbohydrate to be visibly formed in light was in many cases starch. Thus, from a general physiological viewpoint, the relation of photosynthesis to plant nutrition was accepted and it formed the basis for a variety of experimental works, with plants. However there seemed to be little progress as regards the actual biochemical mechanism of photosynthesis until the beginning of the twentieth century.

Photosynthetic Pigments

Pelletier and Caventou (1818) named green leaf pigment as *chlorophyll.* Tswett, at the beginning of this century, discovered how to separate the pigments by chromatography. He used the term *chlorophyllin* for the two green pigments in the leaves which originally had been shown to be present by Sorby in 1873. However, Willstatter and Stoll in their work on the leaf pigments and on photosynthesis published in 1913 and 1918 used the word chlorophyll for the unmodified leaf pigments. They used chlorophyllin to describe modified chlorophyll; this may have been due to the fact that in the calcium carbonate used for chromatography the green pigments were apt to suffer a chemical alteration. It had been widely held that chlorophyll was the agent in the plant wholly responsible for photosynthesis. Fruitless experiments were directed towards proving this while chlorophyll continued to be a main focus of attention. However it happened that in more recent times between 1930 and 1940 the focus of attention came back again to water.

C.B. van Niel (1949) in the 1930s had carried out the most searching studies with photosynthetic bacteria. These do not produce oxygen, which distinguishes them from the green plants. The bacteria were found to cause oxidation of a substance in their environment, in the absence of atmospheric oxygen. van Niel could express symbolically all the types of photosynthesis in one way. The result of the light absorbed by the pigments in the living cells was to resolve water, represented by HOH, into the parts H and OH. This could give rise to a reduction and a simultaneous oxidation. As a result, the active processes involving the reduction of carbon dioxide and oxygen production could be carried out by means of ferments (enzymes) in the dark. This theory after about 2,400 years (since Thales of Miletus) seemed to be a single thought which has motivated experiments extending even to the present time, leading towards a more complete understanding of the mechanism of photosynthesis.

Ratio of CO_2 Consumed to O_2 evolved

French plant physiologist Boussingault (1864) determined the ratio of CO_2 consumed and O_2 evolved during photosynthesis. He found that the photosynthetic ratio- the volume of O_2 evolved to the volume of CO_2 used up-is almost unity. Sachs (1864), the German botanist, (who also discovered plant respiration) demonstrated the formation of *starch* grains during photosynthesis. Sachs kept some green leaves in the dark for some hours to deplete them of their starch content. He then exposed one half of a starch-depleted leaf to light and left the other half in the dark. After some time the whole leaf was exposed to iodine vapour. The illuminated portion of the leaf turned dark violet due to the formation of starch-iodine complex; the other half did not show any colour change.

German botanist Engelmann (1880) studied a direct connection between oxygen evolution and chloroplasts of green leaves. He also demonstrated the relation between the action spectrum of photosynthesis and the absorption spectrum of chlorophyll. He placed a filament of the green alga *Spirogyra,* with its spirally arranged chloroplasts, on a microscope slide together with a suspension of oxygen-requiring, motile bacteria. The slide was kept in a closed chamber in the absence of air and illuminated. Motile bacteria would move towards regions of greater O_2 concentration. After a period of illumination the slide was examined under a microscope and the bacterial population counted. Engelmann found that the bacteria were concentrated around the green bands of the algal filament.

Engelmann illuminated the alga with a spectrum of light by interposing a prism between the light source and the microscope stage. The largest number of bacteria surrounded those parts of the algal filament that were in the blue and red regions of the spectrum. The chlorophylls present in the alga absorbed blue and red light; since light has to be absorbed to bring about photosynthesis. He concluded that chlorophylls are the active photoreceptive *pigments* for photosynthesis. The state of knowledge of photosynthesis at the beginning of this century could be represented by the equation:

$$(CO_2)_n + H_2O + \text{light green/plant } (O_2)_n \longrightarrow \text{starch + chemical energy}$$

It was not clear whether the source of O_2 was water or CO_2.

Concept of Limiting Factors

Various external environmental factors (quality and quantity of light incident on the leaves or algae, the ambient temperature, and the concentration of CO_2 and O_2 in the surrounding atmosphere and, more generally, the availability of water and nutrients) limit the rate of photosynthesis. Early attempts were made to study the effects of light intensity, temperature, CO_2 concentration, etc. on the overall yields of photosynthesis. The studies were made on unicellular green algae, *Chlorella* and *Scenedesmus,* and the unicellular flagellate *Euglena.* Unicellular algae were found more suitable for quantitative research. They can be grown in all laboratories under fairly standard conditions throughout the year and can suspend uniformly in aqueous buffer solutions. The aliquots of the suspension can be transferred with a pipette as though they were true solutions. Studies related to external factors photosynthetic rate on single cell alga and forest cover have also been conducted by various scientists.

Concept of Light and Dark reactions (Flashing light experiments)

Blackman (1905) interpreted the shape of the light-saturation curves by suggesting that photosynthesis is a two-step mechanism involving a photochemical or *light* reaction and a non-photochemical or *dark* reaction. The dark reaction (enzymatic), is slower than the light reaction and hence at high light intensities the rate of photosynthesis is entirely dependent upon the rate of the dark reaction. The light reaction has a low or zero temperature coefficients, while the dark reaction has a high temperature coefficient, characteristic of enzymatic reactions. Dark reaction could proceed both in light and in dark, with multiple steps.

These two reactions can be separated by using flash illuminations lasting fractions of a second. Light flashes (lasting less than a millisecond - 10^{-3}s) are produced mechanically by placing a slit in a rotating disc in the path of a steady light beam. It may also be produced electrically by loading up a condenser and discharging it through a vacuum tube. Ruby lasers emitting red light at 694 nm are also used as a radiation source.

Emerson and Arnold (1932), illuminated suspensions of *Chlorella* cells with condenser flashes lasting about 10^{-5} s and measured the rate of oxygen evolution in relation to the energy of the flashes, the duration of the dark intervals between the flashes, and the temperature of the cell suspension. Flash saturation occurred in normal cells when one molecule of O_2 evolved from 2500 chlorophyll molecules. They further explained that the maximum yield of photosynthesis is not determined by the number of chlorophyll molecules capturing the light but by the number of enzyme molecules, which carry out the dark reaction. Emerson and Arnold also studied that for dark time intervals (between successive flashes) greater than 0.06 s, the yield of O_2 per flash was independent of the dark time interval. The yield per light flash increased with dark time intervals from 0 to 0.06 s. They concluded that the dark reaction determining the saturation rate of photosynthesis takes about 0.06 s for completion. They calculated the average dark reaction time (about 0.02 s at 25°C).

Until 1930, it was believed that the primary reaction in photosynthesis was splitting of CO_2 by light to carbon and O_2. The carbon so produced is subsequently reduced to carbohydrates by water after passing through various series of reactions. In 1930, after discoveries of two important aspects of photosynthesis, changed above viewpoint.

Firstly, various bacterial cells can assimilate CO_2 and produce carbohydrates without the use of light energy. Later, van Niel (Dutch microbiologist), while making comparative studies of plant and bacterial photosynthesis, showed that some bacteria could assimilate CO_2 in light without evolving O_2. Such bacteria cannot grow photosynthetically unless they were supplied with a suitable hydrogen donor substrate. van Niel represented photosynthesis as the following general reaction:

$$\underset{\text{(oxidizable substrate)}}{CO_2 + 2H_2A} \xrightarrow{\textit{Light and Chlorophyll}} (CH_2O) + H_2O + 2A$$

van Niel suggested that during photosynthesis of green plants and algae H_2A is H_2O and 2A is O_2. During photosynthesis, primary photochemical process is the splitting of water and yield an oxidant (OH) and a reductant (H). The primary reductant (H) could then bring about the reduction of CO_2 to cell materials, and the primary oxidant (OH) could be eliminated through a reaction to liberate O_2 and reform H_2O. The overall equation of photosynthesis for green plants, after van Niel, is:

$$CO_2 + 4H_2O \xrightarrow{\text{Light and Chlorophyll}} (CH_2O) + 3H_2O + O_2$$

The above reaction is the sum of three individual steps:

$$4H_2O \xrightarrow{\text{Light and Chlorophyll}} 4(OH) + 4H$$

$$4H + CO_2 \rightarrow (CH_2O) + H_2O$$

$$4(OH) \rightarrow 2H_2O + O_2$$

The above reaction sequences indicate that the oxygen is evolved from water and not from CO_2.

Robort Hill's Contribution – Hill Reaction

Robert Hill (1937) made another important observations. He separated the photosynthesizing particles (chloroplasts) of green leaves from the respiratory particles by differential centrifugation of a homogenate of leaf tissues. The separated chloroplasts did not evolve O_2 after illumination. This may be due to possible damage of the chloroplasts during isolation. However, after adding suitable electron acceptors (oxidants) like potassium ferrioxalate or potassium ferricyanide in illuminated suspension, he observed O_2 evolution. He observed that one molecule of O_2 is evolved for every four equivalents of oxidant reduced photochemically. Quinines and dyes were found to be reduced by illuminated chloroplasts. This phenomenon, now known as the **Hill reaction**. *It is a light-driven transfer of electrons from water to non-physiological oxidants* (**Hill reagents**) against the chemical potential gradient. Hill reaction demonstrated that that photochemical O_2 evolution could be separated from CO_2 reduction in photosynthesis.

Ruben and Kamen (1941) demonstrated that the decomposition of water, and O_2 liberation during photosynthesis. They exposed photosynthesizing cells to water enriched in the recently available oxygen isotope of mass 18 (^{18}O). The isotopic composition of the O_2 evolved was the same as that of the water and not that of CO_2

used. They also discovered the radioactive isotope ^{14}C, which was successfully used by later workers to study photosynthetic CO_2 assimilation.

The chloroplasts (O_2-evolving and natural electron acceptor) prepared by Hill unable to reduce CO_2 in presence of light. Therefore, it was then thought that CO_2 assimilation is not possible in a cell-free system. This was the view of physiologists until the early 1950s.

In Arnon's laboratory, a series of (43 consecutive!) experiments were conducted to disprove this hypothesis. In 1951, three different laboratories, including Arnon's, reported the photosynthetic reduction of NADP to NADPH by isolated chloroplasts with simultaneous evolution of O_2. Then Arnon, Allen and Whatley (1954) demonstrated photophosphorylation (i.e. light-induced synthesis of ATP from ADP and Pi), and also the simultaneous assimilation of $^{14}CO_2$ and evolution of O_2 by isolated spinach chloroplasts; cell-free photosynthesis was thus accomplished!

Frenkel (1954), observed the formation of ATP from ADP and Pi on illumination of membrane preparations from photosynthetic bacteria.

Arnon and his associates (including Trebst, Losada and Tsujimoto) separated the chloroplasts into granal and stromal fractions and able to distinguish the *light* and *dark* reactions of chloroplasts. They demonstrated ATP and NADPH formation in the grana in light, and the dark enzymatic CO_2 reduction by enzymes of the stroma. They further discovered cyclic photophosphorylation which produced only ATP, and suggested its role for providing additional ATP for CO_2 fixation in chloroplasts.

Simultaneously, Bassham, Benson and Calvin (California, USA) were studying the incorporation of $^{14}CO_2$ into photosynthesizing algae. They demonstrated that the reduction of CO_2 to sugars proceeded by dark enzymic reactions and also those two molecules of NADPH and three molecules of ATP were required for the reduction of every molecule of CO_2. For elucidating the path of photosynthetic carbon reduction. For this contribution, Calvin was awarded the Nobel Prize in 1961.

Emerson and co-workers measured quantum yields of photosynthesis in algae. They observed that the average quantum yield obtained by using two superimposed light beams of different wavelengths was higher than the average quantum yield obtained by using the two beams separately. This was explained as *enhancement of quantum yield.* Emerson and Rabinowitch (1960) suggested the existence of *two light reactions* in photosynthesis. Hill and Bendall (1960) put forward the Z *scheme* of photosynthesis showing the operation of two photosystems in series in photosynthetic electron transport and phosphorylation. Mitchell (1961), suggested *chemiosmotic theory* to explain ATP formation in biological membranes., Kortschak and co-workers (1965) studied the incorporation of $^{14}CO_2$ into sugar cane leaves. They reported that in sugar cane the first stable compounds formed in photosynthesis are *malic* and *aspartic* acids. These acids are subsequently converted to sucrose via 3-phosphoglyceric acid and hexose phosphates. These findings were extended and confirmed by Hatch and Slack (1966). They observed that after exposure of sugar cane leaf segments to $^{14}CO_2$ for 1 s, more than 93% of the fixed radioactivity was located in the C4 dicarboxylic acids malic, aspartic and oxaloacetic. Thus the C_4 pathway of CO_2 fixation was discovered, which is also referred to as the **Kortschak, Hatch-Slack pathway**.

In recent years, Deisenhofer, Hüber and Michel made most notable achievement in photosynthesis, is the crystallization of the reaction center from a purple photosynthetic bacterium and the determination of its structure by X-ray crystallographic analysis. For this contribution they were awarded the Nobel Prize in 1988.

Crystallization of the light harvesting complex and photosynthetic reaction centers from selected plants and cyanobacteria have been achieved. High resolution X-ray analysis of these crystals in the *dark* and *light* phases would help to unravel the structure-function relationships of the reaction center components and to understand the molecular processes associated with O_2 evolution.

SUMMARIZED HISTORICAL DEVELOPMENTS AND CONCEPTS OF PHOTOSYNTHESIS RESEARCHES

Light energy and Chloroplast

- Light has characteristics of both of a particle and a wave.
- When molecules absorb or emit light, they change their electronic state.
- The quantum yield gives information about the fate of the excited state.
- Photosynthetic pigments absorb the light that powers photosynthesis.
- The chemical reaction of photosynthesis is driven by light.
- Photosynthesis is light-driven redox process.
- Oxygen-evolving organisms have two photosystems that operate in series.
- The chloroplast is the site of Photosynthesis
- Thylakoids contain integral membrane proteins.
- Photosystems I and II are spatically separated in thylakoid membrane.
- The structures of two bacterial reaction centers have been determined at very high resolution.
- The antenna funnels energy to the reaction center.
- Many antenna complexes have a common structural motif.
- Four thylakoid protein complexes carry out electron and proton transport.

- Energy is captured when an excited chlorophyll reduces an electron acceptor molecules.
- The reaction center chlorophylls of the two photosystems absorb at different wavelengths.
- The photosystem II reaction center is a multisubunit pigment-protein complex.
- Water is oxidized by Photosystem II.
- Pheophytin and two quinines are early electron acceptors of Photosystem II.
- Electron flow through the cytochrome b_6f complex transports protons to the thylakoid lumen.
- Plastoquinone and plastocyanin are putative electron carriers between Photosystems II and I.
- The photosystem I reaction center reduces $NADP^+$.
- A chemiosmotic mechanism converts the energy stored and electric potentials to ATP.

Regulation and Repair of the Photosynthetic Apparatus

- Carotenoids serve as both accessory pigments and photoprotective agents.
- Thylakoid stacking permits energy partitioning between the photosystems.
- Some xanthophylls also participate in energy dissipation.
- The photosystem II reaction center is easily damaged.
- Photosystem I is protected from active oxygen species.

Genetics, Assembly and Evolution of Photosynthetic Systems

- Chloroplast genes exhibit non-Mendelian pattern of inheritance.
- Many chloroplast proteins are imported in cytoplasm.
- The biostnthesis and breakdown of chlorophyll are complex pathways.
- Complex photosynthetic organisms have evolved from simpler forms.

Calvin Cycle

- The Calvin cycle induces carboxylation, reduction and regeneration.
- The carboxylation of Ribulose bisphosphate is catalyzed by the enxyme Ribisco.
- Triose phosphate are formed in the reduction step of the Calvin cycle.
- Activity of the Calvin cycle requires the regeneration of Ribulose-1,5-bisphosphate.

- The Calvin cycle was elucidated by the use of radioactive isotopes.
- The Calvin cycle regenerates its own biochemical components.
- The Calvin cycle is regulated by several mechanisms.
- Light-dependent enzyme activation regulates the Calvin cycle.
- Light-dependent ion movements regulate Calvin cycle enzymes.
- Light-dependent membrane transport regulates the Calvin cycle.

The Photorespiratory Carbon Oxidation Cycle

- Photosynthetic CO_2 fixation and photorespiratory oxygenation are competing reactions.
- Competition between carboxylation and oxygenation decreases the efficiency if photosynthesis.
- Carboxylation and oxygenation are closely interlocked in the intact leaf.
- The biological function of photorespiration is unknown.

CO_2 Concentrating Mechanisms I: Algal and Cyanobacterial Pump

- Many plants either do not photorespire at all.

CO_2 Concentrating Mechanism II: The C_4 Carbon Cycle

- Malate and aspartate are carboxylation products of the C_4 cycle.
- The C_4 cycle concentrates CO_2 in buldle sheeth cells.
- The concentration of CO_2 in bundle sheet cells has energy cost.
- Light regulates the activity of the key C_4 enzymes.
- In Hot, dry climates, the C_4 cycle reduces photorespiration photorespiration.
- The stomata of CAM plants open at night and close during the day.
- CAM is regulated by phosphorylation of PEP carboxylase.

Synthesis of Starch and Sucrose

- Starch is synthesized in the chloroplast, sucrose in the cytosol.
- The synthesis of sucrose and starch are competing reactions.

CHAPTER 2

Visible Radiations and Photosynthesis

BASIC CONCEPTS

Visible Radiations Have Characteristics of Both a Particle and a Wave

The radiation is the propagation of electromagnetic energy through space. A triumph of physics in the early twentieth century was the realization that light has properties of both particles and waves. A wave is characterized by a **wavelength**, denoted by the Greek letter lambda (λ), which is the distance between successive wave crests. The **frequency**, represented by the Greek letter *nu* (ν), is the number of wave crests that pass an observer in a given time. A simple equation relates the wavelength, the frequency, and the speed of any wave:

$$c = \lambda \nu$$

where c is the speed of the wave-in the present case, the speed of light (3.0×10^8 m s^{-1}). The light wave is a transverse (side-to-side) electromagnetic wave, in which both electric and magnetic fields oscillate perpendicular to the direction of propagation of the wave and at 90° with respect to each other.

Light is also a particle, which we call a **photon**. Each photon contains an amount of energy that is called a **quantum** (plural quanta). The energy content of light is not continuous but rather is delivered in these discrete packets, the quanta. The energy (E) of a photon depends on the frequency of the light according to a relation known as Planck's law:

$$E = hv$$

Where h is Planck's constant (1.58×10^{-34}cal/s or 6.63×10^{34}J/s).

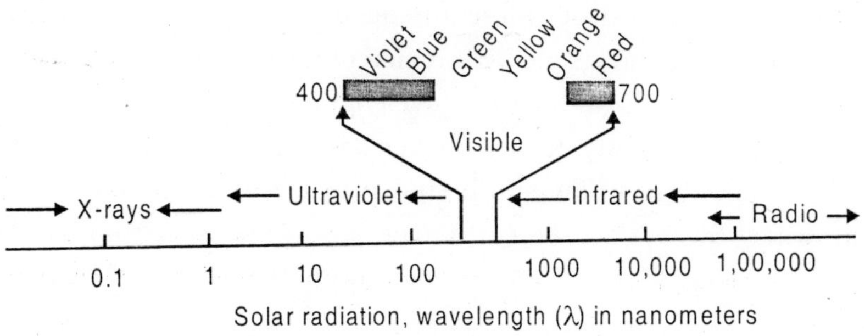

FIG. 2.1: *Solar radiation spectrum (λ in μm).*

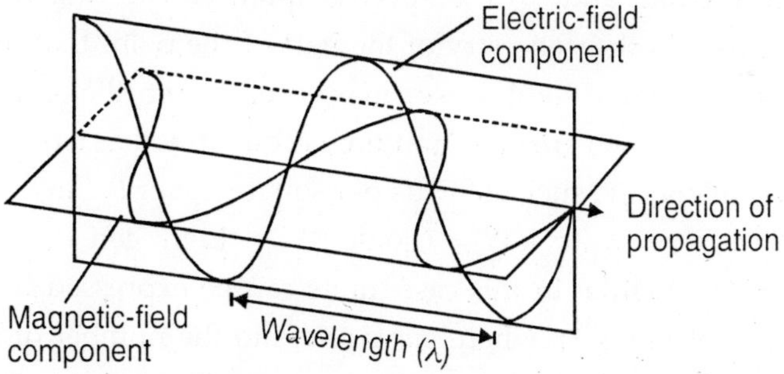

FIG. 2. 2: *Light is a transverse electromagnetic wave, consisting of oscillating electric and magnetic fields that are perpendicular to each other and perpendicular to the direction of propagation of the light. Light moves at a speed of 3×10^8 m s^{-1}. The wavelength (λ) is the distance between successive crests of the wave.*

It is customary in biology to refer to the wavelength of the light wave λ, rather than the frequency κ; the two are related by the simple equation: $\kappa = c/\lambda$ Where c is the velocity of light (3×10^{10} cm/s in a vacuum). Therefore, we can consider the energy in a mole of photons, $E = N\kappa\lambda$, where N is Avogrado's number [6.02×10^{23} molecules (or photons) per mole]. Thus $E = Nhv = Nhc/\lambda$.

TABLE 2.1: Energy of light quanta at different wavelengths

Wavelength of light (nm)	Colour of light kcl/mol	Energy of quanta E_0'v
350 Ultraviolet	80	35
450 Blue		
550 Yellow	602.6	
650 Red		
750 Infrared	40	1.7

Sunlight is like a *rain of photons* of different frequencies. Our eyes are sensitive to only a small range of frequencies-the visible-light region of the electromagnetic spectrum. Light of slightly higher frequencies (or shorter wavelengths) is in the ultraviolet region of the spectrum, and light of slightly lower frequencies (or longer wavelengths) is in the infrared region. The output of the sun is shown in Figure 2.1 , along with the energy density that strikes the surface of Earth. The absorption spectrum of chlorophyll *a* indicates approximately the portion of the solar output that is utilized by plants.

Working With Light

Three light parameters are especially important when working with light: (1) the **amount**, (2) the **direction**, and (3) the **spectral quality**. The first two parameters are important with respect to the geometry of the part of the plant that intercepts the light. For a flat leaf, a flat or planar light sensor makes sense, and the amount of energy that falls on a flat sensor of known area per unit time is quantified as **irradiance**. Units can be expressed in terms of energy, such as watts per square meter ($W\ m^{-2}$). Time (seconds) is contained within the term watt: 1 W = 1 joule (J) s^{-1}. Light can also be thought of as a stream of particles, or **quanta**. In this case, units can be expressed in moles per square meter per second ($mol\ m^{-2}\ s^{-1}$), where *moles* refers to the number of photons (1 mol of light = 6.02×10^{23} photons, Avogadro's number). This measure is called **photon irradiance**.

Quanta and energy units can be inter-converted relatively easily, provided the wavelength of the light is known. The energy of a photon is related to its wavelength as follows:

$$E = \frac{Hc}{\lambda}$$

where *c* is the speed of light ($3 \times 10^8\ m\ s^{-1}$), *h* is Planck's constant (6.63×10^{-34} J s) and λ is the wavelength of light, usually expressed in nm (1 nm = 10^{-9} m). From this equation it can be shown that a photon at 400 nm has twice the energy of a photon at 800 nm. For simplicity, above can be reduced to the following:

$$E = \frac{1988 X 10^{-19} J}{\lambda}$$

where λ is expressed in nanometers. From this calculation, we can determine that a photon of 400 nm light contains 4.97×10^{-19} *J*.

Now, suppose we have 3 µmol of 400 nm light falling on 1 m^2 every second, or a photon irradiance of 3 µmol m^{-2} s^{-1}. This quantity is approximately the amount of blue light at 400 nm that strikes the surface of Earth on a sunny day. If we want to convert photon irradiance to energy irradiance, we must first convert micromoles to moles (1 µmol = 10^{-6} mol): (3 µmol m^{-2} s^{-1}) × (10^{-6} mol μmol^{-1}) = 3×10^{-6} mol m^{-2} s^{-1} of photons, or quanta. Calculating the number of quanta from Avogadro's number gives: (3×10^{-6} mol quanta m^{-2} s^{-1}) x (6.02×10^{23} quanta mol^{-1}) = 1.8×10^{18} quanta m^{-2} s^{-1}. As we already calculated, each photon, or quantum, of 400 nm light contains 4.97×10^{-19} *J*. Thus (4.97×10^{-19} J $quantum^{-1}$) x (1.8×10^{18} quanta m^{-2} s^{-1}) = 0.9 J s^{-1} m¯2. Since 1 J s^{-1} = 1 W, therefore, the irradiance will be of 0.9 W m^{-2}.

Turning our attention to the direction of light, light can strike a flat surface directly from above or it can strike the surface obliquely. When light deviates from perpendicular, irradiance is proportional to the cosine of the angle at which the light rays hit the sensor. Thus, irradiance is maximal when light strikes a surface directly from above, and it decreases as light becomes more oblique- similar to the situation with a typical leaf. Sensors that correct for the angle of incidence of light are said to be *cosine corrected*. But there are many examples in nature in which the light-intercepting object is not flat (e.g., complex shoots, whole plants, chloroplasts). In addition, in some situations light can come from many directions simultaneously (e.g., direct light from the sun plus the light that is reflected upward from sand, soil, or snow). In these situations it makes more sense to measure light with a spherical sensor that measures light omnidirectionally (from all directions). The term for this omnidirectional measurement is **fluence rate** (Rupert and Letarjet 1978), and this quantity can be expressed in watts per square meter (W m^{-2}) or moles per square meter per second (mol m^{-2} s^{-1}). It is obvious from the units whether light is being measured as energy (W) or as photons (mol). In contrast to a flat sensor, a spherical sensor is equally sensitive to light, independent of direction. Depending on whether the light is collimated (rays are parallel) or diffuse (rays travel in random directions), values for fluence rate versus irradiance can be quite different from one another. They are equivalent only under special conditions.

Photosynthetically Active Radiation (PAR)

Photosynthetically active radiation (PAR, 400 to 700 nm) may also be expressed on the basis of energy (W m^{-2}) or quanta (mol m^{-2} s^{-1}) (McCree 1981). It is important to note that PAR is an irradiance-type measurement. In research on photosynthesis, when PAR is expressed on a quantum basis, it is given the special term **photosynthetic photon flux density (PPED)**. However, it has been suggested that the term *density* be discontinued (Holmes et al. 1985) because within the International System of Units (*Systeme Internationale d'Unites*, or SI units) *density* can mean area or volume. Moreover, area is contained within the term flux. PPED has in some cases been shortened to PPF, but it is not clear whether this abbreviation represents an irradiance-type or a spherical measurement.

When light moves through water such as lakes or oceans, it scatters in such a way that its direction of travel is randomized. Thus, aquatic photosynthetic organisms can receive light simultaneously from above, below, and the sides. In oceanography and limnology, the light that travels from the water surface toward the bottom is called *downwelling irradiance*, and the light that travels in the reverse direction is called *upwelling irradiance.*

Before the advent of specialized light meters suitable for plant physiology, light was measured in **lux** or **foot-candles**. These measurements are based on the perception of light by the human eye, which is maximally sensitive to light within the green at 555 nm. Sensitivity falls off on both sides of this wavelength and approaches zero within the blue and red, wavelength regions that are important for photosynthesis. Thus, lux and foot-candles are of little use in plant physiology. These units can be converted to the units of irradiance and fluence rate, but only if detailed knowledge about the spectral quality of the light is available, something that can be obtained only by use of an instrument called a *spectroradiometer*, which measures the amount of light at each wavelength.

Therefore, when choosing how to quantify light, it is important to match sensor geometry and spectral response with that of the plant. Flat, cosine-corrected sensors are ideally suited to measure the amount of light that strikes the surface of a leaf; spherical sensors are more appropriate in other situations, such as when studying a chloroplast suspension or a branch from a tree.

How much light is there on a sunny day and what is the relationship between PAR irradiance and PAR fluence rate? Under direct sunlight, PAR irradiance and fluence rate are both about 2000 μmol m^{-2} s^{-1}, though higher values can be measured at high altitudes. The corresponding value in energy units is about 400 W m^{-2}. When light is completely diffuse, irradiance is only 0.25 times the fluence rate.

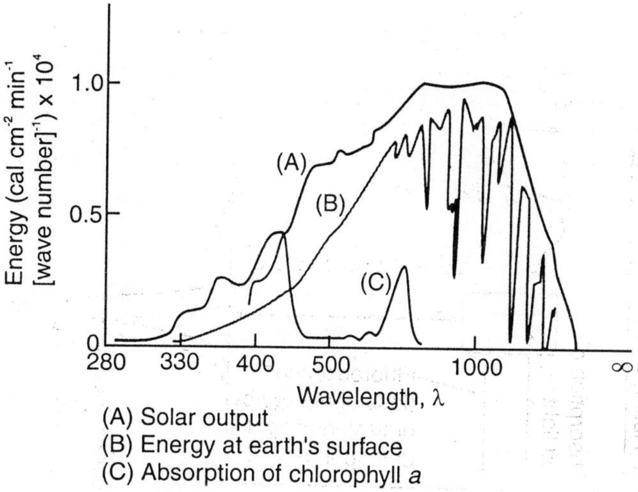

FIG. 2. 3: *The solar spectrum and its relation to the absorption spectrum of chlorophyll. Curve A is the energy output of the sun as a function of wavelength. Curve B is the energy that strikes the surface of Earth. The sharp valleys in the infrared region beyond 700 nm represent the absorption of solar energy by molecules in the atmosphere, chiefly water vapor. Curve C is the absorption spectrum of chlorophyll a, which absorbs strongly in the blue (about 430 nm) and the red (about 660 nm) portions of the spectrum. Since the green light in the middle of the visible region is absorbed only weakly, much of it is therefore reflected into our eyes and gives plants their characteristic green color. Note that curve C has absorbance rather than energy units.*

When Molecules Absorb or Emit Light, They Change Their Electronic State

Chlorophyll appears green to our eyes because it absorbs light in the red and blue parts of the spectrum, so only some of the light enriched in green wavelengths (about 550 nm) is reflected into our eyes. The absorption of light is represented by following, in which chlorophyll (Chl) in its lowest-energy, or ground, state absorbs a photon (represented by *hv*) and makes a transition to a higher-energy, or excited, state (Chl*):

$$\text{Chl} + hv \longrightarrow \text{Chl}^*$$

The distribution of electrons in the excited molecule is somewhat different from the distribution in the ground-state molecule. Figure 2.3 describes the absorption and emission of light by chlorophyll molecules. Absorption of blue light excites the chlorophyll to a higher energy state than absorption of red light, because the energy of photons is higher when their wavelength is shorter.

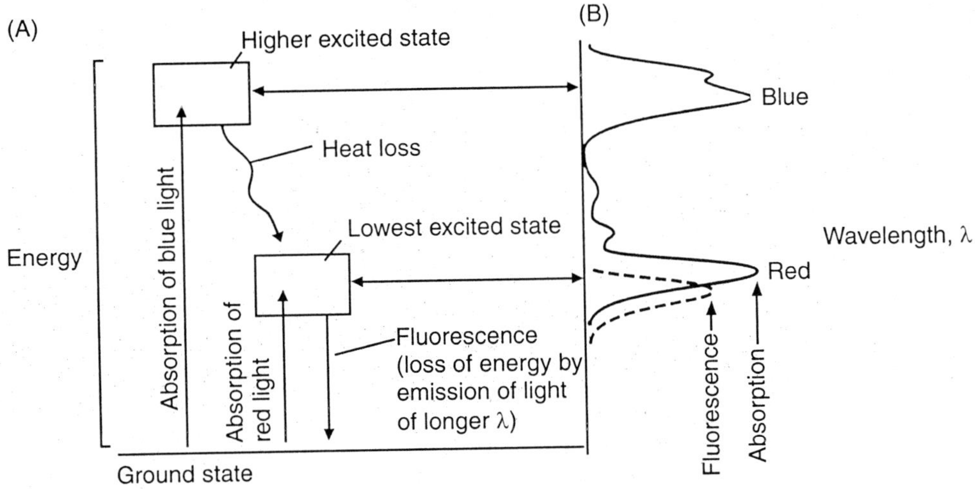

FIG. 2.4: *Light absorption and emission by chlorophyll. (A) Energy level diagram. Absorption or emission of light is indicated by vertical lines that connect the ground state with excited electron states. The blue and red absorption bands of chlorophyll (which absorb blue and red photons, respectively) correspond to the upward vertical arrows, signifying that energy absorbed from light causes the molecule to change from the ground state to an excited state. The downward-pointing arrow indicates fluorescence, in which the molecule goes from the lowest excited state to the ground state while re-emitting energy as a photon. (B) The spectra of absorption and fluorescence. The long-wavelength absorption band of chlorophyll corresponds to light that has the energy required to cause the transition from the ground state to the first excited state. The short-wavelength absorption band corresponds to a transition to a higher excited state. (After Sauer 1975).*

In the higher excited state, chlorophyll is extremely unstable, very rapidly gives up some of its energy to the surroundings as heat, and enters the lowest excited state, where it can be stable for a maximum of several nanoseconds (10^{-9} s). Because of this inherent instability of the excited state, any process that captures its energy must be extremely rapid.

In the lowest excited state, the excited chlorophyll has several possible pathways for disposing of its available energy. It can re-emit a photon and thereby return to its ground state-a process known as **fluorescence**. When it does so, the wavelength of fluorescence is almost always slightly longer than the wavelength of absorption of the same electron state, because a portion of the excitation energy is converted into heat before the fluorescent photon is emitted. Conservation of energy therefore requires that the energy of the fluorescent photon be lower than that of the excitation photon-hence the shift to longer wavelength. Chlorophylls fluoresce in the red region of the spectrum.

Alternatively, the excited chlorophyll can return to its ground state by directly converting its excitation energy into heat, with no emission of a photon. A third process that deactivates excited chlorophyll is **energy transfer**, in which an excited chlorophyll transfers its energy to another molecule. A fourth process is **photochemistry**, in which the energy of the excited state causes chemical reactions to occur. The rate of the earliest steps in the photosynthetic energy storage process is among the fastest known chemical reactions. This extreme speed is necessary for photochemistry to compete with the other possible reactions of the excited state.

The Quantum Yield Gives Information About the Fate of the Excited State

The process with the highest rate will be the one most likely to deactivate excited chlorophyll; slower processes only occasionally win out in the race to dispose of the energy of the excited state. This concept is expressed quantitatively by way of the quantum yield. The **quantum yield** (ϕ) of a process in which molecules give up their excitation energy (or *decay*) is the fraction of excited molecules that decay via that pathway. Mathematically, the quantum yield of a process, such as photochemistry, is defined as follows:

$$\phi = \frac{\text{yield of photochemical products}}{\text{total number of quanta absorbed}} \quad (7.4)$$

The quantum yield of the other processes is defined analogously. The value of ϕ for a particular process can range from 0 (if that process is never involved in the decay of the excited state) to 1.0 (if that process always deactivates the excited state). The sum of the quantum yields of all possible processes is 1.0.

In functional chloroplasts kept in dim light, the quantum yield of photochemistry is approximately 0.95, the quantum yield of fluorescence is 0.05 or lower, and the quantum yields of other processes are negligible. The vast majority of excited chlorophyll molecules therefore lead to photochemistry.

The quantum yield of formation of the products of photosynthesis, such as O_2, can be measured quite accurately. In this case the quantum yield is substantially lower than the value for photochemistry, because several photochemical events must take place before any O_2 molecules form. For O_2 production the measured maximum quantum yield is approximately 0.1, meaning that 10 quanta are absorbed for each O_2 molecule released. The reciprocal of the quantum yield is called the **quantum requirement**. The minimum quantum requirement for O_2 evolution is therefore about 10. Quantitative measurements of the absorption of light and the fate of the energy contained in the light are essential to an understanding of photosynthesis.

Photosynthetic Pigments Absorb the Light That Powers Photosynthesis

The pigments of the plant first absorb the energy of sunlight. All pigments active in photosynthesis are found in the chloroplast. The **chlorophylls** and **bacteriochlorophylls** (pigments found in certain bacteria) are the typical pigments of photosynthetic organisms, but all organisms contain a mixture of more than one kind of pigment, each serving a specific function. All chlorophylls have a complex ring structure that is chemically related to the porphyrin-like groups found in hemoglobin and cytochromes. In addition, a long hydrocarbon tail is almost always attached to the ring structure. The tail anchors the chlorophyll to the ring structure. The tail anchors the chlorophyll to the hydrophobic portion of its environment. The ring structure contains some loosely bound electrons and is the part of the molecule involved in electron transitions and redox reactions.

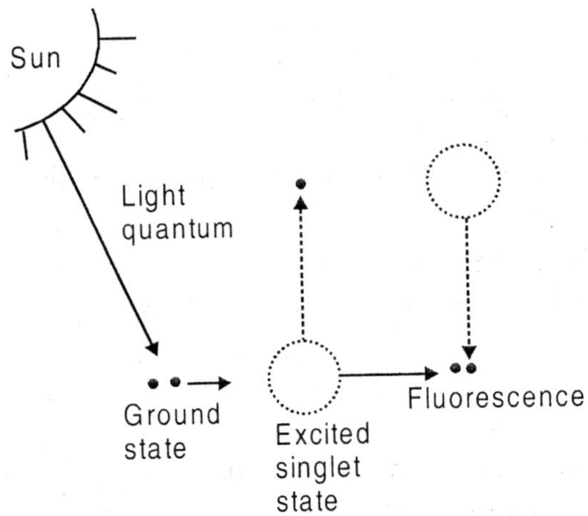

FIG. 2.5: *Absorption of photon by chlorophyll atoms.*

Photosynthesis Takes Place in Complexes Containing Light-Gathering Antennas and Photochemical Reaction Centers

A portion of the light energy absorbed by chlorophylls and other pigments is eventually stored as chemical energy via the formation of chemical bonds. This conversion of energy from one form to another is a complex process that depends on cooperation between many pigment molecules and a group of electron transfers proteins. The majority of the pigments serve as an **antenna**, collecting light and transferring the energy to the **reaction center**, where the chemical reactions leading to long-term energy storage take place.

How does the plant benefit from this division of labor between antenna and reaction center pigments? Even in bright sunlight, a chlorophyll molecule absorbs only a few photons each second. If every chlorophyll had a complete reaction center associated with it, the enzymes that make up this system would be idle most of the time, only occasionally being activated by photon absorption. However, if many pigments can send energy into a common reaction center, the system is kept active a large fraction of the time.

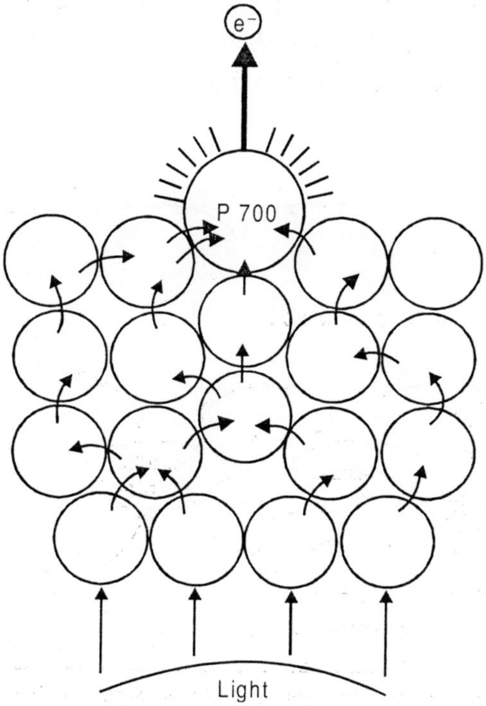

FIG. 2.6: *Harvesting of solar light by chlorophyll* a *molecule.*

In 1932, Robert Emerson and William Arnold performed a classic experiment that provided the first evidence for the cooperation of many chlorophyll molecules in energy conversion during photosynthesis (Emerson and Arnold 1932). They gave very brief (10^{-5} s) flashes of light to a suspension of the green alga *Chlorella pyrenoidosa* and measured the amount of oxygen produced. The flashes were spaced about 0.1 s apart, a time that they had determined in earlier work was long enough for the enzymatic steps of the process to be completed before the arrival of the flashes and found that at high energies the oxygen production did not increase when a more intense flash was given: The photosynthetic system was saturated with light.

The relationship of oxygen production to flash energy measured by Emerson and Arnold. They were surprised to find that under saturating conditions, only one molecule of oxygen was produced for each 2500 chlorophyll molecules in the sample. We know that several hundred pigments are associated with each reaction center and that each reaction center must operate four times to produce one molecule of oxygen; hence the value of 2500 chlorophylls per O_2.

The reaction centers and most of the antenna complexes are integral components of the photosynthetic membrane. In eukaryotic photosynthetic organisms, these membranes are found within the chloroplast; in photosynthetic prokaryotes, the site of photosynthesis is the plasma membrane or membranes derived from it.

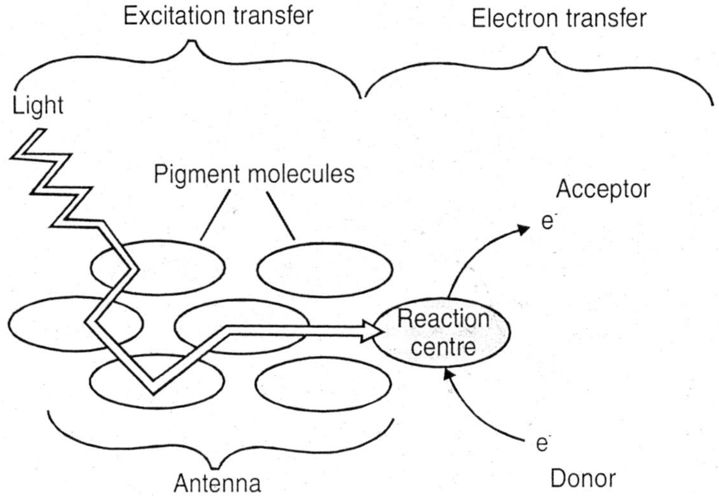

FIG. 2.7: *The basic concept of energy transfer during photosynthesis. Many pigments together serveas an antenna, collecting light and transferring its energy to the reaction center, where chemical reactions store some of the energy by transferring electrons from a chlorophyll pigment to an electron acceptor molecule. An electron donor then reduces the chlorophyll again. The transfer of excitation in the antenna is a purely physical phenomenon and involves no chemical changes. The first chemical reaction after photon absorption take place in the reaction center. Subsequent reactions stabilize the unstable products of the initial chemical reaction.*

Many pigments together serve as an antenna, collecting light and transferring its energy to the reaction center, where chemical reactions store some of the energy by transferring electrons from a chlorophyll pigment to an electron acceptor molecule. An electron donor then reduces the chlorophyll again. The transfer of excitation in the antenna is a purely physical phenomenon and involves no chemical changes. The first chemical reactions after photon absorption take place in the reaction center. Subsequent reactions stabilize the unstable products of the initial chemical reaction.

Action and Absorption Spectra

The visible radiation produced by the sun is a form of energy. In fact it is the ultimate source of virtually all the biological energy on the earth. However, there is a catch to understanding this because, in order for light energy to be useful in any biological process (photosynthesis most certainly included) it must first be absorbed and converted to a form of chemical energy useful to the cell.

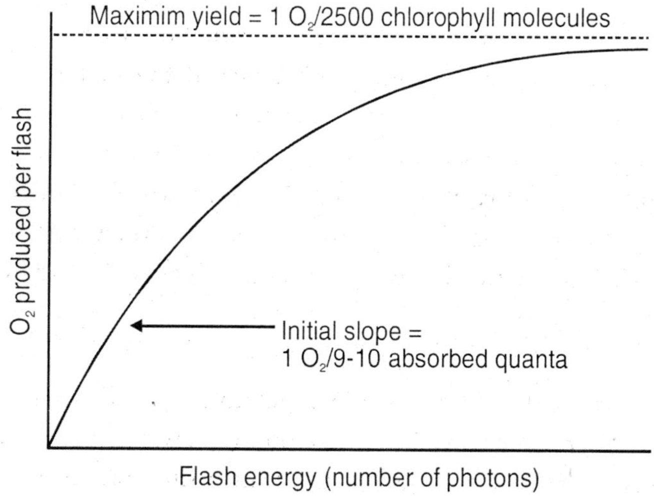

FIG. 2.8: *The relationship of oxygen production to flash energy, the first evidence of the interaction between the antenna pigments and the reaction center. The light saturation curve, obtained in 1932 by Robert Emerson and William Arnold, showed that at saturating energies, the maximum amount of* O_2 *produced is 1 molecule per 2500 chlorophyll molecules. The slope of the low-intensity part of the curve gives the quantum requirement of oxygen production.*

Many molecules and atoms within cells have the ability to absorb units of light energy. But a given molecule or atom will not absorb light of just any wavelength of energy level. Rather, light is absorbed only when the photon is of just the right energy level to *move an electron* from one energy level to another, more energetic level. Different molecules vary drastically with respect to what is *just right energy level.* Some will absorb light of the visible spectrum, while others require different wavelengths. There are various wavelengths of sunlight, therefore, what portion of the sun's total spectrum is essential for photosynthesis?. This question is perfectly testable, assuming that we can separate various portions of the sun's spectrum- and such separation is indeed possible. That is, we can expose a plant to ray or visible rays or any other part of the solar radiation spectrum and note whether the plant responds by fixing CO_2 into food molecules. The result obtained would show that *visible radiation* was the effective form of the spectrum for inducing photosynthesis.

The visible radiation is really a mixture of light of various colours. What portions of the visible spectrum (e.g., red, orange, green, blue) are most efficient in supporting carbon fixation? Simply supply various colours of light to plants and measure the amount of photosynthesis that occurs in response to each type of light.

Absorption Spectrum

It is the image recorder when electromagnetic radiation from a source emitting a continuous spectrum is passed through a substance. If the material is in the gaseous phase, bands will appear in the same position as the lines that occur in the characteristic emission spectrum of that substance in solids or liquids. These bands broadened and can be used to identify or quantify the material. The term **absorption** refers to complete retention of light, without reflection or transmission. **Absorbance** is the measure of the extent to which light is attenuated during passage through a coloured liquid or solid.

Leaf Reflectance

Leaves are the primary scattering units in a vegetative canopy. Radiant energy incident on a leaf can be scattered from the leaf surface on to or away from the leaf, giving rise to secularly **reflected** and **diffusely reflected** and **transmitted radiation**. The leaf also will absorb radiant energy. Partitioning of radiation is a function of leaf cellular structure, leaf surface coating and roughness, and morphological parameters. Leaf anatomical structure and biochemical composition are critical to the amount of radiation absorbed, transmitted or reflected.

Relatively little incident visible radiation (350-700 nm) is reflected or transmitted in healthy leaves due to absorption of chlorophyll and other leaf pigments. Decreased visible reflectance with leaf development and maturation in cotton (*Gossypium hirsutum* L.), white oak (*Quercus alba* L.) and apple (*Malus domestica* Borkh) has been attributed to pigment accumulation. In contrast leaf absorbs relatively little incident near-infra-red (NIR) radiation (75-135 nm) because NIR scattering by the leaf mesophyll result in high reflectance and transmittance.

Action Spectra

The use of action spectra has been central to the development of our current understanding of photosynthesis. An action spectrum is a graph of the magnitude of the biological effect observed as a function of wavelength. Examples of effects measured by action spectra are oxygen evolution and hormonal growth responses due to the action of phytochrome. Often an action spectrum can identify the chromophore responsible for a particular light-induced phenomenon. Action spectra were instrumental in the discovery of the existence of the two photosystems in O_2-evolving photosynthetic organisms.

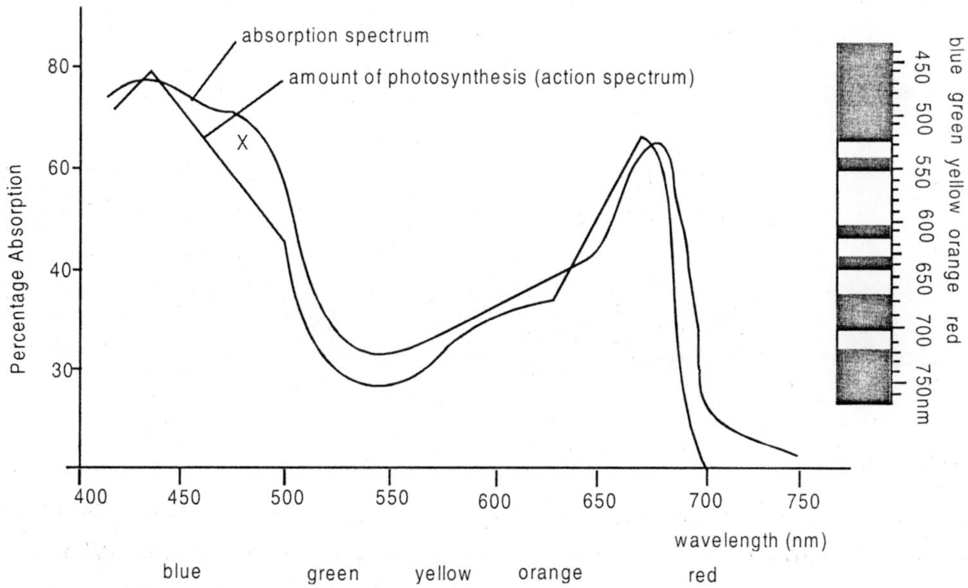

FIG. 2.9: *(A) Absorption and action spectra of chlorophyll (B) Wavelenghts absorbed by chlorophyll shown by black areas.*

Some of the first action spectra were measured by T.W. Engelmann in the late 1800s. Engelmann used a prism to disperse sunlight into a rainbow that was allowed to fall on an aquatic algal filament. A population of O_2-seeking bacteria was introduced into the system. The bacteria congregated in the regions of the filaments that evolved the most O_2. These were the regions illuminated by blue light and red light, which are strongly absorbed by chlorophyll. Today, action spectra can be measured in room-sized spectrographs in which the scientist enters a huge monochromator and places samples for irradiation in a large area of the room bathed by monochromatic light. But the principle of the experiment is the same as that of Engelmann's experiments.

Difference Spectra

An important technique in studies of photosynthesis is light-induced **difference spectroscopy**, which measures changes in absorbance. In this technique, bright light, often called **actinic** light, is used to illuminate a sample, while a dim beam of light is used to measure the absorbance of the sample at wavelengths other than that of the actinic beam. In this way a difference spectrum is obtained, which represents the changes in the absorption spectrum of the sample induced by illumination with the actinic light. Absorption bands that disappear upon illumination appear as negative peaks; new bands that appear upon illumination appear as positive peaks. Difference spectra give important clues to the identity of molecular species participating in the photoreactions of photosynthesis. The difference spectrum of the photooxidation of P700 (a chlorophyll that absorbs light of wavelength 700 nm, as we'll discuss later in the chapter) is shown in Figure 2.6 (Ke 1973).

By the use of special flash techniques, it is possible to record the difference spectrum at a given time after flash excitation. Multiple difference spectra recorded at different times after flash excitation can be used to measure the kinetics of the chemical reactions that follow photon excitation of a reaction center. These techniques can have extraordinary time resolution, in some cases less than a picosecond (10^{-12} s), and have provided great insights into the earliest events in the photosynthetic energy storage process.

De-excitation

After absorption of photon, the chlorophyll molecule becomes excited. The process of absorption is extremely rapid and takes 10^{-15} sec to complete absorption. **What happens to the excited molecule?** The excited molecule remains in its excited state for a while and excess energy (absorbed by photon) comes back to its ground state. The phenomenon of coming back to the ground state from an excited state is called **de-excitation** and during this process energy is liberated in mainly four following possible pathways:

(i) Heat emission, (ii) Light emission (fluorescence, phosphorescence, delayed light emission), (iii) Energy transfer to another molecule, and (iv) photochemistry (or photoionization).

Chlorophyll Fluorescence

Pigments that are involved in photosynthesis are capable of carrying on certain reactions even after they are extracted from the chloroplasts. They can absorb light and *fluorescen* that is, remit the light energy absorbed as light, but necessarily of longer wavelength (that is, of lower energy). The fluorescence photon has wavelength generally higher than that of the photon absorbed. In other words the energy of the emitted photon is lower than the energy of the absorbed photon. This can be explained by considering the substrates. The absorbed photon can lead the molecule to any substrate of S_1 depending on the energy of the absorbed photon. However, the molecule comes down quickly (10-12 sec) to the lowest substrate of S_1 in the stepwise fashion by heat emission (wavy vertical line). The molecule comes down to one of the substrates of S_0 by light emission (fluorescence). The molecule further goes down to one of the lower substrates of S_0 by heat emission. Thus, a part of the energy of the absorbed photon is emitted (de-excitation) as heat and the rest is emitted as fluorescence photon. Hence the energy of the photon emitted is less than the energy of the photon absorbed.

Chlorophyll solutions fluorescen dark red. If the reactions of photosynthesis are blocked by poisons, fluorescence will occur in vivo because the energy absorbed cannot be used. The fluorescence spectrum is characteristic of the pigment, so it is possible to tell which pigment is fluorescing, hence, which was activated. In addition, energy acceptors (that is, oxidized substrates capable of being reduced) can quench the fluorescence, demonstrating the ability of the pigment to transfer its energy to these acceptors instead of reemitting it as light.

Blue photons (blue light), carrying higher energy, raise electrons to the excited singlet state 2, red photons raise electrons to excited state 1. These, of course, are the photons, which are preferentially absorbed. Green photons are mostly not absorbed and it is the green light which is reflected (from and transmitted through) leaves which gives them their colour. Excited state 2 decays very rapidly to excited state 1 as electrons cascade down the energy gradient, losing energy by radiation less de-excitation. None of this energy is available for photosynthesis. It is the red, more long-lived, excited state created directly by a red photon (or indirectly by a blue photon) which is the starting point of photosynthesis and, therefore, of most of biological energy transducation. In addition the creation of the red excited state is accompanied by the generation of the massive oxidising potential, the positively charged *holes* within PSII, which are refilled by electrons drawn from water. Otto Warburg believed photosynthetic energy transduction to be perfect. It is not. Oxidation is inevitably associated with reduction. Chlorophyll fluorescence however, is only one minor channel of energy dissipation. Moreover, there are many factors, mostly still imperfectly understood, which can switch dissipation from one channel to another.

In functioning chloroplasts most of the incident energy is used photochemically. Therefore, in vivo we find only a small fluorescence yield (about 1%) in contrast to solutions in which photosynthesis does not take place. The fluorescence yield also increases in intact chloroplasts when photosynthesis is blocked by inhibitors such as diuron. In contrast to chlorophyll a solutions, intact chloroplasts show several fluorescence emission maxima which correspond to the various chlorophyll forms of photosystem I and II. At a temperature of - 196° C isolated chloroplasts show fluorescence emission maxima at 685 nm (F-685), 695 nm (F-695) and 735 nm (F-735). F-685 is associated with the antenna chlorophylls of LHCII, F-695 with the RC chlorophylls and F-735 with photosystem I. In vivo, virtually all the fluorescence is emitted by photosystem II. After previous dark adaptation intact leaves or chloroplasts show characteristic, temporal chlorophyll fluorescence changes upon irradiation, which were first, observed by Kautsky and are called variable fluorescence.

Measurement of the variable chlorophyll fluorescence is an important tool in photosynthesis research, which can be used to analyze partial reactions of the electron transport chain and the site of inhibitor activity. This technique has been used successfully to reveal the mechanisms of PSII herbicides, such as ureas and triazines. Fluorescence methods in general are highly advantageous due to their non-invasive nature, their high sensitivity and the fact that they can be applied easily to highly scattering samples like, e.g., intact leaves, needles, etc. Steady state fluorescence and/or quasi-steady state techniques like fluorescence induction have been used extensively in the past to probe a large variety of phenomena related to the energy absorption, energy distribution, etc.

Another pathway of de-excitation leads to **phosphorescence** which is also a light emission but having characteristics different from fluorescence. In this case the excited molecule first goes from S_1 state to the triplet state (T_1) by heat emission, and then comes down the ground state by light emission, called phosphorescence.

De-excitation can occur through another pathway leading to light emission known as **delayed light emission** or **delayed fluorescence.** Some of the molecules populated at the triplet state can come back to the S_1 state by absorbing the required energy from neighbouring molecules and then come down to the ground state emitting visible photon called delayed light.

Energy Transfer

An excited molecule can, under suitable condition transfer its excess of energy to another molecule at the ground state. As a consequence the first molecule comes down to its ground state and the second molecule is raised to its excited state. This is called **energy transfer**. In this case no light emission takes place. As expected, the energy gap between S_1 and S_0 of one molecule should match with that of the other for energy transfer to occur. *The fluorescence wavelength of the donor molecule match with the absorption wavelength of the acceptor molecule.* Energy transfer takes place among *identical molecules* also, and molecules must be very close to each other (within several nanometers). During photosynthesis, chlorophyll molecule of the light-harvesting class absorbs photon, gets excited and the transfer its excited energy to another chlorophyll molecule of same type, and the event may be repeated for several times until excitation energy reaches the site where photochemistry (formation of chemical energy) takes place.

Photochemistry

In photosynthesis a molecule in its excited state can utilize its excess energy to drive a chemical reaction. The event involves other molecules apart from the one which absorbs a photon gets excited.

Through this reaction the energy of the photon is utilized to create stable reducing and oxidizing agents. This occurs through electron transfer and charge separation. The reaction is as follows:

$$XYZ \xrightarrow{\text{Photon}} XY^{*}Z \longrightarrow XY^{+}Z \longrightarrow X^{+}YZ$$

In above reaction there are three molecules (X, Y and Z) at ground states. For example Y absorb is a quantum of energy and is raised to an excited state. In an excited state the potential of the molecule becomes more negative (this happens because when an electron goes to an higher electronic level, the electron become strongly bound to the molecule, the molecule tends to lose it; in other words the affinity of the molecule for the electron is reduced). The affinity of the three molecules is in such a way that in the more negative state of potential, Y* transfers an electron spontaneously to Z. This makes C positively charged (its potential becomes more positive than in the ground state, and state drawing electron spontaneously from X) and Z becomes negatively charged. Thereby Y comes back to the ground state, but X is now positively charged. This causes a separation of charges in the system with the formation of reductant (A^-) and an oxidant (X^+) as a result of light absorption by Y. There is no net change in Y. This phenomenon is called **sensitizer of photochemical reaction**. At this stage, now X^+ has the property to oxidize and Z^- has the property to reduce. This example explains the basis of conversion of light energy to chemical energy.

In chloroplast, Y may be considered equivalent to specialized chlorophyll molecule which is known as **trap** or **reaction centre**. The specialized chlorophyll is different from light-harvesting chlorophyll because in this chlorophyll there occurs no electron transfer (photochemistry) when it absorbs a photon; it only transfers its excitation energy to another molecule. In green plant photosynthesis, the electron from Z^- is transferred through several redox couples eventually to NADP and X^+ becomes sufficiently positive to oxidize water producing protons, electron and oxygen. In this process water is the ultimate source of electrons needed for the reduction of NADP.

Photochemical Quenching

Linear photosynthetic electron transport moves electrons from water through the two photosystems to ferredoxin. From ferredoxin these are passed to CO_2, via NADP, or to O_2 (the Mehler reaction or pseudocyclic electron transport.) If CO_2 is removed, NADP becomes rapidly reduced and the lack of NADP as a terminal electron acceptor affects other carriers in the chain. Once Q_A (the primary electron acceptor in PSII) is reduced (i.e., once its open electron traps are closed) it no longer **quenches** fluorescence and the fluorescence signal rises. Although Q_A is a quinone "Q" stands for **quencher** in this context. This was the name that Duysen and Swears (1963) first applied to a then unknown component in PSII which quenches fluorescence when it is oxidised but not when it is reduced. Such quenching is now referred to as q_Q or *photochemical quenching*.

CHAPTER 3

Proteomics and Genomics of Photosynthetic Apparatus

PLASTIDS

Plastids are self replicating organelles surrounded by an envelope composed of two membranes. They are found only in photosynthetic cells of eukaryotes but exceptionally found in generative and sperm cells of certain angiosperms too. Plastids are varied in size, shape and colour. Green coloured plastids are chloroplast, that contain the chlorophyll pigment and are site of photosynthesis. In green-leaf cells particularly in the parenchyma cells, the chloroplasts lie in the cytoplasm. The chloroplasts occupy more than 90 per cent of the volume of the cytoplasm in leaf and are typically lense-shaped (in mosses, fern and seed plants), appearing biconvex, plano-convex or concavo-convex in profile. In algae they are heteromorph. In higher plants they are usually 4-8 μm in diameter and 2-3 μm thick. Changes in shape and size can also be induced osmotically both in the leaf cells and chloroplast isolated from the cell. They often move in the cell under the influence of illumination and as a result of cytoplasm streaming.

Other coloured plastids are found in flowers and fruits. On the basis of the storage products colorless plastids or leukoplasts are termed amyloplasts (starch), proteinoplasts (protein) or elaioplasts (lipids). Plastids have three compartments: (i) The intramembrane space between the outer and inner membrane, (ii) The stroma space and (iii) The thylakoid space.

THE MULTIFUNCTIONAL CHLOROPLAST

Chloroplasts in green algae and higher plants contain photosynthetic thylakoid membranes. Four multi-subunit protein complexes, Photosystem I (PSI), PSII, the ATP-synthase complex, and the cytochrome b_6f complex, which together comprise 75 to 100 proteins, perform the photosynthetic reactions. It can be postulated that the thylakoid membranes contain a large number of other proteins that are involved in the biogenesis and regulation of the four multi-subunit complexes. These additional proteins are also involved in processes such as biosynthesis and ligation of cofactors, insertion of proteins into membranes, and folding and degradation of proteins. Several such proteins have been identified in the thylakoid membrane. They include the proteases DegP and FtsH, protein translocation components, such as SecY, SecE, SecA, Alb3 and Hcf106, the PSII assembly factor Hcf B6 and a lumenal isomerase TLP40. Kinases (Synders and Kohorn, 1999) phosphatases (Vener et al., 1998), and possibly other signal transducers are present in the thylakoid membrane. Several proteins without obvious function have been identified in the intra-thylakoid space, the thylakoid lumen.

Although best known for their role in Photosynthesis, chloroplasts also synthesize many essential compounds, including plant hormones (Lange, 1998) fatty acids and lipids (Essingmann et al., 1998), amino acids (Ho et al., 1999), Vitamins (B1, K1 and E, Belanger et al., 1995), Purine and pyramidine nucleotides (Smith et al., 1998) and secondary metabolites such as alkaloids and isoprenoids (Keller et al., 1998). In addition to these, chloroplasts also are required for nitrate and sulfur assimilation (Heldt, 1997). Several of the enzymes in these biosynthetic pathways have been identified by analyzing mutants of *Arabiodopsis* and other plant species. However, more proteins and possible new pathways remain to be discovered, and it is not unlikely that several of these components are located in or at the surface of the thylakoid membrane.

Number

Their number in a cell varies greatly among plant species. The assimilatory parenchyma (palisade and spongy parenchyma) of leaves contains a especially large number of chloroplasts. In spinach leaves, for instance, 300-400 chloroplasts have been found per palisade parenchyma cell and 200-300 per spongy parenchyma. The dry weight of spinach chloroplasts is about 9×10^{-12} g and that of corn chloroplasts about 38×10^{-12} g.

OTHER INCLUSIONS

Starch grains

These are often single, largest inclusion found in the stroma of chloroplasts. They appear in profile as ovule structures generally electron translucent but occasionally they are electron opaque. Starch generally accumulates in the bundle sheath chloroplasts of C4 plants and only to a smaller extent in the mesophyll chloroplasts.

Crystalline inclusions

Crystalline inclusions or bodies are of 1 mm in size and lie in stroma. They are proteinaceous in nature and their function is unknown.

Phytoferritin

It is an iron protein complex which appears as an electron opaque aggregation of particles in the plastid stroma. Algae possess **pyrenoids**, which organize the deposition of reserve substances. The microfilaments, 5-6 nm in diameter have been occasionally seen in electron micrographs are thought to run lengthwise in the stroma near the plastid envelope. The function of these microtuble-like structures is not well understood.

MEMBRANE SYSTEM OF CHLOROPLAST

Literature on chloroplast ultrastructure from past 40-50 years is dominated by electron microscopic studies. This type of observation involves the fixation of the specimen by harsh chemical processes. Electron micrographs show the chloroplast as being enclosed by a double membrane and transversed by the thylakoid membrane system. In the many mesophyll cells the thylakoids are spatially segregated into appressed grana regions and non-appressed stroma thylakoids. The thylakoid membranes contain large amounts of chlorophyll pigments which autofluoresce in the orange-red wave-lengths. These pigments are structurally organized into protein-pigment complexes, the photosystems, which are embedded within the thylakoid membrane. It is generally believed that the grana regions are enriched in photosystem II (PSII) with a fluorescence maximum at 685 nm at 20°C. Photosystem I (PSI) emits a much weaker fluorescence signal at room temperature at around 730-740 nm and is thought to be located in the stroma thylakoids, with a small amount occurring in the grana margins and the stroma exposed outer grana.

In comparison with the large amount of electron microscopic pictures of chloroplast available, relatively little is known about the ultrastructure or even the gross anatomy of live, photosynthetically-active chloroplasts. Recently, Kohler et al. (1997) reported connecting tubules, so called **stromules** between chloroplasts when studying live plastids labelled with green fluorescent protein. Sarafis (1998) observed emerging lobes in confocal studies of *Ophiopogon nigrescence.* Mehta et al. (1999) studied *Pisum* chloroplasts and showed one chloroplast with one protruding and one forming stromule.

EM studies have shown that C4 bundle sheath chloroplasts differ structurally from mesophyll chloroplasts. From these and biochemical studies three types of bundle sheath chloroplasts have been identified. Bundle sheath chloroplasts from species belonging to NAD-ME type and PEP-ME type are generally granal. NADP-ME type chloroplast are agranal or have very few rudimentary grana. Variations in the amount of cyclic and non-cyclic electron flow between the bundle sheath chloroplast of different C4 species has been observed. The agranal type lacks non-cyclic electron flow and oxygen-evolving activity, the rudimentary type has low non-cyclic electron flow and oxygen evolution rates and granal type bundle sheath chloroplasts exhibit comparatively high non-cyclic electron flow and O_2 evolution rates.

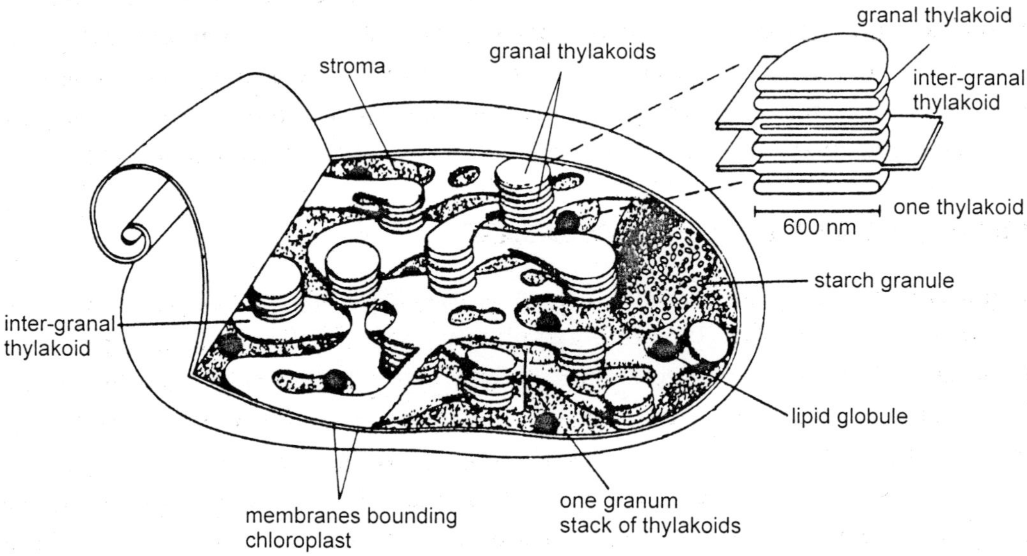

FIG. 3.1: *A three dimensional structure of chloroplast.*

The surrounding material, which does not contain chlorophyll, is called *stroma.* It is hydrophilic protenaceous matrix. CO_2-incorporating apparatus is present within the stroma. The stroma along with thylakoids also contains particulate components such as *fraction portion I,* ribosomes and polysomes, and the fibrils of DNA, nucleoids (DNA), plastoglobuli (lipid reserve) and starch grains. Robert Hüber, Johann Deisenhofer and Hartmut Michel were awarded the 1988 Nobel Prize in Chemistry for work in the photosynthetic process, for succeeding in making a structural analysis of a membrane linked to protein complex which opened up completely new possibilities in chloroplast structural research.

TRANSPORT SYSTEM OF CHLOROPLAST MEMBRANE

The majority of the protein components of chloroplasts are encoded in the DNA. They are synthesized on cytoplasmic ribosomes outside the organelle and subsequently imported into the chloroplasts. *In-vitro* studies using the protein-synthesizing machinery of a wheat-germ extract coupled to polyadenyl messenger RNAs (poly A^+mRNA) isolated from algal or higher plants have demonstrated the transfer of a number of nuclearly encoded polypeptides into the chloroplast; examples are ferredoxin, plastocyanin, ferredoxin-NADP reductase, polypeptides associated with antenna pigment complexes, the small subunit of ribulose-bisphosphate carboxylase, and fructose 1,6-bisphosphatase. These proteins are assembled on the cytoplasmic ribosomes as a precursor of much higher molecular weight protein referred to as the transit peptide, with sequence extension, at the aminoterminal. The precursor of plastocyanin (molecular mass = 10.5 kDa), for example, is assembled with a transit peptide of molecular mass 15 kDa. The precursors are bound to the outer chloroplast membrane, translocated through the chloroplast envelope, processed to the mature polypeptide (mainly by proteolytic cleavage of the transit peptide), and then either assembled as membrane constituents or incorporated into the stroma. The import of precursors into the chloroplast is an energy-dependent process and consumes ATP synthesized by photophosphorylation. The requirements for binding of precursors to the envelope membrane, the nature of the receptors, and the mechanisms regulating the translocation and assembly of polypeptides are still being actively investigated.

There are two classes of precursor proteins based on their thylakoid assembly requirements. The integral membrane proteins (e.g. Chl *a/b* protein of LHC II) must cross the plastid envelope membranes, the aqueous envelope periplasm, the stromal compartment and then integrate into the thylakoid lipid bilayer. These precursor proteins contain a hydrophobic N-terminal extension called the *transit sequence* that contains topogenic information for targeting the precursor to the chloroplast and localizing the matured protein on the thylakoid. Lumenal proteins (e.g. plastocyanin) must cross all the plastid compartments and in addition must traverse the thylakoid bilayer membrane. The plastocyanin precursor contains two transit sequences as chloroplast-targeting components. A chloroplast-targeting domain (at the N-terminal) directs the pre-protein towards and into the chloroplasts; this domain is cleft by a stromal protease. A second, strongly hydrophobic, domain at the C-terminal directs the partially cleft precursor to the lumen where it is split off by a thylakoid protease.

All the chloroplast proteins enter plastids through the general import pathway. Several components that constitute the translocation machinery have been identified. Toc 33, a small GTP binding protein is the latest of a family of outer envelope (Toc) proteins identified. *Arabidopsis* mutants with defective Toc 33 show defective protein import (Jarvis et al., 1998). Tic 22, a peripheral membrane protein belonging to the family of inner envelope (Tic) proteins, has been projected to connect inner and outer membrane protein complexes (Kouranov et al., 1998). Proteins destined for the thylakoid membranes are transported into or across the thylakoid membranes by different pathways: ΔpH, thylakoid secretory (Sec) pathway, chloroplast signal recognition particle (cpSRP) and by spontaneous insertion. Components involved in these different mechanisms are being characterized.

The soluble protein cpSecA is involved in translocation of a number of nuclear-encoded lumenal protein through the thylakoid membrane. Analysis of maize mutants, defective in cpSecY, a chloroplast localized component of the translocon, shows significant reduction in thylakoid membrane accumulation (Roy and Barkan, 1998). The severe reduction in chloroplast translation in these mutants suggests that cpSecY goes beyond its involvement with the SecA-dependent protein translocation. Roy and Barkan (1998) proposed that cpSecY plays a role in the interaction between chloroplast translation and membrane biogenesis.

The chloroplast homolog of SRP, cpSRP54 forms a transit complex with cpSRP43 and LHCP. The formation of this complex is essential for translocation and integration of LHCP into the thylakoid membrane (Schuenemann et al., 1999). Nilsson et al. (1999) have shown that cpSRP54 interacts tightly and specifically with ribosome nascent chain complexes of D1 protein, implicating a role for cpSRP54 in D1 biogenesis. An *Arabidopsis* mutant, *chaos*, has been identified as a mutation in the gene encoding cpSRP43. The *chaos* mutant phenotype differs distinctly from mutants of cpSRP54, suggesting that the functions of two proteins do not overlap. Pilgrim et al. (1998) have shown that a chloroplast homolog of SRP54 (cpSRP54) plays an important role in chloroplast biogenesis. Mutants with reduced levels of cpSRP54 display delayed maturation of proplastids and aberrant light induced movement. Analysis of these mutants suggests a role for cpSRP54 in the biogenesis of several proteins, co-translationally and post-translationally. *Arabidopsis* mutants that lack cpSRP54 have defective plastid biogenesis whereas chaos shows only defects in LHCP targeting. FtsY, a bacterial homolog of the SRP receptor protein has been identified from Arabidopsis and shown to play a role in the SRP pathway.

Maize mutants of hcf 106 are defective in the ΔpH pathway. Hcf 106 encodes a membranes protein that may play a critical role. Another gene, tha4, that has similarity to hcf 106 in topology and structure has been identified from maize. It is suggested that these two gene products may function in common pathway. Whether these different translocation pathways converge is question that awaits additional evidence. The necessity for different pathways for protein targeting into thylakoid membranes may be specified by the nature and sub-organelle location of the transported proteins.

EXCHANGE OF IONS AND METABOLITES

Chloroplasts isolated from higher plants are surrounded by two distinct membranes which are together called the chloroplast envelope. These envelopes are the site of synthesis of a number of membrane constituents. These are involved in the transport of many metabolites.

The chloroplast envelope is the site of synthesis and assembly of membrane galactolipids, α-tocopherol and carotenoids. The polypeptide composition of the envelope is different from that of the thylakoids and of the stroma. The chloroplast envelope functions both as a barrier separating the stroma from the cytosol and as a bridge allowing the transport of specific metabolites between these two compartments. The chloroplast envelope is permeable to various molecules. The **outer envelope** membrane acts as a molecular sieve and is permeable to all molecules up to a molecular mass of 10 kDa. The **inner membrane** (osmotic barrier) is almost impermeable to charged species such as ferredoxin, NAD(P), NADPH + H^+, acetyl CoA, pyrophosphate, many sugar phosphates, small cations (Na^+, Mg^{2+}), and to sucrose and sorbitol. CO_2, O_2, acetic acid, pyruvic acid, glycerol are uncharged molecules. They can diffuse through the inner membrane.

Pi and triose phosphates, ATP, and dicarboxylate anions (malate, aspartate, oxaloacetate, etc.) pass through the inner membrane in the presence of specific carriers present in the envelope called *translocators*. Of these translocators, the proteinaceous factor (phosphate or C_3 translocator), which allows the transport of triose phosphates from the stroma to the cytosol in exchange for the import of Pi into the stroma, has been studied in most detail.

The chloroplast envelope also mediates in the export of fatty acids synthesized in the chloroplast to other organelles (for lipid synthesis). The exact nature of the polar lipid interactions with plastid membrane proteins is not yet clear. It is also not clear that how the membrane components which are synthesized in the envelope are subsequently transferred and assembled in the thylakoids, how do the polyunsaturated fatty acids in the chloroplasts originate, what are the specific functions of various envelope proteins?

CHLOROPLAST DNA

Plastid genomes of algae and land plants exist as circular DNA molecules with sizes in the range of 50-400 kbp. In *Chlamydomonas,* there are 80 copies of the genome whereas in mesophyll cells of higher plants there are up to 10,000 copies. Most of the genomes examined contain two large inverted repeats harbouring the ribosomal RNA (rRNA) genes and these are separated by two single-copy regions. The complete nucleotide sequences of plastid genomes are now available for six vascular plants, (*Epifagus virginiana, Marchantia polymorphs,* tobacco, rice, *Pinus thunbergii* and maize) and six unicellular eukaryotes (*Euglena gracilis, Chlorella ellipsoidea, Cyanophora paradoxa, Porphyra purpurea* and *Odontella sinensis*). The sequences of plastids from land plants and green algae reveal approximately 120 genes, which are mostly conserved among these organisms and can be divided into three major groups.

- The first group includes sequence required for plastid gene expression with approximately 50 genes encoding subunits of RNA polymerase, rRNAs, tRNAs, ribosomal proteins and additional factors (e.g., elongation factor, EF-Tu and initiation factor).
- The second group comprises approximately 40 genes, encoding components of the photosynthetic apparatus mostly subunits of the thylakoid-associated complexes PSI, PSII, the Cytbf complex, the ATP synthase and the large subunit of the stromal enzyme Rubisco.
- A third group includes open reading frames whose function is still largely unknown. Some of these are conserved between species and are designated ycf (hypothetical chloroplast open reading frame); others appear to be species-specific.

The plastid genome of *Porphyra purpurea* contains twice as many genes as land plants. Although not at all the additional genetic material has been characterized, it includes genes involved in photosynthesis that are encoded by the nucleus in plants, and genes for plastid biosynthesis functions (e.g. for the synthesis of fatty acids, amino acids, pigments and thiamine), other genes encode proteins involved in transport.

CHLOROPLAST RIBOSOMES

Stroma show presence of single ribosomes, clusters of ribosomes and presence of polysomes after staining with uranyl acetate. The single chloroplast ribosomes are smaller than cytoplasmic ones measuring 17 nm in diameter (25 nm for the cytoplasmic ribosomes). It has sedimentation coefficient of 70S and can be split into two subunits sedimentating at 55-60S and 40S respectively. They resemble bacterial ribosomes more closely than eukaryotic cytoplasmic ribosomes.

PROPLASTIDS

Proplastids, the progenitors of other plastids are found in root and shoot meristems, embryos, endosperms and in young developing leaves. They are not structurally complex. They consist of an envelope, a nucleiod with DNA fibrils, a small number of ribosome, thylakoids, sometimes small starch molecule. Proplastids are varied in shape. They are only slightly larger than mitochondria. In light grown plants the proplastids in stems and leaves usually differentiate into chloroplasts. Proplastids may also have important biochemical functions such as the synthesis and storage of starch and other metabolites.

PLASTID INHERITANCE

In most angiosperms plastid inheritance is maternal because the sperm cells lack plastids. This condition may arise during the first pollen mitosis when all plastids go into vegetative cell resulting in an aplastidic generative cell. In some species e.g. *Beta vulgaris, Solanum tuberosum* almost all plastids go into the vegetative cell during first pollen mitosis; a few plastids pass into the generative cell. During the development of the sperm cells in these plants the plastids are lost and as a result plastid inheritance remains maternal. In a few angiosperm genera, e.g. *Oenothera* and *Pelargonium*, plastid inheritance is biparental because plastid distribution is equal during the first pollen mitosis and the plastids are not eliminated resulting in plastids being present in the sperm cells.

CHLOROPLAST BIOGENESIS

In angiosperms (flowering plants), chloroplast biogenesis is usually studied by germinating seeds in total darkness and then exposing the seedlings to light. The changes occurring during illumination are followed by examination of the ultrastructure of the plastids by electron microscopy, the assay of photosynthetic electron transport and CO_2 fixation rates, and determination of the chromophore-containing components, e.g. cytochromes and ferredoxins, by spectroscopy. Such studies have shown that chloroplasts develop from relatively undifferentiated structures known as *proplastids* via simple colourless organelles termed *etioplasts*.

The most important change occurring during illumination (greening) is the reduction of protochlorophyllide to chlorophyllide (chlorophyll without the phytyl side chain). The response of leaves to light involves changes in the expression of specific genes at the

RNA level. RNAs complementary to many segments of the plastid genome are expressed more abundantly in light. Substantial increases in the content of carotenoids, cytochromes, ferredoxin, plastocyanin, ferredoxin-NADP reductase, etc. have been measured during the greening process. Photosystem I activity is expressed at an earlier stage than PSII activity. The enzymes of the CO_2 fixation pathway are present in the etioplasts, albeit at very low concentrations compared to mature chloroplasts.

In many plants, differentiation of protoplastids or etioplast into chloroplast occurs only upon perception of a light signal. This light induced development involves rapid accumulation of chlorophyll, photosynthetic membranes and associated photosynthetic proteins. The transition is mediated by two classes of photoreceptors, phytochromes and cryptochromes. Mutational analysis had led to the identification of nuclear encoded regulating elements that affect biogenesis of the chloroplast in the variety of ways. The DET, COP (both *Arabidopsis*) and RegA (*Volvox*) proteins have been shown to repress transcription of genes required for plastid biogenesis (Kirk et al., 1999). Another class of mutants has been described (CAB underexpression or cue) that show defects in chloroplast development and light induction of nuclear encoded Cab mRNA transcription, providing additional evidence that photoreceptor signal transduction pathways and plastid signaling pathways share common factors in an intricate network.

Genetic analysis in *Arabidopsis* has identified a number of genes which influence chloroplast development. These mutants include those with defects in differentiation and greening (dag), chloroplast and leaf development (dcl) and chlorophyll accumulation (pac). Each of these mutant suggests that the nucleus can sense the physiological status of the developing plastid and respond accordingly. PAC is a nuclear encoded protein involved in processing of chloroplast mRNA (Meurer et al., 1998). Nuclear gene expression is unaffected in *pac* mutants whereas chloroplast transcript such as psbA-D, petB and several genes from the ndh cluster are dramatically reduced in abundance indicating that PAC may function in the recognition and processing of maturation signals found in chloroplast mRNAs. A number of mutants with defects in pigment biosynthesis have been shown to have pleiotropic effects on chloroplast biogenesis (Leon and Arroyo, 1998). Disruption of the im gene in *Arabidopsis* results in a variegation mutant (IMMUTANS) which is similar to *iojap* and *albostrians* mutant (Carol et al., 1999). The IM protein product has an alternative oxidase activity that indicates a role as phytoene desaturase, an enzyme critical for preventing photo-oxidative damage during early chloroplast biogenesis.

Expression of a number of nuclear genes is influenced by the state of plastid development. Studies on IMMUTANTS have shown that disruption of carotenoid biosynthesis leads to photo-oxidation, which in turn inhibits the expression of a set of nuclear encoded photosynthesis related proteins (Wu et al., 1999). This effect is most noticeable on the expression of the light-harvesting complex a/b (CAB) proteins. This regulation is partly due to change in nuclear gene transcription. There is evidence for both positive and negative plastid signaling to the nucleus but the nature of the signal remains elusive. In *Chlamydomonas*, at least some of the properties of the plastid signal can be achieved by the addition of chloroplast precursors to cells lacking chlorophyll biosynthesis. The plastid signals in higher plants have been speculated to be metabolities, secondary messengers or macro molecules. Although the nature of the signal is not known, analysis of *cue* mutants lead to the proposal that the plastid signal follows at least part of the pathway through which phytochrome mediated signaling operates (Lopez-Jeuz et al., 1998).

INTEGRATION OF PLASTID AND CELL FUNCTION

Plastid isolated from higher plants does not replicate and survive only a short period of time in vitro. This indicates that these organelles are not autonomous and are dependent on cytoplasmic functions and nuclear gene products. The inter-dependence of plastids and the cell in which they reside is relatively specific and the transfer of plastids into foreign plant cells results in impaired function and loss of the transferred organelles. It is clear that extensive changes have occurred in the photosynthetic endosymbiont during the integration of process. For example, plastids of higher plant do not have peptidylglycan walls which are found in Bacteria. Presumably, the genes required for wall biosynthesis have been lost or inactivated. Interestingly, the cyanelles of *Cyanophora paradoxa* which resemble plastids and perform a photosynthetic function in this organism retain a partial peptidylglycan wall (Herdman and Denman, 1977). Some plastid functions are not found in prokaryotes and must have been acquired during the process of the host endosymbiont coevolution. The involvement of plastid in plant growth regulator biosynthesis is an example of this type of change. Biosynthetic pathway of host and endosymbiont have also become integrated like some steps of lipid biosynthesis in higher plants that occur on ER membrane while few other steps are localized on plastid envelope.

The integration of plastid and cell function must have involved in the transfer of many genes from the plastid to the nucleus. The fact that all known enzymes in chlorophyll and carotenoid biosynthesis are encoded by nuclear genes underscores this point. The expression of genes involved in photosynthesis in higher plants is regulated developmentally and shows organ, tissue, and cell specificity. Furthermore, plastid populations in higher plants acquire specialized functions such as starch storage (amyloplast) carotenoid (chromoplasts) or lipid (leucoplasts) biosynthesis and storage.

Plastids with specialized functions are usually localized in specific organs or tissues such as tubers (amyloplasts) or epidermal layers of fruit and petals (chromoplasts). This implies that the host regulates the expression of specific biosynthetic pathways, a control that was once localized in plastids. The discussion above concerning the origin of plastids provides a useful perspective when trying to understand the organization and regulation of genes encoding plastid function in higher plants. First, we should anticipate that plastid gene organization and decoding mechanisms will have prokaryotic features. Second, the transfer of genetic information from the plastid to the nucleus raises several questions. Why were some genes transferred but not others? What signals' are needed to direct nuclear gene products to the plastid compartment?

PROTEOMICS

The term proteomics or more appropriately functional proteomics describes the ability to apply global experimental approaches to assess protein function. Proteomics has emerged as a new experimental approach in part because mass spectrometry has simplified protein analysis and characterization, and several important and recent innovations have extended the capability of mass spectrometry. The ionization techniques, MALDI (Matrix-assisted laser desorption ionization) and ESI (Electrospray ionization), have stimulated development in mass spectrometers to enhance the production of two different types of information. The first type of information is the accurate measurement of molecular weight. To measure molecular weight to the low ppm level, MALDI is used typically in conjunction with T-OF (time-of-flight) mass analyzers. The second type of information, produced by tandem mass spectrometers (MS/MS), is diagnostic of amino acid sequence. These approach require efficient methods to digest proteins in the gel, extract the resulting peptides and transfer them to the mass spectrometer. By employing robotics to excise spots from the gel or to slice the gel into series of small cubes and to digest the proteins, whole-gel analysis might be possible. Identification of proteins that have purified by 1-D or 2-D SDS-PAGE can now be performed with low narogram quantities of protein (Figeys et al., 1996). McCormack et al., (1997) have used the method with immunoaffinity precipitation, protein affinity-interaction chromatography and to identify proteins that interact with large protein complexes.

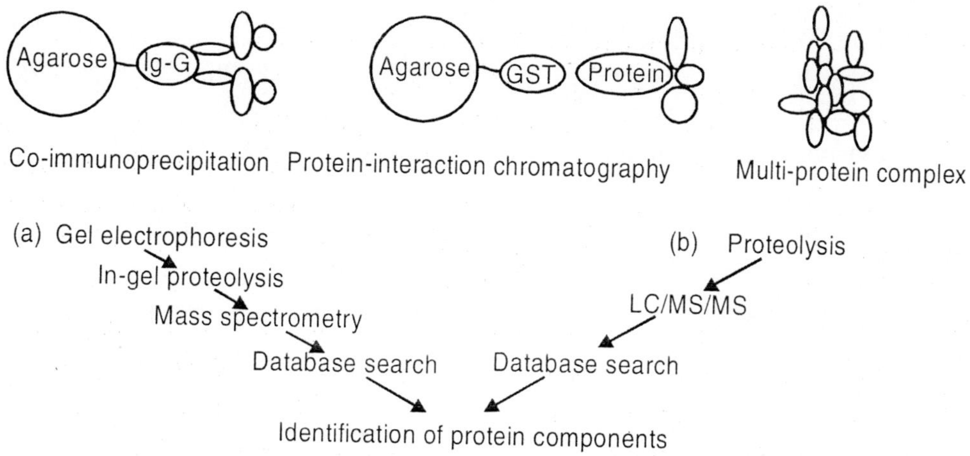

FIG. 3. 2: *The components of protein complexes can be determined using mass spectrometry from co-precipitation reactions, protein affinity interaction chromatography or isolation. One of two approaches can be used to isolate and/or identify the components. (a) The collections of proteins can be separated using sodium dodecyl sulfate polyacrylamide gel electrophoresis (SDS-PAGE) and then bands can be removed for proteolysis. Each protein from the gel can be subjected to mass spectrometry. Data is then searched through databases to identify the protein. A second approach (b) subjects the collections of proteins to proteolysis directly. The complicated collection of peptides is then analyzed directly by high-performance liquid chromatography (HPLC), combined with tandem mass spectrometry (LC/MS/MS), and the resulting spectra searched through a database. Software is used to assign peptides to their respective, proteins, and thus identify the proteins present.*

Recently, Peltier et al. (2000) studied proteomics of the chloroplast. The soluble and peripheral proteins in the thylakoids of pea have been systematically analyzed by using two-dimensional electrophoresis, mass sectrometery, and N-terminal Edman sequencing, followed by database searching. After correcting to eliminated possible isoforms and post-translational modifications, they estimated that there are atleast 200 to 230 different lumenal and peripheral proteins. Sixtyone proteins were identified, for 33 of these proteins, a clear function or functional domain could be identified, whereas for 10 proteins, no function could be assigned. For 18 proteins, no expressed sequence tag or full-length gene could be identified in the databases, despite experimental determination of a significant amount of amino acid sequence. Nine previously unidentified proteins with lumenal transit peptides are presented along with their full-length genes; seven of these proteins possess the twin arginine motif that is characteristic for substrates of the TAT pathway. Logoplots were used to provide a detailed analysis of the lumenal targeting signals, and all nuclear-encoded proteins identified on the 2-D gels were used to test predictions for chloroplast localization and transit peptides made by the software programs Chlorop, PSORT, and Signal P.

The 58 proteins (excluding three stromal proteins) were classified in a pie diagram according to their function, following the categories used in the analysis of a 1.9-mB contiguous sequence of chromosome IV of *Arabidopsis* (Bevan et al., 1998). This is a helpful tool to visualize the more general role of the lumenal and peripheral membrane proteins. For 17% of the proteins, no clear function could be predicted, whereas for 31% of the proteins, no expressed sequence tag or full-length gene could be assigned. The remaining 52% could be classified according to their function.

Naturally, a significant fraction of the proteins is involved in energy production (21%), either in photosynthetic electron transport or in ATP production. So far, only one transporter, Brittle-1, was found on the two-dimensional electrophoresis maps of both the lumenal and peripheral fractions. Brittle-1 has an adenlyate translocator function in the transfer of ADP-glucose in amyloplasts. Brittle-1 was cloned using a transposon-tagged maize mutant (Sullivan et al., 1991) and was found to be localized in amyloplast membranes (Sullivan and Kaneko, 1995). Because starch also accumulates in chloroplasts, Brittle-1 is likely to have a similar function as in amyloplasts. Indeed, in vitro import assays showed that Brittle-1 could be targeted to the chloroplast inner envelope membranes, but the protein had not been found previously in chloroplasts in vivo (Li et al., 1992).

Transmembrane prediction programs indicate that Brittle-1 could be an integral membrane protein, but the predicted transmembrane domains are shorter and less hydrophobic than are observed usually. Any integral membrane protein was not detected on the two-dimensional electrophoresis maps, which is expected because the protein fractions were centrifuged at high gravity values to remove any membranes. Thus, it was surprising to find Brittle-1 on both the lumenal and peripheral maps. Brittle-1 is an integral membrane protein. Because the protein had not been detected previously in chloroplasts, further experimentation is needed to more precisely define the localization of Brittle-1.

The two-dimensional electrophoresis maps also revealed a lipid storage protein (PG1), which was classified in the cell growth and division category. PG1 recently was identified in plastoglobules extracted from pea chloroplast membranes, and in vitro import assays confirmed its localization in the chloroplast (Kessler et al., 1999). Plastoglobules have been observed in different types of plastids and are thought to serve as lipid reservoirs for thylakoid membranes. In addition to PC1, plastoglobules possess several other proteins associated with the outer surface, such as the 30-kD plastid lipid-associated protein (Pozueta-Romero et al., 1997). It is quite likely that several of those proteins are present on the two-dimensional electrophoresis maps, and they should be identified in the near future.

Other proteins involved in cell growth include a set of four histone H4-like proteins in the 10-to 14-kD size range. A large fragment of histone H4 is extremely well conserved among many eukaryotes, varying little in monocotyledons and dicotyledons, green algae, insects, and mammals. The identified peptides of the four proteins all corresponded to this conserved region. Two chloroplast proteins that are electrophoretically similar to histones H2A, H2B, and H3 of pea cell nuclei were part of the chloroplast nucleoids. The amino acid composition of these proteins demonstrates high similarity with the HU proteins of *E. coli* (Yurina et al., 1995), but the identified peptides of the histone-like proteins in this study did not match those with HU proteins. In total, ~ 30 polypeptides from 94 to 12 kD in size were detected in nucleoids. These nucleoids have been found on the thylakoid surface (Liu and Rose, 1992) and chloroplast inner envelopes (e.g., Sato et al., 1993). The histone-like proteins were found exclusively in two-dimensional electrophoresis maps of the peripheral proteins, which would be agreement with a fairly tight binding (i.e., sonication resistant) of plastid DNA to the thylakoid surface. An alternative explanation for these histone-like proteins is that a minor contamination with nuclei has occurred. To avoid such a contamination the intact chloroplasts purified on linear Percoll gradients, in which nuclei are known to sediment to the bottom, whereas intact chloroplasts can be collected between ~50 to 70% Percoll (Galagher and Ellis, 1982).

The finding of a known chloroplast-localized RNA binding protein points to the role of the thylakoid as a surface for transcription and translation, in agreement with observations of membrane-associated polysomes (Hattorie and Margulies, 1986; Jagendorf and Michaels, 1990) and plastid DNA.

One enzyme involved in fatty acid biosynthesis was identified and was the only protein classified under metabolism. It is indeed well known that fatty acid and lipid metabolism takes place both in the plastid as well as in the endoplasmic reticulum (Miquel and Browse, 1992). Further studies are required to determine where in the chloroplast these enzymes are localized or not.

Approximately 12% of the proteins were classified under protein destioation and storage. They include the protease DegP, the assembly factor Hcf136, Ferritin, three chaperones, and a *cis-trans* isomerase that possesses a lumenal transit peptide. Ferritin concentrates and stores cellular iron to ~10^{11} times the solubility of the free ion and earlier had been determined to be a plastid protein located at the stromal side of thylakoid membranes (e.g., Waldo et al., 1995). Ferritin accumulation is positively correlated to iron loading of the plant, is regulated at a transcriptional level (Gaymard et al., 1996), and is under developmental control (Lobreaux and Briat, 1991). The FKBP isomerase is only the second isomerase identified in the thylakoid; the other one was named TLP40 (Fulgosi et al., 1998).

The presence of fairly high amounts of the chaperones Hsp70, Cpn60, and Cpn21 on the lumen map but not on the peripheral map is intriguing. These ATP-dependent chaperones have been demonstrated to be functionally important for protein import across the envelope and for assembly of protein complexes (e.g., Lubeck et al., 1997; Keegstra and Cline, 19990. The MALDI-TOF or Edman sequencing data matched best to the genes for stromal chaperones.

The conclusion is that either (1) the genes for the lumneal chaperones have not been sequenced or (2) the gene products have a dual location due either to dual targeting to the stroma and lumen or to alternative splicing, as observed for stromal and thylakoid ascorbate peroxidases (Mano et al., 1997; Yoshimura et al., 1999). ChloroP predicts alternate stromal cleavage sites for the three chaperones if the program is provided with only the N-terminal sequences truncated before the experimentally determined (Hsp70) or postulated N terminus of the mature protein (Cpn60 and Cpn21; prediction for alternate cleavage site). This often is not the case for other substrates (data not shown). When these alternatively cleaved precursors then are analyzed by SignalP, lumenal cleavage sites are detected within the first 100 amino acid residues of the precursor proteins. Thus, if the actual stromal transit peptide is cleaved more toward the N terminus, it might be possible that these chaperones are targeted to the lumen. At this point, note that different stromal-processing peptidases may exist (Su and Boshchetti, 1994; Koussevitzky et al., 1998) and that a similar alternative cleavage behavior and dual targeting have been proposed for polyphenol oxidase (Koussevitzky et al., 1998).

A third possible explanation for the concentration of chaperones on the lumen map is that the proteins are bound to the stromal side of the thylakoid membrane. They might be released together with lumenal proteins upon sonication, just like $CF_1\alpha$ and $CF_1\beta$. It is likely that the stromal chaperones are involved in protein assembly at the thylakoid surface. Schlichter and Soll (1996) searched for lumenal chaperones and protease-treated thylakoid before release of the lumenal content; they found shorter isoforms of the stromal chaperones, based on immunodetection with antisera generated against the stromal chaperones. Their results might be explained by only partial digestion of the stromal chaperones when they are assembled in membrane-bound complexes at the thylakoid surface. Clearly, additional experiments are required to draw any final conclusions.

Consensus Sequences and Prediction of Localization and Cleavage Sites

The neural network program ChloroP was most successful in localizing the newly discovered chloroplast proteins, whereas PSORT assigned 52% of the identified proteins to the chloroplast. What does this mean for the use of these programs to confirm localization of chloroplast proteins? If the proteins indeed are localized in the chloroplast, ChloroP recognizes ~94% of them, whereas PSORT probably identifies ~52%. ChloroP was trained with a positive test set of 75 known transit peptide containing chloroplast proteins (excluding lumenal proteins) and with a negative test set (75 proteins from other nonchloroplast localizations; Emanuelsson et al., 1999). When the authors tested 715 *Arabidopsis* entries in the SWISS-Prot database, they observed that 96% of those annotated in the database as being chloroplast localized were indeed predicted to be the chloroplast. By contrast, 11% of proteins annotated as nonchloroplast proteins were predicted to be in the chloroplast, which would result in ~ 2500 false positive for the full genome. In addition, the program is somewhat based toward the known chloroplast proteins, as exemplified by the 100% score for the 12 well-known photosynthetic proteins on our maps. The PSORT program is much more ambitious in that it predicts proteins to 17 different locations (three within the chloroplast), in the plant cell. Judging from information available on the PSORT the program has not been as rigorously tested as ChloroP has been (Nakai and Horton, 1999).

It is also possible that PSORT is more conservative in its prediction of proteins to be localized in the chloroplast. Testing newly identified proteins by both programs is probably the best option at the moment to get additional support/confirmation for chloroplast localization. It has been noted that in many cases, an additional amino acid residue needs to be removed to obtain the mature protein sequence (Richter and Lamppa, 1998; Emanuelsson et al., 1999).

Protein Stability and Degradation

Steady-state levels of different chloroplast proteins represent a balance between their synthesis and assembly on the one hand, and their degradation on the other. Most chloroplast proteins are very stable once they have been properly assembled and there is little turnover of most photosynthesis proteins. For example, there is little or no degradation of Rubisco in barley leaves placed in darkness for 7 days. Massive protein degradation is implicated in the transition of plastids from one phase to another, as well as in the final deterioration of chloroplast during senescence. But, proteolysis is also important in maintaining a functional organelle under both optimal and stress conditions. Thus, protein degradation processes are important in determining the overall function of the organelle.

The expression of chloroplast protein complexes often requires coordination between the organellar and the nuclear genomes. How this "cross-talk" is achieved is largely unknown but it is thought that non-stoichiometric amounts of multi-subunit protein complex components are proteolytically removed. Schmidt and Mishkind (1983) demonstrated that when the synthesis of chloroplast proteins in chlamydomonas, including that of the large subunit of Rubisco (LSU), is inhibited by chloramphenicol, the nuclear-encoded small subunit (SSU) is synthesized and imported into the plastid but rapidly degraded. This degradation is carried out by a constitutive mechanism probably by a nuclear-encoded protease, since degradation also occurs in cells deficient in chloroplast ribosome. Degradation of subunits of another, chloroplast protein complex was observed in a Lemna mutant. In this mutant the translationally active levels of Rieske Fe-S protein mRNA were greatly reduced as compared to those of other components of the Cytb6f complex. These subunits were expressed but their products failed to accumulate due to increased rate of degradation. Another example of rapid degradation of unassembled subunit is that of the chloroplast-encoded 48 and 34.5 KDa polypeptides of PSII due to nuclear mutation. In the case of the PsaD subunit of PSI, when the precursor protein is bound to the core complex, it is sensitive to exogenous protease treatment. Only after the precursor is processed to its mature form by the stromal peptidase does it become resistant to this treatment.

The structural complexity of the chloroplast dictates active sorting of proteins to the different compartments during biogenesis. Since the mechanisms for protein sorting in the chloroplast are complex, they are prone to failures and errors, which may result in the mistargeting of proteins. A mistargeted protein is assumed to undergo degradation. This was recently demonstrated when the intentional mistargeting of OEE33, a component of the lumenal oxygen-evolving enzyme, to the stroma resulted in its rapid degradation. This degradation is mediated by an ATP-dependent, serine-type, stromal protease. These characteristics are similar to those of c/p protease, which is found constitutively in the stroma. This suggests the involvement of c/p protease in the degradation of this mistargeted protein, although its direct role in this process still needs to be demonstrated.

It has been demonstrated that the D1 protein of the PSII reaction center is rapidly degraded under photoinhibitory conditions. Similar light-mediated turnover of D_2 and CP43 proteins of PSII was also observed. Degradation of early light-inducible protein (ELIP), which is structurally related to LHC proteins, was also observed. This integral thylakoid-membrane protein is stable under high-light conditions but degrades rapidly upon transition to low light. Recently, the specific degradation of ELIP was shown to be mediated by a serine-type protease which is peripherally bound to the thylakoid membrane, is independent of ATP and has an estimated molecular mass of 65 KDa.

Environmental effects on soluble chloroplast proteins have also been demonstrated. Rubisco, subjected to oxidative stress either *in vitro* or *in vivo*, lost its activity and became susceptible to either exogenous or endogenous proteases respectively. Inactivation of Rubisco is thought to be mediated by intermolecular cross-linking involving Cys-247 of LSU. Accordingly, treatment of the inactive enzyme with DTT can partially restore its activity and resistance to proteolysis. The cross-linked protein was found to be associated with chloroplast membranes which may be indicative of a major conformational change that leads to increased sensitivity to proteases. Oxygen and hydroxyl radicals have also been shown to be involved in initiating the degradation of a number of unidentified proteins in isolated chloroplast.

Proteolytic degradation of a protein often requires not only an active proteases but also a signal within the target protein itself. Such a signal can be either internal, within the amino acid sequence of the target protein or added post-translationally to the protein. The best characterized mechanism for signaling proteins for degradation is that of the ubiquitin system which is highly conserved among all eukaryotes. However, the existence of the ubiquitin system in chloroplast is unlikely because chloroplast proteins do not react specifically with anti-ubiquitin antibodies. Internal signals have been suggested to target proteins for rapid degradation. The PEST hypothesis correlates the existence of regions enriched in Pro, Glu, Ser and Thr with short half-lives of proteins. The rapidly degraded D1 protein of PSII contains a PEST region in one of its stromal loops and this region was thought to be involved in the rapid turnover of this protein. However, deletions or substitutions in this region did not stabilize the protein. This is consistent with the inconclusiveness of the PEST hypothesis, while PEST regions do play a destabilizing role in some proteins; they have no effect on stability of others.

Another signal for rapid degradation is known as the "N- end rule" which relates the metabolic stability of a protein to the identity of its N-terminal amino acid residue. Although in eukaryotes the degradation of "N-end" substrates is mediated by the ubiquitin system, the N-end rule has been shown to operate in bacteria as well where the ubiquitin system does not exist. In E. coli the degradation component of the N-end rule pathway is mediated by c/p protease which has a homologue in chloroplast. This observation makes it tempting to speculate that the N-end rule operates in chloroplast as well.

Another protein-tagging system has been recently discovered in bacteria. This system is responsible for the degradation of proteins synthesized from incomplete mRNA in E. coli. In the absence of a stop codon on the mRNA, the ribosome reads through into the ssrA RNA, resulting in the addition of a peptide tag at the C-terminus of the protein. This tagged protein is then degraded by a C-terminus-specific protease. Whereas, the ssrA RNA is absent from the chloroplast genome, another component of the system does exist in chloroplast. The protease involved in the degradation of the tagged protein, the Tsp protease, is homologous to the processing peptidase of D1 protein-and the Tsp-recognizable sequence LAA is encoded by D1's own mRNA. Whether the homology is limited to the processing of D1 or can be extended to the complete degradation of proteins in the thylakoid lumen by the C-terminal-tag-recognizing protease needs to be explored.

GENOMICS

The term functional genomics can referred to as development and application of global experimental approaches to assess gene function by making use of the information and reagents provided by structural genomics (Hieter and Boguski, 1997). It involves the use of high-throughput methods for the study of large numbers of genes in parallel. Having identified a new sequence, the comparison with sequence databases is the simplest way to obtain functional information. EST (expressed sequence tag) information is present in public databases for a variety of species including a number of plants. A recent release of the EST database of the National Centre for Biotechnology Information lists more than, 39,000 Arabidopsis ESTs. Powerful programmes for rapid sequence comparisons with databases are accessible via the Internet. SAGE (serial analysis of gene expression) is extension of EST sequencing. An inventory of transcripts is established based on very short cDNA sequence tags from the 3′ part of mRNA molecules, concatenated and cloned before sequencing. These short tags have proven to identify corresponding genes in databases. An alternative technology is referred to as DNA chips. It is based on a method to synthesize large amounts of different oligonucleotides in situ on a glass support. Oligonueleotides corresponding to genes of interest are hybridized with fluorescent probes and signals are detected using techniques similar to ones used for the microarrays.

The complete sequence of the tobacco plastid DNA and the identification of gene by numerous investigators makes it possible to estimate that the higher plant plastid genome has the coding capacity for 123 genes (Shinozaki et al., 1986). Following table shows genes coded by plastid DNA. All these genes fall into 3 categories (refer to chloroplast DNA).

The plastid gene is identified for decoding of the plastid genome or -encode proteins involved in photosynthesis. Clearly, these two groups of genes are coupled, because to express genes involved in photosynthesis the plastids transcription and translation apparatus must be synthesized. This later process is no small undertaking which is shown by the fact that mesophyll leaf cells accumulate between 10^6 and 10^7 plastids ribosomes. Plastid ribosomes and mRNAs are present at much reduced levels in non-photosynthetic plastids such as proplastids, chromoplasts or amyloplasts.

TABLE 3.1: Chloroplast genes

I. Genes required to decode plastid DNA transcription		
	(Genes for the genetic system)	(Gene product)
	transcriptions rpo, A.B.C.,	B,B[1] subunits of RNA
Polymerase	Translation r DNA	16S,23S,4.5S,5S Ribosomal RNA
	trn (30 genes)	t RNAs
	rps (12 genes)	30 S ribosomal protein subunits
	rpl (7 genes)	50 S ribosomal protein subunits
	infA	initiation factor 1,
Putative secretory factor	tufA	elongation factor Tu (in algae)
II Genes involved in photosynthesis		
	RbcL	large subunit of Rubisco
	psaA, psaB	80-85 kD PSI chlorophyll
Apoproteins	psaC	9kD PSI Fe S protein
	psbA, psbD	32-34 kD PSII reaction centre
	chlorophyll apoproteins	
	psbB, psbC	47 kD, 43 kD PSII chlorophyll
Apoproteins	psbE – psbF	Cytochrome 559 (PSII)
	psbH	10 kD PSII phosphoprotein
	atpA, B,E,F,H,I	ATP synthetase subunits
	ndh A,B,C,D,E,F,G,H,I,J,K	NADH dehydrogenase
	18>100 codons	unidentified
	'18<100 codons	unidentified

TABLE 3. 2: Genes for the photosynthetic apparatus

Genes	Products	Tobacco	Pea	*Chlamydomonas*	Euglena
Ribulose bisphosphate carboxylase (Rubisco)					
rbcL	Rubisco, large subunit	+	+	+	+
rbcS	Rubisco, small subunit	-	-	-	-
Photosystem I complex (PS I)					
psaA	PS I, p 700apoprotein A1	+	+	*	.
psaB	PS I, p700 apoprotein A2	+	+	+	(+)
psaC	PS 1, 9 kDa polypeptide	+	+		
Photosystem II complex (PS II)					
psbA	PS II, D 1 protein	+	+	**	.
psbB	PS II, 47 kDa polypeptide	+	+	(+)	(+)
psbC	PS II, 43 kDa polypeptide	+	+	(+)	.
psbD	PS II, D2 protein	+	+	+	
psbE	PS II, cytochrome b_{559} (8 kDa)	+	+		

Genes	Products	Tobacco	Pea	*Chlamydomonas*	Euglena
psbF	PS II, cytochrome b_{559} (4 kDA)	+	+		
psbG	PS II, G protein	+			
psbH	PS II, 10 kDa phosphoprotein	+	+		
Cytochrome b/f complex					
petA	b/f complex, cytochrome f	+	+		
petB	b/f complex, cytochrome b_6	*	*		
petD	b/f complex, subunit IV	*	*		
ATP synthetase complex					
atpA	H^+-ATPase, subunit α	+	+	+(+)	
atpB	H^+-ATPase, subunit β	+	+	+	(+)
atpE	H^+-ATPase, subunit e	+	+	(+)	
atpF	H^+-ATPase, subunit I	.	.	(+)	
atpH	H^+-ATPase, subunit III	+	+	(+)	+
atpI	H^+-ATPase, subunit IV	+	+		
Putative NADH dehydrogenase components					
ndhA	NADH dehydrogenase, ND1	*			
ndhB	NADH dehydrogenase, ND2	**			
ndhC	NADH dehydrogenase, ND3	+			
ndhD	NADH dehydrogenase, ND4	+	+		
ndhE	NADH dehydrogenase, ND4L	+			
ndhF	NADH dehydrogenase, ND5	+			

+, continuous genes present; ., split gene present; -, gene absent; (+), genes detected by heterologous hybridization. Numbers of + and . symbols indicate copies per genome.

Furthermore, functions carried out by these non-photosynthetic plastids (plastid replication, carotenoid-biosynthesis, starch accumulation) are not known to require plastid gene expression and most likely involve primarily nuclear genes. The localization of these genes in the nucleus means that differentiation of specialized photosynthetic plastids does not require concomitant build up of plastid transcription and translation capacity. The plastid compartment in non-photosynthetic cells can therefore functions as other non-DNA containing organelles such as glyoxysomes or microbodies. Perhaps the advantages of this situations contributed to the present distribution of genes in the nucleus and plastid.

It has been suggested that some genes have been retained in the plastid because the proteins they encode would be difficult to transport through the chloroplast envelope. However, this idea, although genius, does not explain why some ribosomal proteins remain encoded by plastid DNA but not others of similar size and solubility. Further it has been shown that when psb A, a plastid gene, is moved to the nucleus of a transgenic plant and provided with appropriate expression signals, it will direct a functional psb A gene product into plastids (Cheung et.al., 1988). The data suggest that it is possible to move plastid genes which are migrated to the nucleus without loss of function. Further support for this idea comes from the observation that although EF-Tu is encoded by a plastid gene in *Euglena*, in higher plants it is a nuclear gene. Similarly, rpoA is a nuclear gene in *Geranium* but a plastid gene in other higher plants. This suggests that gene transfer is still occurring, although slowly, possibly because gene transfer disrupts complex regulatory circuits which now coordinate plastid and nuclear gene expression.

DECODING

Transcription of Plastid Gene

The genetic system in the chloroplast is essentially prokaryotic in nature, which has given rise to much discussion as to the possible prokaryotic origin of plastids. Control sequences at the 5′ end of the gene which are involved in initiating transcription are very similar to the *pribnow* box and the '-35′ region characteristic of bacterial genes. In addition, chloroplast genes do not usually contain intervening sequences characteristic of many eukaryotic genes, although some introns have been found, notably in tRNA genes of tobacco and maize. The mRNA produced from chloroplast genes is not usually polyadenylated, although short sequences of upto 20 residues have been reported, and, of course, no transport is required as the mRNA is produced in the same compartment as the ribosome on which it will be translated. There is no evidence that chloroplast mRNA is transported out of the chloroplasts and translated on cytoplasmic ribosomes.

When *E. coli* is transformed with the plastid genes rbcL and psbA, transcription is initiated in *E.coli* at the site similar or identical to one that used by the chloroplast RNA polymerase. This shows that DNA sequences elements which direct transcription initiation of plastid genes (TTGACA and TATTAT) similar to prokaryotic promoter elements. Furthermore, as in bacteria, these promoter elements precede rRNA, tRNA and protein coding genes.

Transcription in the chloroplast resembles that of prokaryotes, particularly in the use of consensus promoter elements. Plastid genomes are transcribed by two different RNA polymerases; a nuclear encoded RNA polymerase (NEP) that transcribes genes required for transcription and translation, and a plastid encoded RNA polymerase (PEP) that transcribes the photosynthetic genes (Hess and Borner, 1999). Use of ribosome deficient mutants, such as *iojap* (Hubschmann and Borner, 1998) and *in vitro* transcription system enabled systematic dissection of the NEP promoters and demonstrated that a single short sequence element (a CRT motif) is enough for transcriptional initiation. Transcription from PEP promoter requires both a –10 and a –35 elements (Weihe and Borner, 1999) which are similar to elements found in the same position of bacterial genes.

Transcription of both polymerases (PEP and NEP) may be enhanced by σ-like factors and by gene specific DNA binding proteins acting as activators or repressors. Two nuclear encoded σ-factor have been identified for PEP in the red alga *Cyanidium* (sigB and sigC) and in maize (sig1 and sig2). A putative σ-like factor, Os-sigA, was isolated from rice (Tozawa et al., 1998) and shown to increase in abundance during light growth, suggesting that this inducible σ-factor may contribute to light dependent transcriptional regulation of plastid genes. Although the correlation between transcription of chloroplast genes and protein expression is generally poor, a recently identified nuclear gene encoding the plastid ribosomal protein (RPL4) has been shown to co-purify with plastid RNA polymerase and transcription factor CDF2, suggesting a possible role in coordinating plastid transcription and translation (Trifa et al., 1998).

Transcription termination of plastid genes has not been studied extensively. In *E. coli,* factor independent transcription terminators typically contain a G-C rich region of dyad symmetry followed by an AT rich stretch that often contains a run of thymidines. *In vitro* experiments indicate that DNA sequences with dyad symmetry located at the 3 end of rbcL do not cause transcription termination. Transcription was terminated with low efficiency by 3′ regions of spinach psbA, rrnB and pet D. In contrast, trn S and trn H caused termination with high efficiency (80%) although these genes are not followed by regions of dyad symmetry. These results suggest that the stem loop structures found at RNA 3′ ends may not play a significant role in factor independent transcription termination.

Chloroplast of higher plants contain gyrase and topoisomerase I activity which could alter the superhelicity of plastid DNA *in vivo.* Chloroplast ribosomes also resemble those from prokaryotes, being 70S particle consisting of 30S and 50S subunits, and ribosomes from *E. coli* and plant chloroplasts are immunologically similar. The tRNA chloroplast population is distinctly different from that in the cytoplasm, as are the aminoacylating enzymes. The prokaryotic nature of the protein synthetic machinery is also illustrated by its sensitivity to inhibitors of prokaryotic gene expression such as chloramphenicol. This antibiotic does not inhibit eukaryotic protein synthesis.

RNA Processing

Many chloroplast genes are transcribed as polycistronic mRNAs and thus require intra and inter-molecular splicing and processing to form mature transcripts. Introns belonging to both group I, where the splicing is initiated by an activated G residue that attacks and breaks the phosphodiester bond at the 5′ splice site, and group II, where the splicing is initiated by a reactive A residue in the intron sequence that attacks the 5′ splice site, forming a lariat intermediate, have been characterized in plastid genes. Although little is known about the splicing machinery of the plastid introns, nuclear mutants such as ac20, crs1 and crs2 (Jenkins et al., 1999) have indicated an essential or direct role for nuclear factors in mRNA splicing.

It has been shown that RNA editing is involved in producing a functional mRNA. The high specificity of RNA editing relies on cis-acting and transacting factors, some of which are nuclear encoded. Inhibition of chloroplast translation blocks RNA editing, suggesting that chloroplast translation products serve as auxiliary factors, perhaps mediating accessibility of the substrate site during editing (Karcher and Bock, 1998).

The 5′ untranslated region (UTR) of psbA has been shown to be processed by the removed of 54 nucleotides including a stem-loop structure. Chloroplast or nuclear mutations blocking psbA translation reveal a correlation between processing and ribosome association. Loss of the ribosome binding site by mutation in the 5′UTR of the message or mutations blocking ribosome association both result in the absence of mRNA processing and translation of the psbA mRNA. These date suggest that 5′mRNA processing of this mRNA may be a consequence of translation but not necessarily a prerequisite for it.

mRNA Stability

Many plastid RNAs have inverted repeats (IR) in the 3′ UTR that are capable of forming hairpin structures required for correct 3′ end formation. The IR plays some role in stabilizing the transcription. Nuclear encoded proteins binding to this 3′ UTR have been identified. Rott et al. (1998) suggest that the binding of these proteins to the 3′ UTR may influence the efficiency of endonucleolytic cleavage or the exonucleolytic trimming in proper 3′ end formation. Some of these 3′ UTR binding proteins also interact with the 5′ UTR or 5′ UTR binding proteins to control RNA degradation, suggesting a connection between RNA processing, stability and degradation. The results suggest that the nuclear gene product, MCD, protects RNA from degradation by interacting with the 5′ UTR (Drager et al., 1998). MCD may provide a specific mechanism protecting the mRNA or it may have another primary role in mRNA translation or processing. It is not yet clear which of these roles MCD plays in influencing the stability of the transcript.

Translation

Many factors required for translation of chloroplast mRNAs are nuclear encoded. Genetic analysis of *Chlamydomonas* has revealed a class of nuclear genes that are required for translation of specific chloroplast mRNAs. Many of these factors directly interact with the 5′ UTR of specific mRNAs. The 5′ UTR of chloroplast mRNAs tend to be A/U-rich and often contain predicted stem-loop structures. Sequences upstream of the chloroplast initiation codon have been proposed to affect translation containing binding sites for activators and repressors.

Studies on photosynthetic mutants of *Chlamydomonas* suggest that some of the nuclear encoded factors are message specific while others may be class specific. Using complementation rescue the gene disrupted in the mutant ac115 has been characterized and shown to encode a novel protein that has been proposed to play a role in stabilizing intermediates of the D2 translation product. Another novel nuclear gene, crp1, the disruption of which blocks translation and mRNA processing of the petA mRNAs in maize has been cloned. This protein shows similarity to yeast proteins involved in translation of mitochondrial mRNAs. The yeast protein is required for association of the mRNA with membrane bound polysomes. Many proteins that bind to the 5′ UTRs of chloroplast transcripts like psbA, psbC and psbD have been shown to be at least partially membrane associated indicating a potential role in the proper localization of these messages to the thylakoid membrane. Of the set of proteins that associate with the psbC mRNA, the 46 KDa protein has been shown to bind an A/U rich region in the 5′UTR. The binding of this 46 KDa protein is light-dependent and can be inhibited by ADP (Zerges and Rochaix, 1998).

A complex of protein has been identified that bind to the 5′UTR of the psbA mRNA. Of these proteins (RB60, RB55, RB47 and RB38), the 47 KDa protein shows homology to poly(A) binding proteins (PABP). Although chloroplast transcripts generally lack poly (A) tails, the psbA mRNA contains a A- rich 5′ UTR that has been identified as the binding site of the chloroplast poly (A) binding protein (cPABP). Identification of two nuclear mutants deficient in psbA translation that also lack the PABP (RB47) suggest that this protein is necessary for translation of the psbA transcript. Modulation of complex binding to the psbA mRNA mediated through cPABP binding shows that ADP-dependent kinase or oxidizing conditions can abolish binding of the complex. Redox potential generated by photosynthetic activity has been proposed to regulate this complex binding through a chloroplast localized protein disulfide isomerase (cPDI). On the basis of the above data, somanchi and Mayfield (1999) suggested a model where chloroplast redox potential regulates binding of message-specific translational activator proteins including RB47, using redox potential generated by photosynthesis and transduced through the cPSI.

Autoregulation of Dytf translation has been shown to occur by an interaction between the 5′UTR of the petA mRNA and the carboxy-terminal domain of the unassembled protein. Cytf that is not incorporated into the cyt b6/f complex attenuates translation of petA mRNA by interacting with the 5′ UTR directly or indirectly. Such attenuation of subunits of a protein complex in the absence of other subunits of the complex is described as control by epistasy of synthesis (CES). The properties of CES have been suggested as way to regulate assembly of multi-subunit complexes within the chloroplast. While a mechanism for translational regulation has not yet completely emerged, the data presented above suggest that this regulation occurs by the interaction of nuclear encoded factors with the 5′UTR of chloroplast mRNA.

Sine-Dalgarno Sequence

Prokaryotic-like Shine-Dalgarno (SD) sequences have been identified in a number of chloroplast transcripts, but not in all. The 5′UTR of many chloroplast messages contain putative SD sequences but few of these have the prokaryotic location in relation to the initiation codon, which must be within ten nucleotides. Fargo et al., (1998) analyzed the function of potential SD sequences of four transcripts, atpB, atpE, rps4 and rps7, by replacement mutagenesis in *Chlamydomonas* and E. coli and showed no effect on the expression of these genes, leading them to propose a SD-independent mechanism for plastid translation. However, studies on SD-like sequences of tobacco rsp14 and *Chlamydomonas* psbA mRNA show a requirement of SD sequences for translation. In *Chlamydomonas* the putative SD sequences in the psbA mRNA is 27 nucleotides upstream of the initiation codon and deletion of this sequence results in a loss of ribosome association and also in a decrease in message stability. This strongly suggests an authentic SD sequence. The existing evidence supports the presence of SD sequences, with non-prokaryotic spacing, for some chloroplast messages,, in others translation initiation may result from mechanism independent of SD sequences. However, these results present evidence for translation initiation of some chloroplast messages from SD sequences with non-prokaryotic spacing, while others show translation initiation independent of SD sequences.

REVERSE GENETICS

A major breakthrough in plastid research occurred in 1998 when Boynton et al. established a biolistic chloroplast transformation method for *C. reinhardtii*. Once the DNA-coated particles have penetrated into the chloroplasts, the transforming DNA is released and becomes incorporated into the plastid genome by homologous recombination. Transformants can be selected in several ways. The most versatile chloroplast selectable maker is the bacterial aadA (amino-glucoside adenyl transferase) gene, which has been engineered into a chloroplast expression cassette and confers resistance to spectinomycin and streptomycin. Because of the homologous plastid recombination system, this cassette allows specific gene disruption and site-directed mutagenesis to be performed on any chloroplast genes. Chloroplast transformation has also been achieved with tobacco plants using similar approaches.

Disruption of chloroplast genes whose function is not essential for cell viability leads to a homoplasmic state in which all plastid gene copies are inactivated. This can occur for genes involved in photosynthesis because photosynthetic function is dispensable in *C. reinhardtii* when the cells are grown with acetate as the carbon source or in tobacco when sucrose is added to the growth medium. Homoplasmic gene disruptions can also be achieved for other dispensable plastid genes that may not necessary have a role in photosynthesis. In contrast, inactivation of an essential plastid gene leads to a heteroplasmic state in which mutant and wild-type gene copies coexist as long as the selective pressure for the disrupted gene is maintained. In addition to gene disruption, it is possible to perform site-directed mutagenesis of chloroplast genes. In this case, the aadA cassette is inserted at a site that is not essential for chloroplast function and is close to the target gene. The construct is introduced into the chloroplast genome via biolistic transformation by selecting for spectinomycin resistance. This reverse genetic approach is a powerful tool for analyzing the function of chloroplast encoded subunits of photosynthetic complexes and for identifying new subunits of these complexes.

Molecular analysis has led to identification of novel functions controlled by chloroplast-encoded genes. Some of these functions are involved with photosynthesis and others essential for cell viability. In the latter case, two genes, clpP and ycf1-like ORF1995 of C. reinhardtii, are notable. Although the role of the ATP-dependent clpP proteases has been studied extensively in *E. coli*, its function in plastids remains to be explored. The function of the membrane-associated Ycf 1 protein is also completely unknown.

Plastid transformation has not only been important for elucidating the function of genes, it has also greatly helped the analysis of the molecular mechanisms underlying chloroplast gene expression. The nuclear control of plastid gene expression is well documented but the influence of the plastid on nuclear gene activity has also been investigated. The nature of the chloroplast signal involved in this process is unknown, but as the number of plastid open reading frames of unknown function is rapidly shrinking, it should soon be possible to determine whether any of them is involved in this signal transduction.

PHOTOSYNTHETIC PIGMENTS

At global level, it has been estimated that each year the total chlorophyll production across the planet exceed 10^9 tonnes with 75 per cent from terrestrial plants, the reminder from aquatic organisms, largely marine phytoplankton. Production of carotenoids, the second most abundant plant pigment, is probably of the order of about 2×10^8 tonnes (Hendry, 1999). The distribution of the various chlorophylls and bacteriochlorophylls is shown in following table:

TABLE 3. 3: The distribution of chlorophylls and bacteriochlorophylls

Pigment	Organism	Light absorbed
Chlorophyll *a*	All plants, algae and the photosynthetic bacteria Cyanophyta and Prochlorophyta	Red and Blue-violet
Chlorophyll *b*	All plants, the algal Chlorophyta and Euglenophyta and bacterial Prochlorophyta	Red and Blue violet
Chlorophyll *c*	Algal Phaeophyta (brown algae), Pyrrophyta (dinoflagellates), Bacillariophyta (diatoms), Chrysophyta, Prasinophyta, Cryptophyta	Red and Blue violet
Chlorophyll *d*	Some Rhodophyta (red algae) and Chrysophyta	Red and Blue-violet
Chlorophyll *e*	Algal Xanthophyta	Near-red and Blue-violet
Bacteriochlorophyll *a* and *b*	Purple bacteria including Chromatiaceae and Rhodospirillaceae	Near-red and Blue-violet
Bacteriochlorophyll *c*, *d* and *e*	Green and brown sulphur bacteria – Chlorobiaceae and Chloropflexaceae	
Bacterioviridin	Green sulphur bacteria	Near-red and Blue-violet
Phycocyanin	Blue-green, red algae	Orange red
Phycoerythrin	Red, blue-green algae	Green
Carotenoids	Most plants, bacteria	Blue, Blue-green

TABLE 3. 4: Chemical composition of intact spinach chloroplasts

Component	Percentage Chloroplast (Dry Weight)
Total Protein	60
Water insoluble protein	27
Water soluble protein	33
Total lipids	20
Chlorophylls	4.3
Carotenoids	0.9
Colourless lipids	14.8
RNA	1.7-3.5
DNA	?
Free Amino acids	2.5
Orthophosphatase	0.7
Ions (K^+, Na^+, Mg^{2+}, Ca^{2+}, (Cl^-)*	3.6
Polysaccharides**	0.6
Hexosamine	0.2
Starch	variable

*Data of *Beta vulgaris* chloroplast, ** Data of *Allium porrum* chloroplast.

TABLE 3. 5: Lipid composition of spinach chloroplasts (based on 100 mol chlorophyll)

Lipid	Mol.
Chlorophylls	100
Chlorophyll *a*	70
Chlorophyll *b*	30
Carotenoids	39
ß-carotene	9
Lutin	14
Viloxanthin	10
Neoxanthin	4
Antheraxanthin	2
Quniones	11
Plastoquinone	4
Lipid	Mol.
α-tocopherol	0.5
Vitamin K_1	0.5
Tocoquinone	0.5
Glycolipids	447
Monoglactosyldiglyceride	270
Digalactosydiglyceride	154
Sulfolipid	23
Phospholipids	43
Phosphatidylglycerine	29
Phosphatididylethanolamine	2
Phosphatidylinosite	2

Chlorophyll a is found in all photosynthetic eukaryotes. The restricted distribution of the other chlorophylls (b, c, d, or e) may have evolutionary significance.

Most biological pigments are relatively large molecules with, typically, molecular masses ranging from about 200 (quinines), 300 (flavonoids), 400 (betalains) and 500 (carotenoid) to 800 (chlorophylls). Most pigments are composed, at least in part of unsaturated hydrocarbons. Many are ring (cyclic) structures. Of the 17 or so elements found in pigments of biological materials, just four : H, C, N and O predominate.

Properties of Chlorophyll Molecule

The empirical formula of the chlorophyll molecule is $C_{55}H_{72}O_5N_4Mg$. The molecule is made up of five small rings, together forming a larger ring. In the centre of the large ring there is a magnesium atom. Together they form a flat square measuring 1.5 × 12.5 nm. The square is equipped with a long *tail* of carbon atoms which can be thought to act as a handle or fixture which keeps the molecule properly positioned in the plant.

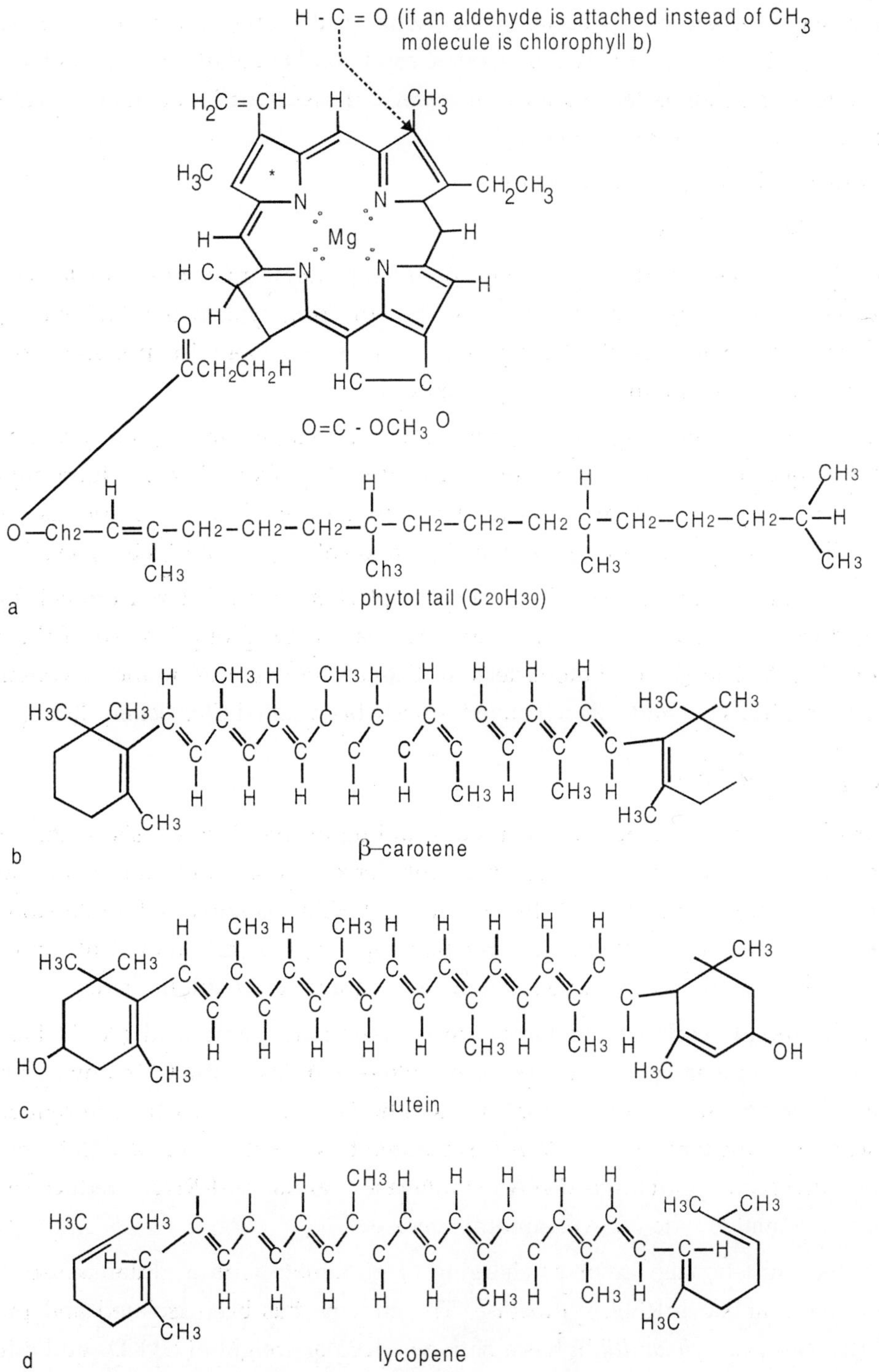

FIG. 3. 3: *Structures of some chlorophyll and carotenoids pigments.*

There are many double bonds between the atoms in the ring system. It is possible to circle the large ring in such a way that every second bond encountered is a double bond. Such a system of bonds is termed a *closed conjugated system*. It is this that provides the molecule with two important properties:

- great light-absorbing power and
- great stability power.

The tail, the magnesium atom and some peripheral atoms may break or loose if the molecule is denatured with heat, acids, etc., but the ring system is intact even in old fossils such as petroleum. A similar ring system as in chlorophyll is present in our red blood pigment, haemoglobin and in the cytochromes.

In addition to the pyrrole rings, *a-b* chlorophyll has a cyclopentanone ring with a methyl ester. A phytol ester is linked by a propionic acid with pyrrole ring D. Chlorophyll *a* is having formyl group at C atom 7 instead of a methyl group. Other chlorophylls, such as chlorophyll C_1 and C_2 are found in brown algae and chlorophyll *d* in red algae.

Photoautotrophic bacteria possess bacteriochlorophyll *a* which differs from chlorophyll in two features: it is hydrated in ring B and carries a vinyl group instead of the acetyl group on ring A. The green photobacterium, *Chlorobium* has been found to contain the bacteriochlorophylls *c, d* and *e* which have formerly been called *Chlorobium* chlorophylls.

Chlorophyll Synthesis

Much progress has been made in our understanding of the biosynthesis of chlorophyll with much of the recent inputs coming from enzymology and molecular biology. Nevertheless, although the main pathway leading to Chl a is fairly well understood, our knowledge of the biosynthetic interrelationship of the different Chl and their intermediates, formation of bChl, regulation and degradation is still limited.

Most of the recent work on the formation of 5-aminolevulinic acid (ALA) has been concerned with regulatory aspects and enzymologic studies of the C5 pathway utilized in plants, algae and some bacteria for ALA synthesis destined for Chl. The sequence of events leading to the formation of ALA is separated by the subsequent action of the three enzymes involved: glutamyl-t-RNA synthetase, glutamyl-t-RNA reductase and glutamate-1-semialdehyde (GSA) aminotransferase.

The first biosynthetic step involves charging of glutamate with a glutamate-specific t-RNA by the aminoacyl-tRNA synthetase. The enzyme has been isolated and purified from *Chlamydomonas reinhardtii*. It has a native molecular weight of 60 KDa and migrates under denaturating conditions as a single band of 32 KDa, indicating an $\propto_2$ homodimer structure. The same t-RNA is utilized for protein biosynthesis and ALA formation.

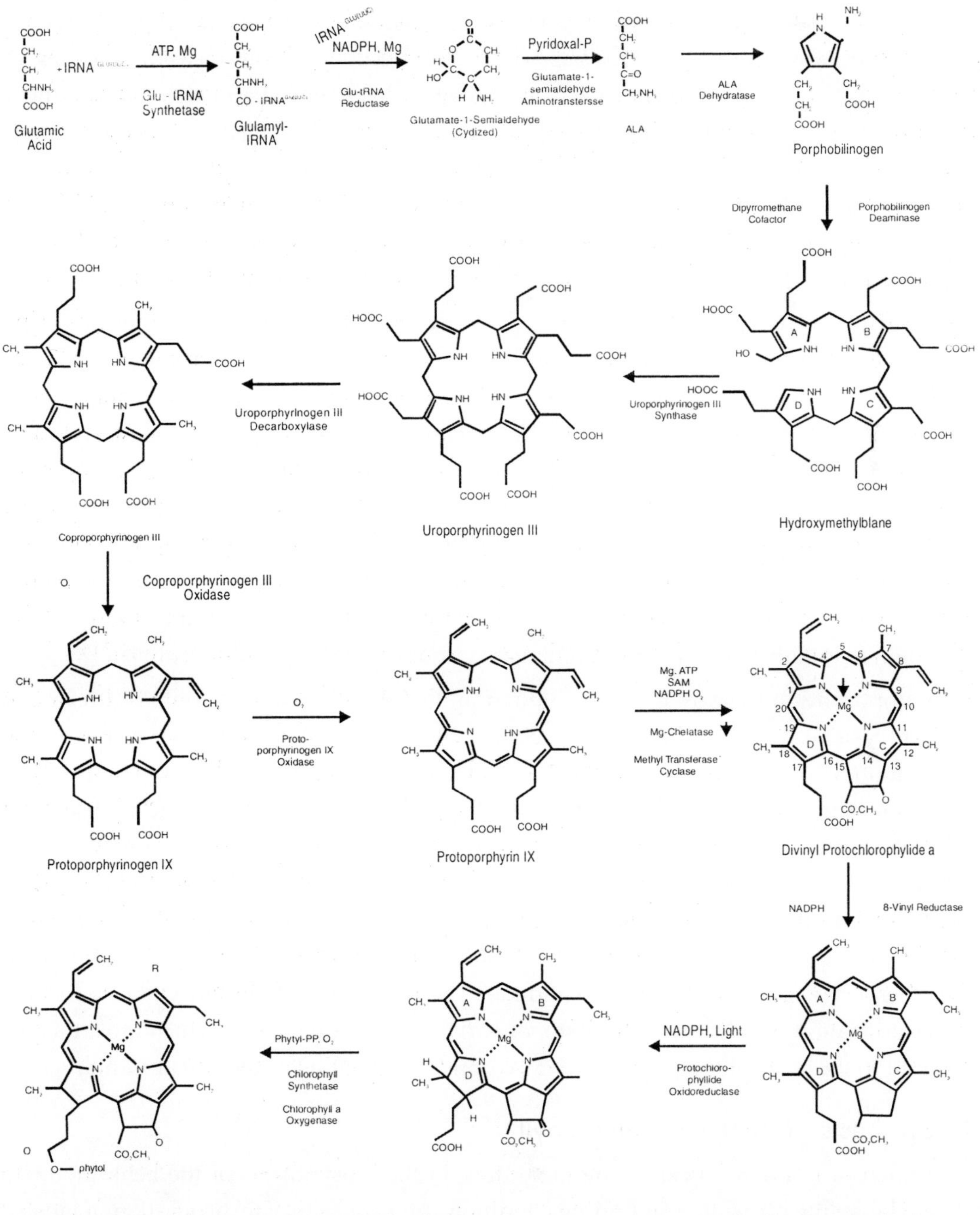

FIG. 3. 4: *The biosynthetic pathway of chlorophyll.*

The charged Glu (glumate) is next reduced to GSA by an NADPH-dependent glutamate-tRNA reductase. The enzyme was isolated from C. reinhardtii and shown to be a monomer of 130 KDa. The final step in ALA formation is the conversion of GSA by aminotransferase-type enzyme that has been isolated from barley and *Synechococcus* as 43 KDa protein. The barley enzyme has been cloned and sequenced and was found to be a dimer of identical subunits with a molecular weight of 46 KDa. The enzymes from barley and *Synechococcus* had 72% homology. The enzyme from *C. reinhardtii* was found to be a monomer of 43 KDa, which in the presence of pyridoxal-5′-phosphate converted GSA to ALA. The involvement of DOVA (4,5-dioxovaleric acid) in the biosynthesis of ALA has also been proposed for *Scenedesmus*, where it is a better precursor of ALA than GSA.

As a next step, two molecules of ALA are condensed to yield porphobilinogen. This reaction is catalyzed by ALA dehydratase. Then, four molecules of porphobilinogen are converted into hydroxymethylbilane by porphobilinogen deaminase. Subsequent reactions include ring closure and simultaneous isomerization of the acetyl and propinoyl groups at the pyrrole ring D, leading to uroporphyrinogen III. This compound is then converted into coproporphyrinogen, followed by two successive oxidations which eventually lead to protoporphyrin IX. In the first reaction, the acetic acid side chains of uroporphyrinogen III are shortened to methyl groups by decarboxylation. Then, two propionic acid side chains are trimmed into vinyl groups, followed by oxidation to establish the system of conjugated double bonds in protoporphyrin IX.

At protoporphyrin IX there is a branch point in the C5 pathway. One route is directed to the synthesis of heme and phytochromobilin whereas the other branch gives rise to the formation for Chl. Ferrochelatase and Mg-chelatase respectively catalyze the first unique reactions in the Fe and Mg branches of the C5 pathway. In the Mg branch subsequent esterification of Mg-protoporphyrin IX leads to Mg-protoporphyrin IX monomethyl ester followed by the formation of the isocyclic ring of the macrocycle. 3,8-Divinyl P chlide a as the final product of these reactions is then reduced at the vinyl group at C8 to yield 3-monovinyl P chlidea. Subsequently, the double bond between C17 and C18 in ring D of the macrocycle is reduced to yield chlide a . The final steps of chl biosynthesis include the esterification of chlide a with phytol or earlier alcohal precursors which must then be reduced to the stage of the phytyl derivative, and oxygenation of the Chla to Chlb.

Biosynthesis of Bacterial Chlorophyll

Not much progress has been made in studies on the biosynthesis of the bchl, the main reason being the difficulties in finding conditions for satisfactory incorporation of labeled precursors and the development of cell-free enzyme systems.

The first bacterial biosynthetic gene to be sequenced was the one coding for 3-desacetyl-3-hydroxyethyl bchlide dehydrogenase from *Rhodobacter capsulatus*. It was shown to be a 33 KDa protein, which is not an integral membrane protein. From the same organism the genes coding for the activity that converts proto to Mg-proto monomethyl ester, pchlide to chlide and Mg-proto monomethyl ester to pchlide were identified by site-directed mutagenesis. The ALA-synthase enzyme was characterized subsequently by cloning and sequencing of the hemA gene from *R. capsulatus*. A pps gene has been isolated from *R. sphaeroides* that suppresses both bchl and carotenoid synthesis in *R. sphaeroids* and *R. capsulatus*, suggesting a coupled regulation of bchl and carotenoid biosynthesis.

Physical Properties of Chlorophylls

The chlorophyll is a dark green waxy solid; but chlorophyll *a* is bluish black while chlorophyll *b* is dark-green solid. Both chlorophylls are soluble in organic solvents giving green solution; chlorophyll *a* melts between 117-129° C and chlorophyll *b* between 120-130° C.

Chlorophyll *b* contains one oxygen atom more and two hydrogen atoms less than the chlorophyll *a*. The difference is due to the presence of -CHO group on position-3 in place of methyl group.

Ordinary pigments preferentially absorb visible light of a certain spectral region and, upon de-excitation release the absorbed energy as heat and, possibly, fluorescence. In contrast, a photopigment can be defined as a chemical compound, capable of light absorption and subsequent conversion of excitation quanta into a form of energy to be used in some physical or chemical process. In the following description only photopigments of biological interest will be considered.

Apart from heat production and fluorescence emission, two phenomena may happen upon excitation: (1) converted energy is stored and used later on for driving energy requiring processes at a given moment, or (2) this energy is instantaneously used for chemical or physical processes. In case (2) there are three possibilities:

(a) energy conversion is used for triggering a reaction,

(b) this conversion results in the transformation of the photopigment into a catalyst and

(c) a product suitable for *use* in chemical reactions is generated by the energy conversion.

Light quantum can be absorbed only if its energy content equals the energy difference between the ground state and one of the excitation levels of the pigment. If absorption occurs, an electron is raised to the level corresponding to its energy. This electron is one of the partners of an electron pair in the ground state. The lowest level is of special interest as regards excited states. Its lifetime is usually at least 103 times longer than that the higher ones. Thus the probability for release of the energy in a form other than heat is much increased as compared with that for higher excitation levels. This phenomenon can be easily demonstrated, e.g. by exciting chlorophyll with red, yellow or blue light. In all cases the fluorescence is the same: the emission spectrum is a mirror image of the red absorption band representing the lowest excited state. This shows that the lowest excitation level can be filled in two ways; either by raising an electron from the ground state to this level, or by electrons *tumbling down* from higher energy states to the lowest excitation one. The energy difference between these levels is released by internal conversion as heat. The lowest excited state is therefore of primary interest with regard to the function of a photopigment.

So far, it is assumed that the law of spin conservation holds for the transfer of an electron from a low-energy orbit to one of higher energy. If, however, for some reason the spin of the excited electron is reversed, the triplet state, a metastable state, is formed. This has a slightly lower energy than the above-mentioned lowest excited singlet state. Owing to spin reversal the direct transition from the triplet to the ground state is forbidden, i.e. the probability of its occurrence is small. Thermal quanta may reverse the spin again and raise the electron to the lowest excited singlet state, S(1), hence it can return to the ground state. Population of the triplet state means that the lifetime of the excited molecule is lengthened considerably, and thus the probability of converting the absorbed quanta to chemical energy is enhanced. Moreover, this triplet state may function as a new ground state and can be excited to higher-energy triplet levels.

Photopigments : Functions

Upon de-excitation, conversion of the electronic energy can occur in various ways. Leaving aside internal conversion, which occurs concomitantly with the other possibilities and results in the production of heat, the following results of energy conversion can be envisaged:

Fluorescence, emitted by the return of an electron from the S_1 level to the ground state

(a) **Delayed fluorescence:** This kind of the fluorescence is emitted if an electron is raised, with concomitant reversal of its spin momentum, from the triplet into the S_1 state by thermal energy, and returns from the latter level to the ground state. In this case the emission spectrum is identical with that of fluorescence.

(b) **Phosphorescence:** This light emission occurs in the case of the forbidden transition from the triplet state to the ground state with reversal of the spin momentum. As the triplet state is of lower energy than the S_1 level, the emission consists of quanta of energy lower than those emitted when the S_1 state is emptied. Therefore, the phosphorescence spectrum is shifted to longer wavelengths as compared with that of fluorescence.

Type of Biological Photoprocess

The major biological processes in which photopigments operate are:

- **Photosynthesis.** Light energy is converted into chemical energy to be used for metabolism.
- **Vision.** Light is used for providing information for the organism.
- **Photomotive processes.** Light causes displacement in the following ways:
- **Phototaxis.** Light induces motion or change of motion of a free organism or a cellular constituent. Two kinds of phototaxis occur:
- **Topophototaxis.** The direction of motion is determined by the direction of the light beam. If the motion proceeds either towards or away from the light source the taxis is called positive and negative respectively. For animals, Kuhn distinguished three kinds of topophototaxis which will not be further considered.
- **Phobophototaxis.** The direction of motion is not determined by the direction of the light beam, but at the boundary of two light fields of different intensities-a shock reaction of the object occurs. This prevents it from crossing the boundary. If it occurs at a transition from strong to dim light, the taxis are called positive, whereas it is called negative at a transition in the reverse sense.
- **Photokinesis.** The motility of an organism is changed by light. Photokinesis is indicated as positive or negative, if the motility is increased or decreased in the light respectively.
- **Photodinesis.** The effect of light on protoplasmic streaming. Light may affect either the rate of streaming, or the amount of protoplasma taking part in it. The effect may be temporary or tonic. In case of acceleration or retardation, photodinesis is called positive or negative respectively.
- **Phototropism.** The light-induced movement of a part of the organism resulting in bending towards the light source (positive), away from it (negative), or occupation of a certain angle to the light beam (diaphototropism), or in a bending not related to the light gradient, but to the structure of the organelle (photonasty). In most cases, the latter process is due to light-induced changes in turgidity, whereas the former ones are based on photic growth reactions.
- **Photomorphogenesis.** Light also control certain formative proceses. It may do so either by itself, or in combination with endogenous rhythms, as occur in photoperiodicity. The reaction may be restricted to activation of secretary activities affecting the behavioural pattern of animals, or it may affect the entire organism as in seed germination.

Since it may not be one of the major photoprocesses, photorespiration, though it differs from dark respiration, is omitted from consideration. The same is true of photodynamic effects. Secondary photo-effects, such as light-induced changes of membrane properties are not considered either.

Photopigments : Properties

Chlorophylls

Chlorophylls are esters of dibasic acids, the chlorophyllins.

These photopigments are active in (a) photosynthesis either by collecting light quanta or by the generation of free electrons leading to the production of reducing compounds and ATP, (b) phototaxis, (c) photokinesis.

Chlorophyll a

It occurs in all green plant's its formula is $C_{55}H_{72}O_5N_4Mg$. Chlorophyll a, like other chlorophylls, contains four pyrrole nuclei (I-IV), arranged in prophyrin ring structure. In addition to these nuclei an isocyclic, or cyclopentanone, ring (V) occurs adjoining to nucleus III. A Magnesium atoms is located in the centre of the porphyrin ring. In most cases, including chlorophyll a, the esterifying alcohols are methanol and phytol.

It is emphasized that the indication of double and single bonds in the Figure does not depict the actual situation. In reality, like other conjugated bond systems, chlorophyll is a resonance hybrid. According to there are major contributions of three resonating structures in this particular case.

A key position in determining various properties of this photopigment is occupied by the central Magnesium atom. Its presence is required for the oxidative process of allomerization. This reaction occurs upon storage of alcoholic solutions of chlorophyll under air. Allomerization consumes molecular oxygen in the molar ratio 1/1 with respect to chlorophyll. The presence of the central magnesium atom is indispensible for this process. The oxidation occurs in the cyclopentanone ring (V) at C-10 a follows: H-C^{10} + O_2-HO-O-C^{10}-, given that allomerization yields at least three products. This conclusion is based on the absence of the Molisch phase test with allomerized chlorophyll. It may be mentioned that this process is believed to be based upon a reversible shift of the hydrogen atom at C-10 to C-9. The sum total of these observation, namely (a) the absence of Mg, and of allomerization, (2) the fact that allomerization means fixation of the H atom at C-10, (3) that it also prevents the phase test, and (4) that this test implies a reversible shift of the H atom in question, suggests that the central magnesium atom is responsible for the mobility of the H atom at C-10.

Still, hydrogen exchange proceeds only very slowly. However, with the proton magnetic resonance technique, showed that hydrogen exchange occurs at C-8 exclusively under their experimental conditions.

Bacteriochlorophyll a

This is the main photosynthetic pigment in both sulphur and non-sulphur purple bacteria. Its formula is $C_{55}H_{74}O_6N_4$ Mg. The structure differs from that of chlorophyll **a** in that C-3 and C-4 are hydrogenated, whereas the side group of C-2 consists of H_3C-C=O. Obtained evidence is that the primary intermediate formed upon excitation is the triplet state of bacteriochlorophyll complexed with the reaction centre.

The following pigments will be mentioned only briefly.

Bacteriochlorophyll b

This pigment was described by it occurs in a *Rhodopseudomonas* species as the main pigment.

Chlorophyll-770

Chlorophyll-770 is found in green sulphur bacteria. It turned out to be identical with bacteriochlorophyll. It is suggested that this chlorophyll rather than the chlorobium chlorophylls functions as a main pigment.

Chlorobium chlorophyll-660

Chlorobium chlorophyll-660 was previously called bacterioviridin. It is found in most of the green sulphur bacteria. It contains neither a methoxyl group at C-10, not phytol. Determined the magnesium content to be 2.92% on a dry weight basis.

Chlorobium chlorophyll-650

Chlorobium chlorophyll-650 occurs less commonly than the previous pigment. The methoxyl group at C-10 and phytol are absent. Its magnesium content are 2.59% on a dry weight basis.

In green plants, including algae, chlorophyll a is usually accompanied by auxiliary chlorophylls. Such chlorophylls are:

Chlorophyll b

Chlorophyll b, $C_{55}H_{70}O_6N_4Mg$, its structure is like that of chlorophyll **a**, except for a replacement of the methyl group at C-3 by a formyl group.

Chlorophyll c

It is synonyms of chlorophyll g and chlorofucine and occurs in diatoms and other brown algae. It does not contain a phytol group, two kinds, namely, chlorophyll c_1 and c_2 have been observed.

Chlorophyll d

It is found in most of the red algae in rather small amount. It was suggested that this pigment is 2-desvinyl-2-formyl-chlorophyll a.

Chlorophyll e

It was observed in yellow-green algae in very small amounts.

Precursors of chlorophylls might also be considered photopigments. Such may be the case with protochlorophyll, as this precursor needs light for conversion to chlorophyll by hydrogenation at C-7 and C-8. In vivo chlorophylls are complexed with lipoprotein carriers. Absorption spectra of chlorophylls are characterized by four independent electronic transitions responsible for bands between 350 and 900 nm.

Carotenoids

These compounds act as photopigments; in photosynthesis as auxiliary pigments; the absorbed energy is transferred to chlorophyll. It has been suggested that part of the carotenoids, the xanthophylls, act as oxygen transporters in photosynthesis. However, such function seems highly improbable. Their possible protective function in preventing photodestruction of chlorophyll will not be considered. Carotenoids function possibly as photoreceptors in phototaxis, photokinesis, and photomorphogenesis. The structure of xanthophyll lutein carotenoid provides an example of carotenoid structure. There are two classes of carotenoid: polyene hydrocarbons, called carotenes, and their oxygen derivatives, the xanthophylls. All are lipochromes; they are soluble in lipids and fats. The pigments are composed of isoprene units. At one or both ends of the isoprene chain carotenoids carry an ionone ring which may be closed b or open d Carotenoids usually occur in the all-trans configuration, whereas cis isomers are encountered less frequently. Carotenoids participating in photosynthesis are as follows:

Carotenes

α-carotene, $C_{40}H_{56}$, in green plants and algae, β-carotene, $C_{40}H_{56}$, in green plants and algae, γ-carotene, $C_{40}H_{56}$, in green plants and algae, and lycopene, C_{40}H56, in purple bacteria are present.

Xanthophylls

Lutein, $C_{40}H_{56}O_2$, in green plants and algae, violaxanthin, $C_{40}H_{56}O_4$, in green plants and algae, neoxanthin, $C_{40}H_{56}O_4$, in green plants and algae, cryptoxanthin, $C_{40}H_{56}O$, in green plants and algae, zeaxanthin, $C_{40}H_{56}O_2$, in green plants and algae, fucoxanthin, in diatoms and other brown algae, (the exact structure is not yet established), spirilloxanthin, $C_{40}H_{54}$ $(OCH_3)_2$, in purple bacterial and rhodopin, $C_{40}H_{58}O$ or $C40H_{56}O$, occurring in purple bacteria.

The absorption spectra of these pigments generally show two or three bands located at wavelengths in between 350 and 550 nm. Carotenoids occur in complexes with lipoproteins in the vivo state.

Phycobiliproteins

These pigments act as photoreceptors in connection with (a) photosynthesis as auxiliary pigments by transfer of their excitation energy to chlorophyll, (b) phototaxis, (c) photokinesis (to a minor degree), and (d) photomorphogenesis as regards the special biliprotein phytochrome, capable of phototransformation. As their names indicate, these pigments are proteins with a chromophoric group. This consists of four pyrrole nuclei linked together by C atoms in a linear or open-cyclic arrangement. There are two major kinds of phycobiliproteins: the blue phycocyanins and the red phycoerythrins. Their chromophoric groups are phycocyanobilins and phycoerythrobilins, respectively. Both pigments coexist in blue-green algae, the Cyanophyceae, and red ones, the Rhodophyceae, but in considerably different proportions. Since phycocyanin as well as phycoerythrin from either blue-green or red algae show different absorption properties, the prefixes C or R are used to indicate their origin. Phycobilins are also observed in Cryptomonads.

Phycobiliproteins are complexes which leach readily out of damaged cells. A complex unit is known to be composed of two or more subunits. A detailed analysis of C-phycocyanin. This phycocyanin from *Oscillatoria agardhii* consists of two polypeptide chains, each of which carries at least one chromophoric group. The chains, of which the amino acid compositions are given below, are connected by a disulphide bridge. Assuming that each chain carries a single chromophoric group, the molecular weights for the chains were found to be 12,100 and 14,100 respectively.

The major phycobiliproteins are: C-phycocyanin, C-phycoerythrin, R-phycocyanin and R-phycoerythrin

B-phycoerythrin, occur in the red alga *Porphyra naiadum* allophycocyanin, and is also present in many red and blue-green algae, whereas phytochrome occurs in plants, including algae. It is active in photomorphogenesis and phototaxis.

Flavoproteins

Probably flavoproteins are the photopigments of phototropism. Flavoproteins may also be responsible for the photosensitivity of fungi. It is less certain that these pigments function as photoreceptors for photomotion in microorganisms. The difficulty in assigning the role of biological photopigment to flavoproteins unambiguously is due to the fact that their absorption spectra showing bands around 450 nm, overlap those of the carotenoids. Action spectra, therefore, are hard to interpret. It thus remains a challenge to try to obtain conclusive evidence about the proposed functions of flavoproteins in photobiology.

CHAPTER 4

Photosystems in Chloroplast

Significant advances have been made towards the understanding of the molecular structure of the thylakoid membrane, i.e. the proteins and lipids and their orientation, the polypeptides associated with light-harvesting pigment-protein complexes and their distribution in the core and peripheral parts of the antennae, etc. Membrane complexes can be isolated from higher plant chloroplasts, and their photochemical and biochemical properties studied *in vitro*. There is strong evidence of a lateral heterogeneity in the distribution of these complexes between stacked and unstacked regions of the thylakoid membrane. Photosystems I and II are now found to exist as two structurally (and even functionally) independent types, designated as PSI_{α}, and PSI_{β}, and $PSII_{\alpha}$ and $PSII_{\beta}$. $PSII_{\alpha}$ which are located in the grana membranes and $PSII_{\beta}$ are exclusively located in the stroma-exposed thylakoids. The light-harvesting capacity of $PSII_{\alpha}$ is greater (because of the presence of more antenna chlorophylls) than that of $PSII_{\beta}$. Little is understood about the mechanism of membrane stacking into grana and the biochemical and physiological significance of this phenomenon, although we know that the degree of stacking is influenced by the cation concentration and by the phosphorylation of light-harvesting proteins.

EMERSON ENHANCEMENT EFFECT

In photosynthesis, two photochemical systems are associated with two pigment system (i.e., pigment system I and Pigment system II). Chlorophyll *a* is found in two forms: (i) with an absorption maxima at short red wave length (673 nm - Chl *a* 673) and (ii) at long red wave length (683 nm - Chl *a* 683). Another form of chlorophyll *a* has also been reported which has an absorption maxima at 700 nm (called p 700), present in much smaller amounts than that of chl *a* 673 and chl *a* 683. PS I contains long wave length absorbing forms (chl a 683 + p 700 in addition to carotenoids) for collection of energy; while PS II contains chlorophyll b in addition to the short red wave length form of chl *a* 673 (in higher plants and green algae). In blue green and red algae, phycobilins and other chl *a* are present as energy collecting centre (PS II). Emerson effect shows that when 653 nm and 700 nm wave lengths of radiations are given, the relative rate of photosynthesis is increased in comparison when 653 nm and 700 nm wave length of light is given successively, relative rate of photosynthesis is reduced.

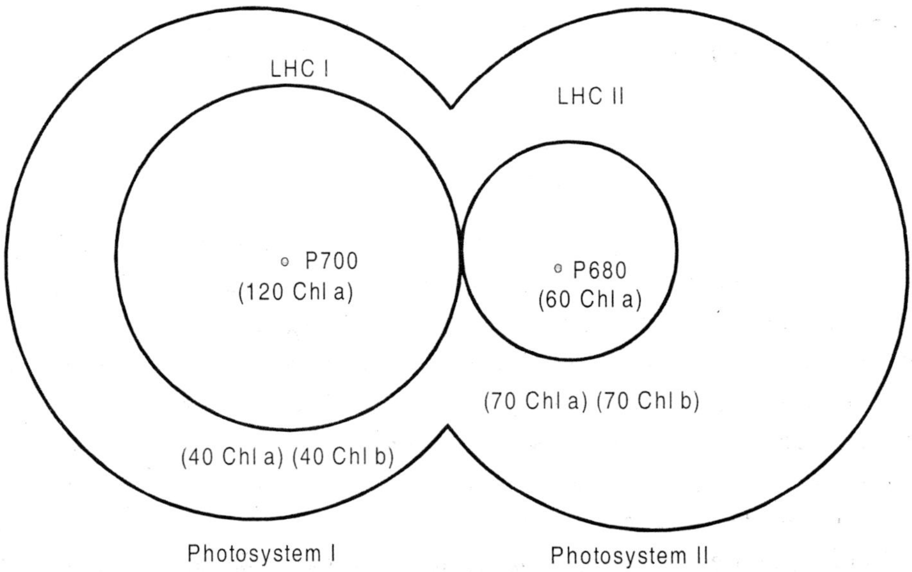

FIG. 4.1: *Presentation of granum thylakoids, light harvesting center (LHC) with position of various components of the light reactions.*

RED DROP EFFECT

Robert Emerson while working on *Chlorella* observed that wavelengths beyond 680 nm decreased the rate of photosynthesis even though chlorophyll *a* could absorb them. This decrease in photosynthetic efficiency was said to be **red drop** effect. It is proposed that at this wavelength only one photosystem remains in operation.

CHLOROPLAST THYLAKOID MEMBRANE

The thylakoid membrane of oxygenic photosynthetic organisms is the most abundant biological membrane on our plant. This membrane contains the molecular machinery that intercepts light energy and convert it into chemical potential. These early stages in the photosynthetic process involve the breakdown of water to molecular oxygen and the generation of reduced nicotinamide adenine dinucleotide phosphate (NADPH) and adenosine triphosphate (ATP). The two latter products act as an energy store, which is used to reduce CO_2 to carbohydrate.

Like other biological membrane, the thylakoid membrane consists of protein and lipids. The proteins of the thylakoid membrane are associated with four major multipeptide complexes, which are well characterized. Two of these complexes bind a large number of chlorophyll molecules and are therefore, responsible for intercepting light energy. They also contain the photochemically active reaction centers that act as traps for migrating excitons. One complex produces the strong oxidant necessary to oxidize water and is known as photosystem II (PS II). The second complex is called phtosystem I (PSI) and produces a reductant sufficiently strong to reduce NADP. The other product of charge separation in PSII is a weak reductant, which is stabilized as reduced plastoquinone (PQ). In the case of PSI, the other product is a weak oxidant that oxidizes the copper-containing protein plastocyanin (PC).

The cooperation between PSII and PSI occurs when the reduced PQ interacts with the oxidized PC. This redox exchange is catalyzed by one of the other membrane protein complexes known as the cytochrome b/f complex. The transport of reducing equivalents from water to NADP via PSII, cytb/f and PSI complexes occur in a such a way that protons are absorbed from the outer surface and deposited at the inner surface of the membrane. Since the thylakoid membrane forms a closed visible system, the pumping of H^+ across the membrane establishes an electrochemical potential gradient between the outer and inner compartments. This energy store is utilized by the fourth multipeptide complex contained in the membrane, known as the CF0-CF1 complex, required to bring about the synthesis of ATP from adenosine disphosphate (ADP) and inorganic phosphate.

Protein Complexes: Distribution

Fractionation of isolated thylakoids into unstacked and stacked fragments have shown that the majority of PSII complexes are localized in the appressed regions of the stacks, while PSI and CF0-CF1 complexes are in the unstacked membranes, including those membrane regions that constitute the end of the grana. The cytb/f complex is evenly distributed between both regions (Fig. 4.2). The light-harvesting chlorophyll a/b complex (LHCII) that acts an antenna for PSII is located mainly in the appressed lamellae. This lateral separation of the various complexes into two distinct domains seems to represent a thermodynamically favourable configuration, partly, as result of surface electrical charge differences between the complexes.

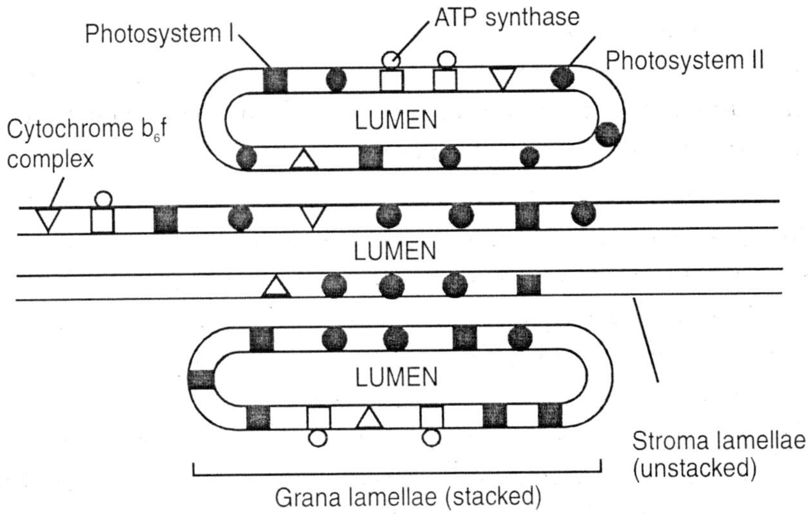

FIG. 4.2: *Organization of the protein complexes of the thylakoid membrane. Photosystem II is located predominantly in the stacked regions of the thylakoid membrane; photosystem I and ATP synthase are found in the unstacked regions protruding in to the stroma. Cytochrome b_6f complexes are evenly distributed. This lateral separation of the two photosystems requires that electrons and protons produced by photosystem II be transported a considerable distance before they can be acted on by photosystem I and the ATP-coupling enzyme.*

PHOTOSYSTEMS

The light reactions in thylakoids are driven primarily by two photosystems namely Photosystem I (PS I) and Photosystem II (PS II) which are interdependent and are linked by membrane bound cyt b/*f* complex and other mobile electron transport carriers. Both PS I and PS II consist of characteristic polypeptides of reaction center core complex, light harvesting antenna system and specific metal ions. PS II commissions the first step in the primary process of the photosynthesis. The light harvesting system of PS II absorbs photon, channels it to P680, the photon trap, where the photosystem converts the light energy to the chemical energy by photooxidation of water to oxygen and reduction of quinone pool to quinol, two major events of thylakoid membrane. Subsequently, the energy flows to produce ATP and NADPH via cyt b/f complex and PS I. These substances are utilised in the Calvin cycle to fix carbon dioxide into carbohydrate.

Compared to PS I, PS II has been investigated in more details during last two decades. The structure of PS II is not only complex, but also extremely delicate. Secondly, it is one of the important players during electron transfer reactions of thylakoids because of its ability to split water molecule, the ultimate source of electrons. The photosystem is also capable of forming reaction centre oxidants with high oxidising potential (Barber, 1995).

PS II is extremely sensitive to all kinds of changes of its surrounding environment and is capable of creating oxidative stress in green leaves.

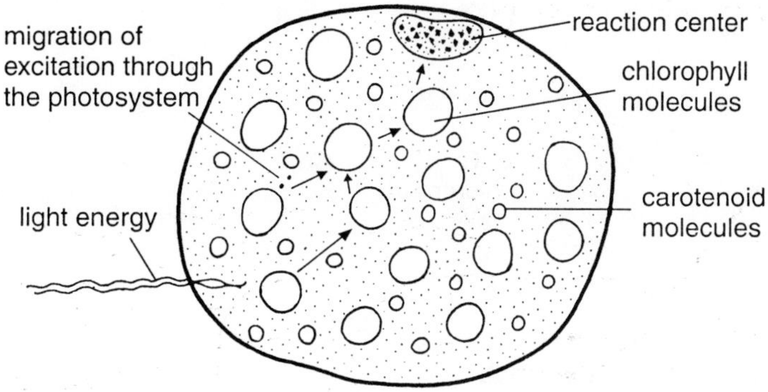

FIG. 4.3: *Diagrammatic presentation of a photosystem showing transfer of energy from antenna molecules to reaction center.*

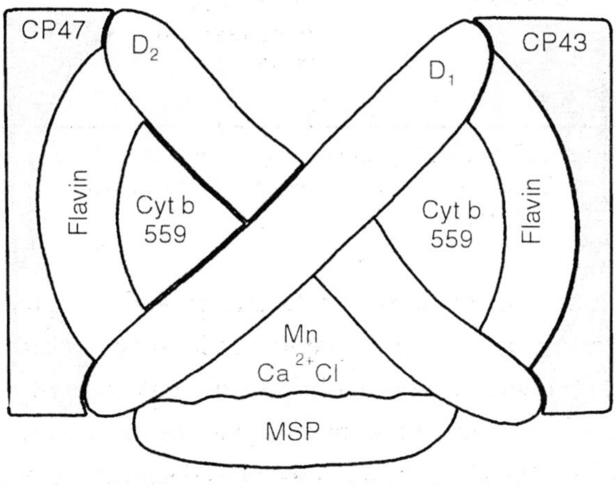

FIG. 4.4: *A diagrammatic presentation of PS II showing its structural organization. CP = Chlorophyll-protein Complex, MSP = manganese-Stabilizing Protein.*

Light Harvesting Complex (LHC)

Each photosystem (PS I and PS II) contains several chlorophyll molecules; yet only one chlorophyll per complex is capable of undergoing light-driven electron transfer. These specialized *reaction centre* chlorophyll *a* molecules are termed as P_{680} (for pigment absorbing light at 680 nm) in photosystem II and P_{700} in Photosystem I. These specialized chlorophyll molecules are bound to an integral membrane protein which has molecular weight of 110,000 and contains the reaction centre for photosystem I. The molecular weight of protein-chlorophyll complex of the reaction centre for photosystem II is 47,000. Associated with each photosystem is a structure the **light harvesting complex** (LHC), or **antennae complex**.

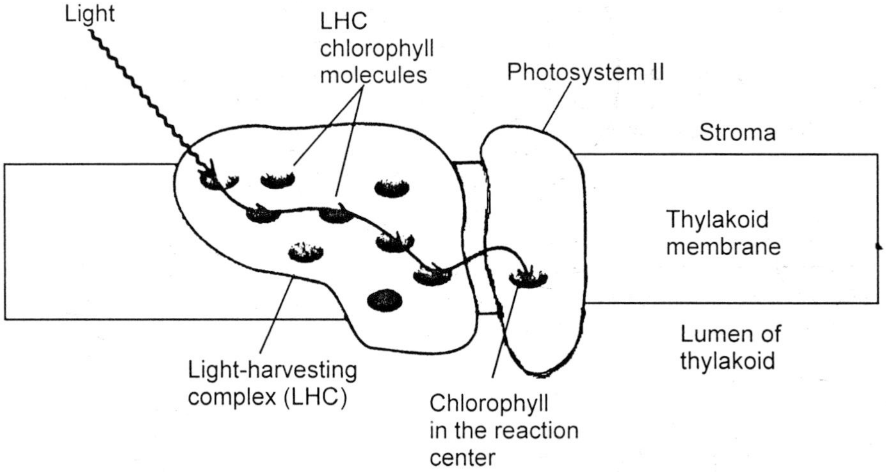

FIG. 4.5: *Light harvesting complex (LHCs) are associated with each of the two photosystem complexes contain dozen of chlorophyll molecules. Any of these can absorb a photon. The energy is transferred to the reaction centre chlorophyll in the photosystem complex.*

The primary function of LHC to capture light energy and to transfer that energy to PS I or PSII. It has no catalytic activity and contains at least two main polypeptides, each of molecular weight 32,000, which bind up half of the chlorophyll *a* and nearly all of the chlorophyll *b*. LHCs are situated on the inner surface of the thylakoid membrane. Dozens of chlorophyll molecules of each LHC absorb photons and transfer it rapidly (in less than 10^{-9} sec) to the specialized chlorophyll molecule in the reaction centres of the two photosystems. LHCs act to absorb light energy and to **funnel** it to a single reaction centre. The function of protein in membrane is to maintain an optimal orientation for energy transfer between chlorophylls. How it maintain this is not still clear.

Photosystem II

Each photosystem consists of a central core complex (CC) and a light harvesting complex (LHC). By the use of detergents the CC can be further split into the pigmented reaction centre subunits (RC) and unpigmented proteins. Analogously, the LHC can be split into the light-harvesting pigment-proteins. In the CC (also called photosystems in some thylakoid models) the photochemical and biochemical responses occur which involve the chlorophylls and redox systems.

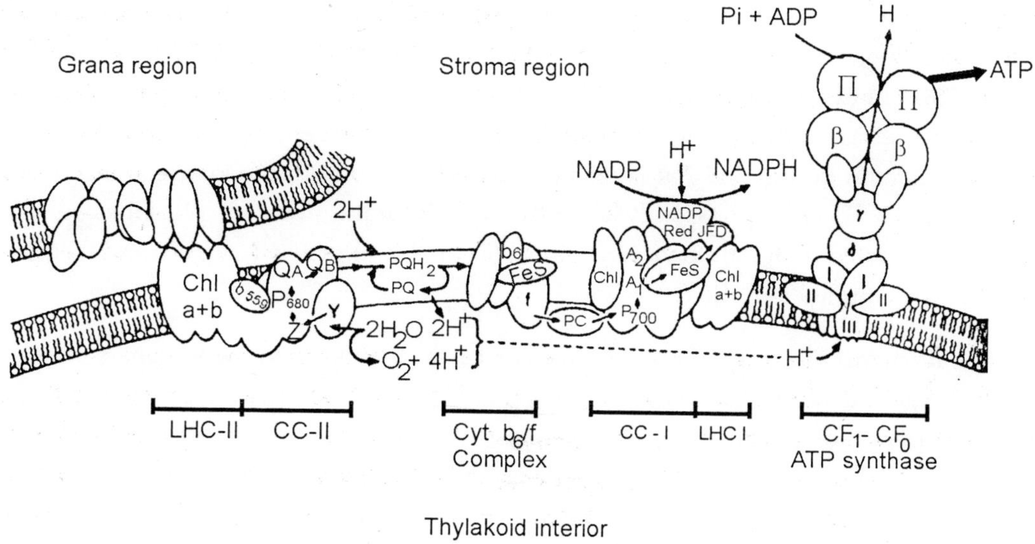

FIG. 4.6: *Arrangement of chlorophyll protein complexes in photosynthetic unit.*

The light-harvesting antenna or complex of PSII can be further subdivided into proximal and distal antennae. The proximal antenna, which surrounds the P680 core and is tightly coupled to it, consists of two pigment-protein complexes, CP47 and CP43 (CP, chlorophyll-protein complex). Both complexes contain Chl *a* and β carotene bound to the apoproteins, but no Chl *b*. Excitation energy captured by the bulk antenna (proximal as well as distal) is transferred to P680 from the proximal antenna. The distal antenna is mainly composed of light-harvesting complexes (LHC II) and contains covalently bound Chl *a*, Chl *b* and xanthophylls.

The three-dimensional structure of LHC II at 6 Å resolution has recently been determined by electron crystallography. This low-resolution structure shows that LHC II forms a trimer, with each monomer containing 15 chlorophylls and several carotenoids. In addition to its function in light harvesting and energy transfer to P680, the LHC II also plays a role in energy distribution between the two photosystems by undergoing reversible phosphorylation-dephosphorylation. This mechanism involves the physical movement of a pool of LHC II between the appressed and nonappressed regions. When PSII is overexcited relative to PSI, the intersystem PQ pool becomes reduced and a membrane-bound kinase is activated. The kinase phosphorylates threonine groups on the exposed surface of a mobile form of LHCII, thus increasing its surface negative charge. The consequence of this is that the phosphorylated complexes repartition from the relatively electroneutral appressed regions to the electrically charged unappressed regions. Here the phospharylated LHCII complexes become sufficiently intimate with PSI so that they can act as an additional antenna for this photosystem. Thus, some of the excess PSII light is redistributed to PSI, so as to improve the balance between them and increase the efficiency of photosynthesis for that particular lighting condition. When the reverse situation occurs and PSI is overexcited relative to PSII, the kinase is inactivated and the LHCII is dephosphorylated by a membrane located phosphatase. The dephosphorylated LHC II then repartitions into the appressed region and becomes functionally coupled to PSII.

A **regulatory cap** made up of polypeptides extrinsically bound to the lumenal surface of the thylakoid membrane. It comprises of a set of hydrophilic extrinsic proteins (EP) that have apparent molecular masses of 33, 24 and 17 kDa and are designated as EP33, EP24 and EP17 respectively. The EP33 acts as a barrier between the water-oxidizing unit attached to the membrane and the aqueous lumenal phase. Spectroscopic and protein cross-linking studies suggest that EP33 is in contact with the reaction centre. It may also be required for the assembly and activation of the Mn complex in eukaryotic algae and plants *in vivo*.

PSII Reaction Centre

A reaction center containing the primary reactants, plastoquinones, and the water-oxidizing complex. The PSII reaction centre can be functionally subdivided into the light-absorption and electron transfer segments and the water-oxidizing complex. The PSII *core* is the minimal unit able to perform light-induced charge separation and electron transport (but not to evolve O_2), and in this respect resembles the purple bacterial reaction centre. This core has been isolated now in many laboratories and comprises 4 to 6 Chl *a*, 2 Pheo, 2 β carotene, 1 D_1-D_2 heterodimer, 1 CP47, 1 CP43, 1 (or 2) cytochrome *b*559 and 1 10 kDa polypeptide (I) per P680. No quinone (Q_A) is present in the PSII core unit. The P680 and Pheo are believed to be bound to the D_2-D_1 (so called because they were first identified as two *diffuse* bands by gel electrophoresis and staining) proteins. In O_2-evolving PSII reaction centre complexes, the quinines (Q_A and Q_B) and ferrous iron are also bound to the D_2-D_1 heterodimer.

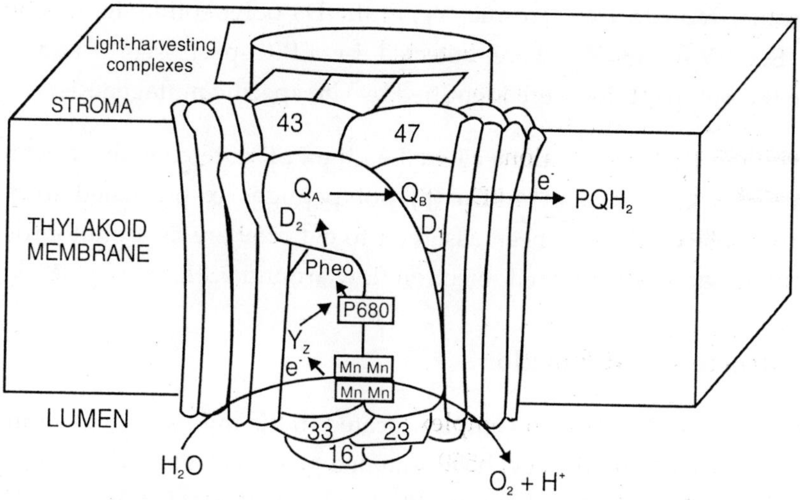

FIG. 4.7: *Schematic model of the PS-II reaction center and the oxygen-evolving complex. P680 is associated with the D1 and D2 proteins, which are analogous to the L and M proteins of the bacterial reaction center. Electrons are transferred from P680 to pheophytin (Pheo) and then to Q_A and Q_B (PQH_2). P680+ (oxidized P680) is rereduced by Y_Z, a tyrosine residue in the D1 protein. The oxygen evolution complex involves Mn, Ca^{2+}, and Cl^- as cofactors.*

The primary charge transfer from excited P680 to Pheo takes place in 3 picoseconds. Recombination of the charged species is prevented by the rapid (250-500 μs) oxidation of Pheo $^+$ by a neighbouring Q_A and the instant reduction of P680 $^-$ by a redox active tyrosine residue Y_Z of the D_1 polypeptide.

$$\text{Pheo}^- + Q_A \longrightarrow \text{Pheo} + Q_A^-$$

Q_A^- is oxidized within 100 to 200 μs by a second plastoquinone, Q_B, forming a Q_B semiquinone. This semiquinone accepts another electron from a different Q_A^- to form Q_B^{--}. (Note that Q_A is a single-electron acceptor while Q_B is a two-electron acceptor).

$$Q_A^- + Q_B \longrightarrow Q_A + Q_B^-$$

$$Q_B + Q_A \longrightarrow Q_A + Q_B^-$$

By analogy with the known sequences of the redox events occurring in purple bacterial reaction centers, it is believed that the fully reduced Q_B—accepts two protons from the stroma, forms Q_BH_2, and leaves the reaction centre as plastoquinol and then transfers the protons to the cyt *bf* complex. The Q_B site in the D_1 protein is then occupied by another Q_B molecule from plastoquinone pool.

There is another redox active tyrosine, Y_D in the D_2 polypeptide. Its function, if any, is unknown. Both Y_Z and Y_D were detected by EPR spectroscopy and their precise locations in the polypeptides were identified by site-specific mutagenesis.

Both O_2-evolving PSII preparations as well as the PSII core contain cytochrome *b*559, a heterodimer of 9-kDa (α) and 4-kDa (β) polypeptides co-ordinated to a heme. The possible functions which have been assigned to cyt *b*559 are as a photoprotective agent and as a proton pump during cyclic electron flow around PSII.

Location, Structure and Function

PS II is a transmembrane protein complex located in the thylakoid membrane. It consists of D1 and D2 intrinsic proteins, cyt b559, antenna proteins, light harvesting proteins type I and type II and extrinsic proteins. Photosystem II is that part of the oxygenic photosynthetic apparatus which catalyzes the light-induced oxidation of water coupled to the reduction of plastoquinone. This function can be represented by the simple equation:

$$2H_2O + 2PQ \xrightarrow[\text{PS II}]{\text{Light}} O_2 + 2PQH_2$$

During this process, four photons are transferred across the thylakoid membrane from the stromal side to the lumenal side which are utilized in ATP synthesis. PSII is located within the thylakoid membrane in close association with lipids.

Major advances have been made over the last decade towards our understanding of the structure and function of PSII. These include following:

- The isolation of almost pure PSII particles, free from other thylakoid components.
- Separation and identification of the polypeptides in the purified complex by gel electrophoresis; determination of the primary structure of these polypeptides by isolating the genes, then cloning and sequencing them.
- Most importantly, the determination of the structure of the reaction centers of photosynthetic purple bacterial reaction centers are structurally and functionally related.

The central part of PS II consists of D1 (32 kD)/D2 (34 kD) heterodimer. The transmembrane helices, hydrophilic loops of D1 and D2 proteins. D1/D2 dimer embeds special pair of chlorophyll a, P680, two additional chlorophyll a, two pheophytins, quinone A(QA), quinone B(QB), non-heme iron and manganese cluster. P680 is the primary electron donor and pheophytin, the primary acceptor. The dimer is enfolded by CP47 (947 kD) and CP43 (43 kD) antenna polypeptides containing chlorophyll a molecules. The dimer is also closely associated with two units of cyt b559 polypeptides. A 33 kD extrinsic protein known as manganese stabilising protein (MSP) is associated with D1-D2-CP43-CP47 complex at donor side. D1-D2-CP43-CP47-cyt b559-33kD unit constitutes the core complex of PS II with 250-300 kD molecular mass, which is the smallest PS II unit that can split water.

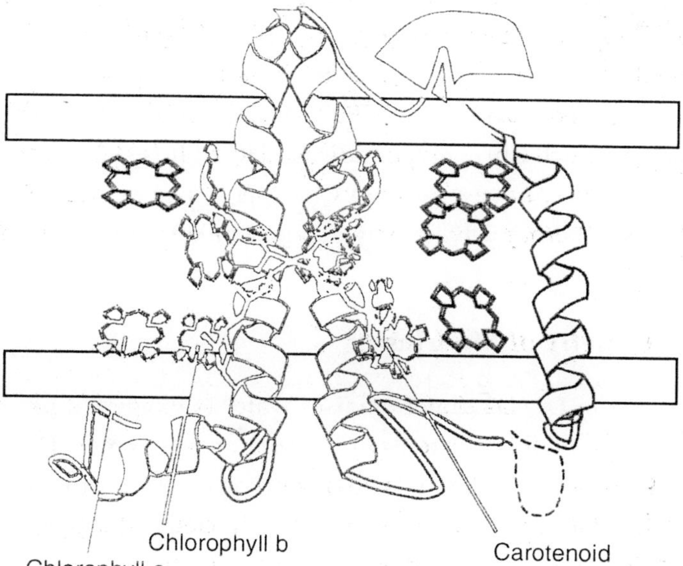

FIG. 4.8: *Structure of the LHCII antenna complex from higher plants. The complex is a transmembrane pigment protein, with three helical regions that cross the nonpolar part of the membrane. Approximately 15 chlorophyll a and b molecules are associated with the complex, as well as several carotenoids. The positions of several of the chlorophylls are shown, and two of the carotenoids form and X in the middle of the complex. In the membrane, the complex is trimeric and aggregated around the periphery of the PS-II reaction center complex.*

Light harvesting chlorophyll a/b binding proteins (LHC II, CP 29, CP 26, CP 24) surround the PS II core complex. CP 29, CP 26 and CP 24 are closely associated with the antenna CP 47 and CP 43. The major light-harvesting complex LHC II is probably organized into trimeric particles that transfer much of their excitation energy to PSII but under certain conditions dissociate from it and migrate independently between stacked and unstacked regions of the thylakoid membrane. CP29 and CP26 are closely associated with the PSII corc, although they can be removed from it without affecting PSII activity. CP24 is removed from PSII along with LHC II. The two LHC I complexes associated with

PSI are called LHC 1-680 and LHC1-730 after their fluorescence emission maxima. The former can be split into two pigmented subcomplexes, LHC1-680A and LHC1-680B, each enriched for a single polypeptide.

It is generally agreed that Chl a/b ratio of LCH II is 1.3 – 1.4 but there is a wide range of reported values for the other Chl a/b proteins. In general, the values approach 3 for the LHC I and minor PSII complexes, with the exception of CP24 which always has a very low Chl a/b ratio. PSII-associated complexes are reported to differ in their Chl-to-protein ratios with values of 12, 8, 9 and 5 for LHC II, CP29, CP 26 and CP 24 respectively. This means that the different Chl a/b ratios could result from all the proteins binding the same amount of Chl a but different amounts of Chl b. All the chlorophyll a/b bind lutein as well as xathophylls neoxanthin and violaxanthin. The latter are enriched in CP29, CP26 and CP24 compared with LHC II. CP22 has a Chl a/b ratio of about 6 and appears to have about half as much Chl/polypeptide as LHC II. It is stable in the absence of Chl, which suggests that it may have a role as a Chl reservoir or carrier rather than as a light-harvesting complex. Alternatively, it may act as a linker between LHC II and the PSII core.

Location and Geometry of Mn Cluster

A Mn cluster located in D1/D2 dimer oxidises water to oxygen. It has not been possible to discover the nature of this cluster, two dimers or a tetramer. Due to lack of direct experimental evidences, several models have been proposed for the Mn-binding site and oxidation of water. Gradually reports of site directed mutagenesis are coming in, identifying the amino acid residues binding to Mn. E69 in cyanobacteria (E 70 in higher plants) of D2 H 332, E 333, H 337 of D1 are reported to be the potential ligands to Mn cluster by mutant studies. D 59, D 61 may ligate to calcium and D59, D61, E65 and H 92 of D1 influence the properties of Mn cluster without significantly affecting its stability or ability to assemble. E 189, H 92 and D 342 influence the catalysis, stability and assembly of Mn. D 342 may bind to Mn or Ca or both. H 190 of D1 strongly influences the redox properties of secondary electron donor Z (y 161, D1) and either ligates to Mn or serves as a H-bond donor.

Studies with wild-type and mutant algae show that the EP33 polypeptide may be required for the assembly and activation of the Mn cluster and to optimize catalytic activity. Chemical modification of various amino acid residues in PSII polypeptides and subsequent measurement of the O_2 evolution rate and spectral properties of the modified PSII suggest that residues in the D_1-D_2 polypeptide (possibly histidine and carboxylate) may provide ligands for the Mn cluster. Similar conclusions were arrived at from site-directed mutations of amino acid residues in D_1 and D_2 polypeptides in the cyanobacterium *Synechocystis*.

Photochemistry

The primary photochemistry of PS II is initiated by charge separation between P680 (special chlorophyll a dimer) and pheophytin (pheo).

$$\text{P680, Pheo} \xrightarrow[\text{3ps}]{h\nu} \text{P680}^{-}\ \text{pheo}\ ^{+}1.2\ \text{ev}$$

Charge recombination is prevented by a series of downhill reductions during electron transfer from Pheo to QA in D2 (200 ps), QB in D1 (0.3 ms) and free plastoquinone (PQ) pool. The orientation of the accessory chlorophyll a molecule determines the preference of path of electron transfer from P680 to pheophytin close to QA (D2 side) over the other route to QB (D1 side) (Heller et al. 1995).

There are two accessory donors YZ (Tyr 161, D1) and YD (Tyr 161, D2, 160 in cyanobacteria) to P680. Z reduces P680 + in 200 ns, half life time of YZ+ is 0.1 -1.0 ms. YZ+/YZ redox potential is between 810 mV and 1000 mv. YD is oxidised by water oxidising complex in S2 states. YD^+/YD has redox potential 720-760 mV. Half life time of YD^+ is about 500 min (Vass and Styring, 1991). Site directed mutants suggest that Q 164, H 189 in Synechocystis 6803, Q 165, H 190 in higher plant D2 and H 190 of D1 are in the vicinity of YD. Saturation-recovery electron spin resonance spectroscopy provides evidence for the symmetric location of YD and YZ (Koulougliotis et al. 1995).

Role of Cyt b_{559}

Cyt b_{559} is closely associated with core complex of **reaction center** (RC II). Heber et al (1979) suggest participation of cyt b_{559} in cyclic electron flow in PS II. Cyt b_{559} exists in two states, low and high potential forms. Thompson et al (1987) propose a structural role of cyt b_{559} in oxygen evolving comblex. Cyt b_{559} is also suggested to have a protective role in photoinhibition. Cyt b_{559} shifts between its high potential and low potential forms, triggered by unknown molecular switch, to act as an electron acceptor (for acceptor side protection) as well as electron donor (donor side protection) see Fig. 4..9 (Barber and De Las Rivas, 1993). Cyt b_{559} may be a structural obligation during evolution of oxygen evolving RC II from bacterial reaction centers with cyclic electron flow. This vestigial polypeptide might have acquired a photoprotective role during evolution.

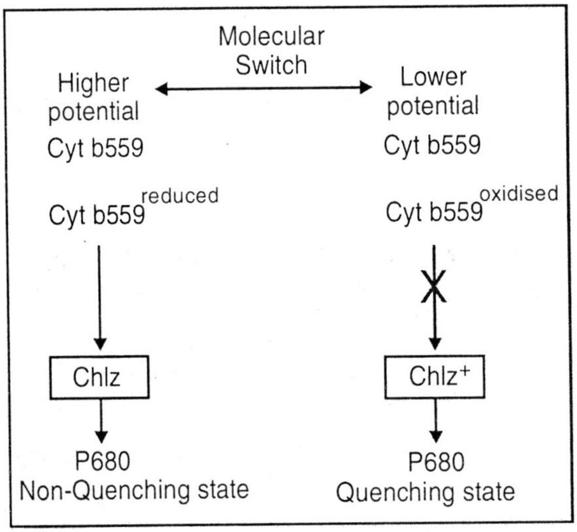

FIG. 4.9: *A model for the role of the cytb559/chlz pathway in protecting against photoinhibition. Quenching of excess excitation energy is regulated by the redox state of chlz during illumination and the redox state of chlz is determined by the redox state of cytb 559.*

Mechanism of Oxygen Evolution in PS II

Though there exists different models regarding organization of Mn-cluster, it is the catalytic site for oxidation of water.

Energetics of Water Oxidation: Free water at pH 7.0 is known to be oxidised in the following steps (see Radmer and Cheniae, 1977).

$$H_2O = .OH + H^+ + e \quad 2.31 V$$
$$H_2O + .OH = H_2O_2 + H^+ + e \quad 0.37 V$$
$$H_2O_2 = .O_2 + 2H^+ + e \quad 1.00 V$$
$$O_2 = O_2 + e \quad -0.45 V$$
$$2H_2O = O_2 + 4H^+ + 4e \quad 3.23 V \quad (0.81 V/e)$$

The energetics of bound water may be modified, but we can have some idea regarding mechanism of oxidation of H_2O to O_2 from the above scheme. The photon absorbed by P680 (680 nm) has energy 1.8 ev. It is sufficient to carry primary charge separation by oxidising P680 to P680$^+$ (Em, 7 = 1.2 V) and P680$^+$/P680 can carry out oxidation of water requiring potential 0.81 V.

Kok's Oxygen Clock

This clock is based on independent chain model of oxygen evolution. The oxygen evolving system comprises five states labelled: S_o, S_1, S_2, S_3, S_4,. Number 0, 1, 2, 3, and 4 indicates the number of positive charges. After absorption of photon (*hv*) the S states advance to the next higher S state and the oxygen is evolved when S_4 reacts with H_2O. , as presented below:

$$S_0 \xrightarrow{hv} S_1 \xrightarrow{hv} S_2 \xrightarrow{hv} S_3 \xrightarrow{hv} S_4 \xrightarrow{hv} S_0 \quad \xrightarrow{2H_2O} O_2 + 4H^+$$

It was previously thought that all four protons are released in the S_4 to S_0 reaction. However, recently it has been the established that charges produced on one centre are independent of all other centres. The scheme is as follows:

$$S_0 \xrightarrow{} S_1 \xrightarrow{} S_2 \xrightarrow{} S_3 \xrightarrow{} S_4 \xrightarrow{} S_0$$

$$IH^+ \quad OH^+ \quad IH^+ \quad 2H^+$$

Another scheme is also proposed which explains that H^+ release pattern is in between 0, 1, 1, 2 and 1,1,1,1 which suggests that water may indeed have begun to chemistry before the last step contrary to the much mentioned belief. NMR analysis also provide evidence that protons could be released in several steps.

Evolution of oxygen per flash is a function of flash number and oscillates with the periodicity of 4. Therefore, any component that oscillates with that period and may be related to this system directly or indirectly. Some of these phenomenon are chlorophyll *a* -fluorescence which shows peaks at the 1st and 5th flashes; delayed light emission (DEL) due to the back reaction of PS II primary photochemistry and **thermoluminiscence.**

The Photosystem II oxygen-evolving reaction

This key reaction of plant, algal and cyanobacterial photosynthesis is the main source of all O_2 on earth but still continues to be one of the great unsolved mysteries of photosynthesis. However, 'light' has begun to get through the 'black box', and recent work has significantly helped to probe the mechanisms involved in the oxidation of water. The problem is being attacked from two main directions:

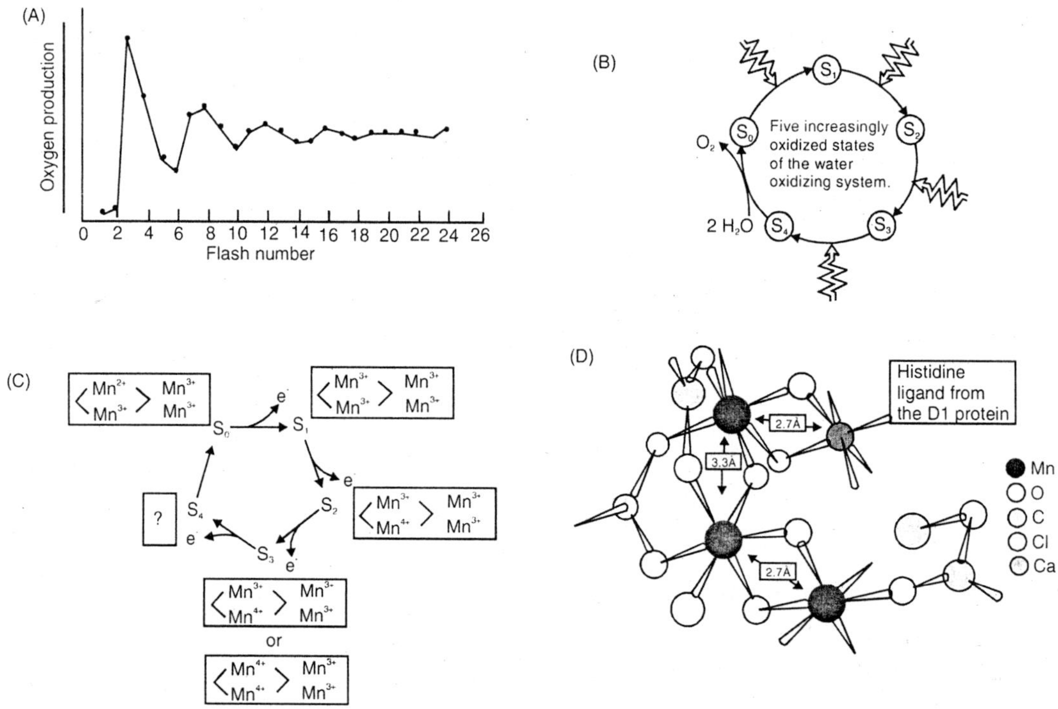

FIG. 4.10: *The pattern of oxygen evolution in flashing light and the S state mechanism for oxygen evolution. (A). sample of dark-adapted chloroplasts is exposed to a series of short light flashes. The maximum production of oxygen occurs on the third flash, secondary maxima occur on even fourth flash, and the yield damps to a steady-state value after about 20 flashes. (B) The S state model developed by Kok and coworkers was advanced to explain the observations of part A. The oxygen-evolving system can exist in states : S_0 to S_4. Successive photons (squiggly arrows) trapped by PS-II advance the system from S_0 to S_1, S_1 to S_2 and so on until state S_4 is reached. S_4 is unstable and read with two water molecules to produce O_2. (C) A structural model for the oxygen-evolving complex based on spectroscopic studies. The four Mn ions are linked to amino acids in the D1 protein , and to oxygen, chloride, and calcium. (One possible scheme for the redox states (S_0 to S_4) of the water-oxidizing Mn complex. Upon the release of oxygen S_4 converts into S_0. No data are yet available for S_4.*

- Isolation of PSII membranes with very good O_2 evolution rates and determination of the valence state and structure of the Mn atoms in the S_n states in the membranes by a variety of spectroscopic techniques.
- Synthesis of Mn clusters with spectroscopic characteristics that resemble the Mn spectra of PSII and which may even catalyze photo-oxidation of water.

Classical spectroscopy such as absorption, infra-red (IR), nuclear magnetic resonance (NMR) and EPR are being supplemented by more advanced techniques of electron nuclear double resonance (ENDOR), electron spin echo, X-ray absorption spectroscopy, etc. to identify the paramagnetic Mn species generated in PSII during laser flash photolysis. Considerable progress has been made in identifying the polypeptides surrounding the Mn cluster of the water-oxidizing complex. The Mn ions are located in a compartment towards the lumenal side of the membrane.

The requirement of manganese in the photoxidation of water has been known since 1970. On the basis of atomic absorption measurements and EPR quantification of Mn ions released by acid treatment of PSII, it is now almost accepted that the active O_2-evolving complex contains four atoms of Mn per P680 (although a few researchers believe there are six Mn per P680). Treatment with 0.8 M Tris buffer pH 8.5 releases most of the Mn from the complex, with the concomitant loss of the three polypeptides EP33, EP24 and EP17 of the regulatory cap. The Mn complex can be reconstituted in light; thekinetics of reassembly of the Mn tetramer appear to be a multistep process.

Joliot, and Kok, proposed the S state model of charge accumulation on the water oxidation catalyst to explain why the yield of O_2 from chloroplasts in response to a sequence of saturated light flashes oscillates with a period of 4. Most of the studies in recent years have tried to correlate the Mn signals from spectroscopic data with the S states of the water-oxidizing complex. Two EPR signals observable at very low temperatures (<35 K) have been attributed to the S_2 state. These are known as the multilane EPR signals with g = 2.0 and the g = 4.1 signal. The signals are perturbed by treating the samples with inhibitors of O_2 evolution, e.g. ammonia ions or NH_2OH. These EPR signals suggest that the Mn ions have a mixed valence in the S_2 state; however, definite assignment of the Mn oxidation state is difficult. The four Mn atoms may be arranged as a tetramer (or a pair of interacting dimers) or as a trimer which gives rise to the two EPR signals and an EPR = silent monomer. In dark-adapted PSII membranes, an EPR signals with g = 4.8 has been detected and has been proposed to arise from a precursor of the multilane signal and assigned to the S_1 state.

Valuable information regarding the ligands of Mn ions and Mn-Mn, Mn-O and Mn-N distances in the cluster has been obtained from X-ray absorption studies. These spectroscopic probes suggest that the atomic environment of Mn ions is heterogeneous. Flash-induced absorption changes in the UV that oscillate with a period of 4 have been assigned to the individual S state transitions. These absorption changes show correlations with the changes in the valence state of Mn from MnII to MnIII to MnIV. Spectroscopic data suggest that the Mn ions undergo oxidation during the $S_0 \rightarrow S_1$ and $S_1 \rightarrow S_2$ transitions, but it is still not clear whether Mn is oxidized on the $S_2 \rightarrow S_3$ transition.

The mechanism of water oxidation is still unknown. It is assumed that the Mn ions form the matrix to which two water molecules co-ordinate to form a dioxygen bridge. There are two possibilities for the chemistry of the oxidation.

- A concerted four-electron oxidation occurs at the $S_3 \rightarrow S_4 \rightarrow S_0$ transitions. It is generally believed, but not unequivocally established, that water oxidation occurs at the $S_4 \rightarrow S_0$ transition.
- A two-step reaction with the formation of a peroxide intermediate prior to the $S_3 \rightarrow S_4 \rightarrow S_0$ transition. Sequential one-electron oxidation of water with the formation of OH^- intermediates is considered to be energetically unfavourable.

Role of Ions Other than Manganese during Evolution of Oxygen

Other ions associated with evolution of oxygen in photosynthesis are Ca^{2+}, Cl and HCO_3 . Ca^{2+} binds to D1/D2 heterodimer and influences the stability and assembly of Mn-cluster as suggested by site directed mutants (Chu et al 1995). Ca^{2+} may play a role in reduction of $P680^+$ by YZ (Satoh and Katoh, 1985). S3 - So transition is also known to be modulated by Ca^{2+}. It may be speculated that Ca^{2+} binding introduces structural modification suitable for Mn binding. Moreover, Ca^{2+} may provide an electrostatic microenvironment to the redox species on the donor side of PS II.

The roles of Ca^{2+} and Cl in the water oxidation are not definitely established. Atomic absorption measurements have shown that PSII membranes contain two to three bound Ca ions per P680. Bound Ca can be released with a loss of O_2 evolution by washing the membranes with 1-2 M NaCl supplemented with 1-5 mM EDTA, in light. Measurement of O_2 evolution of Ca-depleted PSII membranes as a function of added Ca has shown that PSII contains one high-affinity and two low-affinity Ca-binding sites per P680. X-ray absorption spectroscopy has indicated that Ca ions are located very close to the Mn complex. Whether the role of Ca is only to protect the Mn complex or whether it directly participates in the O_2 evolution reaction is still a moot question. The influence of Ca ions in electron transfer and O_2 evolution in cyanobacteria may be different from that in plants.

The molecular mechanism of Cl effect on PS II electron transport is understood to some extent by ^{35}Cl-NMR studies, which suggests a stereo selective binding of Cl to oxygen evolving site itself or very close to it. Coleman and Govindjee (1987) proposed that Cl has two binding sites, intrinsic and extrinsic. Binding of Cl to intrinsic sites (H 332, H 337 of D1 and H 337, D2) enhances the abstraction of H^+ from water by increasing pKa of imidazole (Histidine side chain) rings. Chloride may not act simply a passive counter ion of charged species formed at the active site rather participate actively by affecting pKa and conformation of the polypeptides.

The binding of Cl ions to PSII requires the prior binding of Ca. Binding values were attained by NMR spectroscopy using ^{35}Cl and isotopic labelling with ^{36}Cl. Reported values for the number of Cl^- per P680 required for maximum O_2 evolution have varied from 1 to 5. Some EPR measurements indicate that Cl^- is essential for the $S_2 \rightarrow S_3$ transition. Requirement for Cl^- is not specific; it can be replaced by Br^-.

From the foregoing discussion it is obvious that we still need to learn a lot more about the location, structure, chemical reactions, and the role of cofactors of the water oxidation complex in order to elucidate the mechanism of O_2 evolution.

Non-heme iron of QA - Fe -QB complex, situated on the acceptor side of the reaction center II is identified as HCO_3 binding site by spectroscopic studies (Diner and Petrouleas, 1990; Semin et al, 1990). HCO_3 regulates the electron transfer between QA and QB in PS II and HCO_3 binding is stabilized by positively charged amino acid residues namely arginine and lysine (Govindjee and Van Rensen, 1993). Regulation of electron transport by HCO3 may be a strategy of the system to sense the concentration of carbon dioxide which is converted to glucose as the end product of the process initiated by PS II.

Extrinsic 33 kD and Oxygen Evolving Polypeptide

Importance of 33 kD extrinsic polypeptide in oxygen-evolution has entered into controversy by recent results on mutants. Earlier reports suggest necessity of 33 kD extrinsic polypeptide for oxidation of water (Kuwabara et al., 1985; Kuwabara and Murata, 1982; Miyao and Murata, 1984). Studies on mutants of *Synechocystis* sp. PCC 6803 in which psbO gene encoding 33 KD protein is detected, indicates that 33 kD protein is not essential for oxygen evolution, however, it is required for stability of Mn cluster (Burnap and Sherman, 1991, Philbrick et al., 1991). Thermoluminescence study on IC2 psbO less mutant of *Synechocystis* sp. Pcc 6803 shows damping of S2QB- and S3QB- charge recombination observed by flash induced oscillation of the B-band indicating inhibition of higher oxidation states of Mn-cluster (S3 and S4) in the absence of 33 kD polypeptide (Vass et al., 1992). On treatment with B-mercaptoethanol, inactivation of oxygen evolution is observed (Raval et al., 1994). The 33 kD polypeptide has two cysteine residues (C28, C51) forming a disulfide bridge which is supposed to maintain the conformation of the polypeptide and ability to regind to the PS II preparation (Tanaka and Wada, 1988; Raval et al., 1994). However, site directed mutant (C51A) and double mutant (C28A, C51A) show that both are capable of binding 33 kD to PS II (Betts et al., 1996). Mutant (C51A) is unstable in light where as double mutant (C28A) has properties similar to wild type PS II. This indicates that disulfide bridge does not affect the binding of 33 kD to PS II reaction center. However, sensitivity of mutant (C51A) to light suggests a light dependant interaction that occurs between disulfide bridge and PS II during the redox-cycling of Mn cluster. This may be explained with the model proposed by Raval et al. (1990) where disulfide bridge interacts with the Mn (II) -Mn (III) dimer of manganese electron trap not directly involved in binding and photooxidation of water. Absence of disulfide bridge does not affect oxygen evolution However, it may participate in the redox reaction of Mn (II) - Mn(III) dimer in S3, S4 state.

In spite of the established fact that Mn cluster catalytically oxidises water to oxygen in PS II, geometry of Mn-cluster, its binding site, oxidation state and precise mechanism of catalysis are yet to be deciphered. Molecular modelling and site directed mutagenesis have thrown diffused light to dig out the facts embedded in the black box of oxygen evolving complex of PS II.

PHOTOSYSTEM I

Photossystem I is a multisubunit membrane-protein complex. It catalyzes electron transfer from the reduced plastocyanin in the thylakoid lumen to the oxidized ferredoxin in the chloroplast stroma or cyanobacterial cytoplasm. In cyanobacteria, plastocyanin and ferredoxin are replaced, depending on the nutritional conditions. Purified cyanobacterial PS I complex contains at least eleven different proteins: the core proteins (PsaA and PsaB), three peripheral proteins (PsaC, PsaD, and PsaE) on the stromal side and six integral membrane proteins (PsaF, PsaI, PsaJ, PsaK, PsaL, and PsaM).

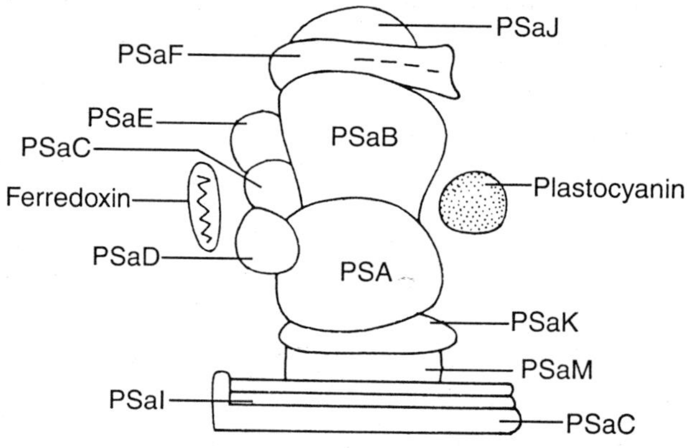

FIG. 4.11: *Diagrammatic presentation of PS I, showing its structural organization.*

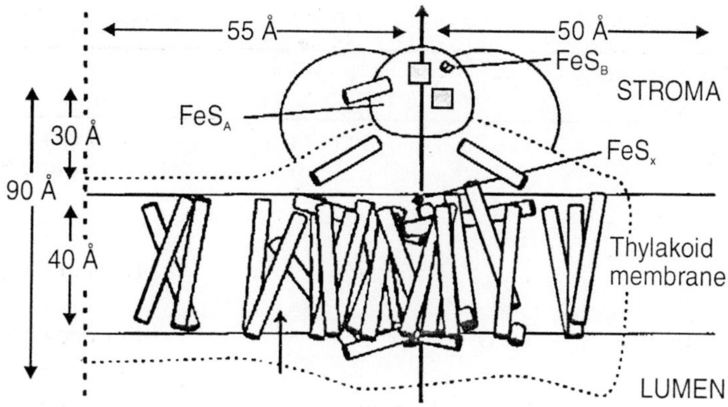

FIG. 4.12: *Structure of Photosystem I from cyanobacteria at 4.0 Å resolution. The figure shows an arrangement of α helices (shown as tubes) viewed parallel to the plane of the membrane. The stromal side of the membrane is at the top, and the lumenal side is at the bottom of the figure. The three FeS centers that are electron acceptors in Photosystem I, FeS_A, FeS_B, and FeS_X are visible on the stromal side as cubes. Pigments are removed for clarity, molecular dimension in Å are shown.*

These PS I proteins, their characteristics and their functions are listed in Table 4..1. In PSI, the reaction center and internal antenna are combined in a single Chl-protein complex that accepts electron from plastocyanin or a cytochrome and delivers them to a small ferredoxin-like molecule, encoded by the psaC gene. This Chl-carotenoid protein consists of two hydrophobic polypeptides of about 82 KDa (PsaA and PsaB gene products) that bind the Chl a and dimer (P700) and the initial electron acceptors A_0 (Chl a), A_1 (Phylloquinone) and A_2 (a 4 Fe-4S center). The two polypeptides together bind 75-100 additional Chl a and 12 – 15 β-carotene molecules. These additional chlorophyll molecules are the "built in" Chl a antenna, filling the same role as the separate antenna complexes CP47 and CP43 in PSII. PS I converts the energy of photons into chemical energy with a remarkable quantum efficiency of 39 to 44%.

In the past 10 years, multidisciplinary research efforts have led to characterization of the components, the electron transfer reactions, and the structure of PS I complex. In this chapter, we will discuss recent progress in the understanding of the function and structure of PS I complex. Detailed discussion of LHCs proteins, the electron transfer and redox cofactors in PS I genes can be found in the recent reviews (Brettel, 1997; Chitnis; 1996; Chitnis, 1997; Fromme, 1996; Green and Durnford, 1996; Ort and Yocum, 1996).

TABLE 4.1: Some subunit proteins of PSI complex (for details refer to Ikeuchi, 1992)

Component	Molecular mass (kDa)	Gene	Location	Function
PSI-A	81-84	psa A©	Transmembrane	Binding site for Chl a, P700, A_0, A_1 and F_X
PSI-B	81-83	psa B©		
PSI-C	9	psa C©	Peripheral	Binding site for F_A and F_B
PSI-D	15-18	psa D		Ferredoxin-docking
PSI-E	8-11	psa E		Ferredoxin- docking
PSI-F	16-19	psa F	Transmembrane and peripheral	Plastocyanin-docking
PSI-I	4-5	psa I©	Transmembrane	?
PSI-J	4-5	psa J©	Transmembrane	?

Notes: The protein protein subunits are usually separated by SDS-PAGE of the PSI complex and identified by protein sequencing and/or gene cloning techniques. Genes denoted by © are coded by the chloroplast, and others by the nuclear genome.

Functions of Photosystem I

- PS I is a unique light-driven generator of the reducing power.
- It functions with PS II in a linear electron transport pathway to generate strong reductants, the reduced ferredoxin and NADPH.
- The products of PS I are among the most electronegative reductants in a biological system. These reductants are used in diverse biochemical processes, including Calvin cycle, nitrate reduction to ammonia, fatty acid desaturation, pyruvate decarboxylation, glutamate synthesis, sulfite reduction, and enzyme regulation through reduced thioredoxin.
- It plays a vital role in cyanobacteria and chloroplast metabolism.

It also participates in the PS II-independent cyclic electron transport pathway which results in proton pumping and ATP synthesis.

Linear Electron Transfer in PS I

In the linear electron transfer, PS I and PS II function according to the Z-scheme. The PS II and PS I reaction centers form the light-driven uphill sections where the energy of a photon is used to increase the reducing ability. The electron transport between the two photosystems involves cytochrome b_6f complex. Among these three integral membrane complexes, the electron transfer is linked by mobile electron carriers: plastoquinone transfers electrons from PS II to the cytochrome bf complex which, in turn, donates electrons to PS I via plastocyanin. In the linear electron transfer, PS I is involved in three steps of electron transfer:

- From plastocyanin to the oxidizing side of PS I.
- Within the PS I complex.
- From the reducing side of PS I to ferredoxin.

Interactions on the oxidizing side

On the lumenal side of thylakoid membranes, PS I accepts electrons from plastocyanin. In cyanobacteria and algae, cytochrome c_6 can serve as an alternative electron donor to the P700 reaction center.

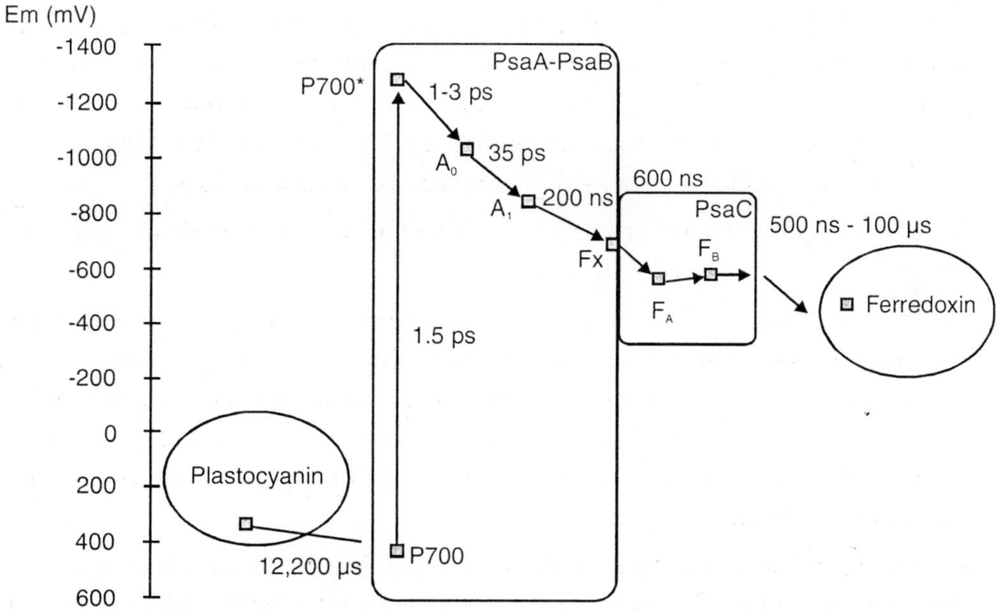

FIG. 4.13: *Electron transfer in PS I. The electron carriers are arranged according to the electron flow on the scale of midpoint redox potential (Em). Typical values for the half-time for the forward reactions are indicated.*

Electron carriers of the PS I complex

The electron transfer within the PS I complex involves a series of electron carriers with increasing electronegative redox potentials E_m from –1290 to –530 mV. The redox centers in the PS I complex are as follows:

1. **P700:** It is a dimer of chlorophyll *a* molecules oriented perpendicular to the thylakoid membranes. Upon transfer of excitation energy from the core antenna system, the primary electron donor P700 undergoes charge separation, producing $P700^+$. The electron lost by P700 is gained by oxidizing plastocyanin.

2. **A_o** : The charge separation of P700 is stabilized by spatial displacement of electrons through a series of low potential redox centers. The primary acceptor A_o is reduced to the anion form upon transfer of an electron from the excited P700. The spectroscopy experiments have supported identity of A_o as a monomeric form of chlorophyll *a*. There are four chlorophyll molecules arranged in two-fold symmetrical positions between P700 chlorophyll pair and the Fe-S cluster. A pair of chlorophyll *a* molecules positioned 16 Å from P700 towards the stroma is approximately half-way between P700 and F_X.

   Another pairs of chlorophyll *a* monomers are located approximately half-way between P700 and the A_o equivalent chlorophyll molecules. They are assigned as accessory chlorophylls. Thus there are two possible branches of electron transfer pathway between P700 and F_X. It is not known if only one or both sides of possible electron transfer pathway are used during the PS I function. It is also not known what is the role of the accessory chlorophylls in the electron transfer.

3. **A_1** : During electron transfer in PS I, A_1 serves as an intermediate electron transfer center that oxidizes A_o and reduces F_X.

4. **F_X** : The next intermediate acceptor in the electron transfer pathway within PS I is the [4Fe-4S] cluster F_X. It is clearly identified in the crystal structure due to its high electron density. It is located near the stromal surface at the edge of membrane plane. F_X is found at the center of the PS I complex from the top view. F_X is 32 Å away from P700 along the membrane normal and 14 Å from the center of the phylloquinone binding pocket.

5. **F_A and F_B** : The F_A and F_B clusters were the first to bound electron acceptors, identified in the PS I complexes. They are bound to the PsaC subunit, unlike other electron carriers which are bound to the core proteins. Like the F_X center, they are also [4Fe-4S] in cluster.

Interactions on the reducing side

On the reducing side of PS I., electrons are transferred from F_A/F_B clusters to ferredoxin. The PS I-ferredoxin complex formation precedes electron transfer and the rate constants for complex formation depend on ionic concentration, suggesting electrostatic interaction between ferredoxin and PS I. Ferredoxin accepts electrons from F_A/F_B clusters of PsaC, implying that these proteins should contact each other.

Therefore, PsaD, PsaE and PsaC are the essential components of the reducing side of PS I. Structure of the reducing side of PS I is not understood at a fine resolution.

Cyclic Electron Transfer

In addition to the PS II-dependent linear electron transfer, PS I also participates in the cyclic electron flow that results in proton pumping and ATP production without the participation of PS II. Cyclic electron flow is essential in the bundle sheath cells of C4 plants that predominantly contain PS I, but its contribution to ATP generation in C3 plants and in cyanobacteria is rather limited.

Cyclic electron transfer has also been proposed be involved in photoprotection and in fine-tuning of the balance between ATP and NADPH production (Bendall and Manasse, 1995).

Overall, the PS I complex utilizes energy of the photons trapped by either its core antenna system or the associated LHCs to function in linear or cyclic pathway producing strong reductants and ATP, and plays an important role in nourishing the whole world.

The Catalytic Subunits

As a light-driven oxidoreductase, PS I catalyses the electron transfer from the reduced plastocyanin to the oxidized ferredoxin. The electron carriers participating the electron transportation are bound to the PsaA, PsaB, and PsaC proteins.

1. **PsaA and PsaB :** These proteins form the hydrophobic core of PS I. The hydropathy analysis of PsaA and PsaB indicated that each protein contains 11 potential transmembrane domains. In higher plants, the genes encoding PsaA and PsaB are present in the chloroplast. The PsaA and PsaB proteins might have evolved from gene duplication and are thought to have similar structural features.The deduced molecular masses of PsaA and PsaB are approximately 83 kDa and 82 kDa, respectively.
2. **PsaC :** It is an acidic peripheral protein with 80 amino acids. It coordinates the terminal electron transfer centers F_A and F_B of PS I. In higher plants, the *psa*C gene is located in the chloroplast genome. PsaC is an extraordinarily conserved protein. PsaC obviously interacts with the core proteins. Chemical cross-linking studies have demonstrated that PsaC can be cross-linked to PsaD and PsaE proteins. So far, many studies have been performed to describe the structure-function relationships in PsaC. However, two questions are still ambiguous:
    - Are both [4Fe-4S] clusters both essential for electron transfer?
    - Which cluster, F_A or F_B, is proximal to F_X?

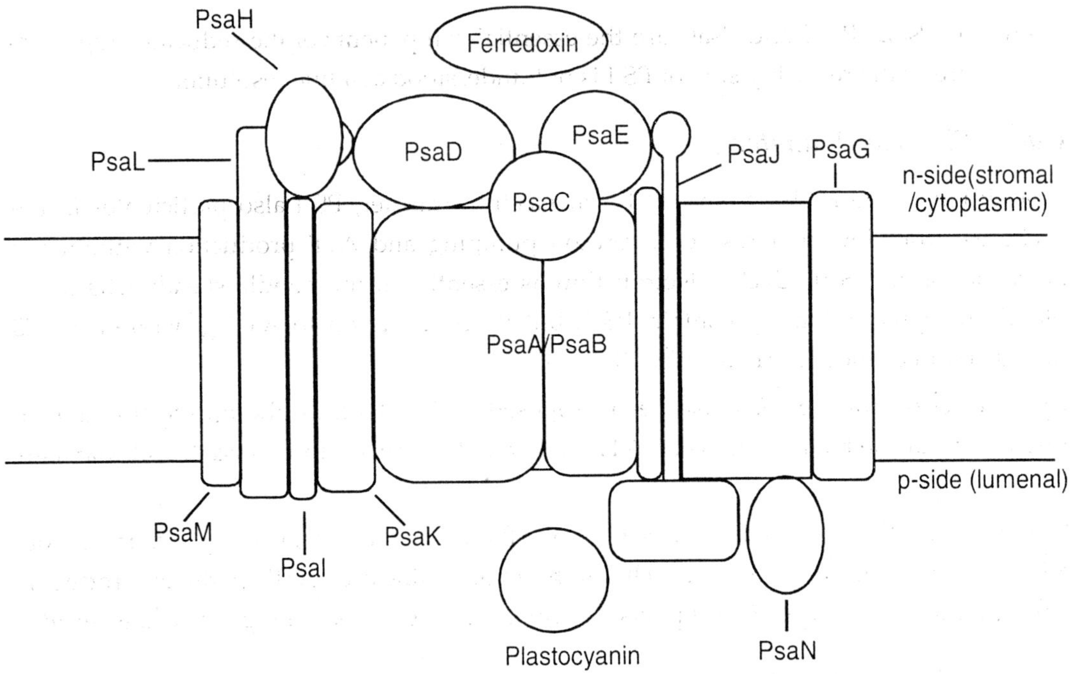

FIG. 4.14: *A model for organization of PS I proteins. A side view of PS I proteins.*

The Docking Subunits

PS I complex accepts electrons from plastocyanin in the thylakoid lumen and donates electrons to ferredoxin in the chloroplast stroma or cyanobacterial cytoplasm. PsaD and PsaE are involved in the docking of ferredoxin, whereas PsaF may interact with plant plastocyanin to form complex for fast electron transfer.

1. **PsaD :** The PsaD subunit of PS I is conserved peripheral protein on the reducing side of PS I. In higher plants, the PsaD protein is encoded by the nuclear gene *psa*D. In general, the cyanobacterial PsaD protein contains 139-144 amino acids while the plant PsaD has about 23 additional residues forming an extension at the N-terminus. This extension of eukaryotic PsaD is accessible to proteases. The insertional inactivation of the *psa*D gene of Synechocystis sp. PCC 6803 indicated that PsaD is essential for efficient function of the cyanobacterial PS I (Chitnis et al., 1989). It interacts with at least three proteins of PS I and stabilizes their organization within the complex and is required for the stable binding of PsaC to the PS I core proteins (Chitnis *et al.*, 1996).

Crystallographic studies at limited resolution have shown that PsaD contains a single, short –helix located near the common interface with PsaC and the integral membrane core subunits which may be important for the stable association of both PsaC and PsaD with the PS I core (Krauß *et al.*, 1996).

2. **PsaE :** PsaE is a moderately conserved peripheral subunit of PS I (Schubert *et al.*, 1997). In higher plants, *psa*E gene is located in the nucleus. Cyanobacterial PsaE contains 70-75 amino acid residues whereas the plant and algal PsaE has an N-terminal extension. Similar to the plant accessible to exogenous proteases (Lagoutte and Vallon, 1992). PsaE facilitates the interaction between PS I and ferredoxin. PsaE may also be required for the cyclic electron transport around PS I (Chitnis *et al.*, 1995).

3. **PsaF :** PsaF is a membrane-integral protein with substantial N-terminal domain exposed on the lumenal side of PS I. The crosslinking product of PsaE and PsaF demonstrated the transmembrane nature of PsaF. Based on biochemical depletion and chemical cross-linking experiments, PsaF has been proposed to function as the plastocyanin or cytochrome c_6 docking site on the oxidizing side of PS I.

Other Proteins of Photosystem I

The other subunits of PS I do not bind electron carriers. They are not involved in docking of soluble electron carriers. However, some of these proteins provide essential structural roles in the organization of PS I complex. Also, these subunits contribute to a roughly continuous ring of transmembrane helices around the core antenna system of PS I. Thus, they may provide a partially enclosed framework to the antenna system, thereby providing protection and stability.

PsaI, PsaJ and PsaM are small hydrophobic proteins in PS I complex, which usually contain one putative transmembrane helix flanked by short hydrophilic, charged domains. These proteins may function as nuts and bolt in the organization of PS I complex. The may assist in the correct organization of other subunits by anchoring peripheral proteins and stabilizing transmembrane helices in lipid bilayer (Xu *et al.*, 1995). The PsaJ-PsaF interaction may involve stabilization of the transmembrane helix of PsaF (Xu *et al.*, 1994). PsaI stabilize the association of PsaL with the PS I stabilize the association of PsaL with the PS I core (Xu *et al.*, 1995). Whether PsaM has the similar role in organization of other subunits has not been identified.

1. **PsaI and PsaL :** PsaL proteins contains a greater diversity than seen in other subunits. The protein contains several conserved regions of homology, most notably in the hydrophobic regions. Function of PsaL was revealed by the inactivation of the *psa*L gene.

PsaI is an integral membrane protein with one transmembrane helix. This presequence of the *Anabaena variabilis* PsaI is predicted to assist the thylakoid translocation and consequently the N-terminus of PsaI may be located in the lumen. PsaI in intact thylakoids is not accesible to external proteases.

2. **PsaJ :** PsaJ is a 4.4-kDa hydrophobic protein. It has been identified in PS I preparations from cyanobacteria and higher plants. Analysis of the deduced amino acid sequence reveals a highly conserved sequence of 22 amino acids which may span the membrane. The hydrophilic N-terminal domain of PsaJ is only 4-5 amino acids long while the C-terminal domain has overall negative charge and more than 15 amino acids.
3. **PsaK :** It is an integral membrane protein that is tightly associated with the PsaA-PsaB complex.
4. **PsaM :** PsaM is a very small protein of about 29 amino acids and has been detected only in cyanobacterial PS I. It helps in stabilizing the trimer in co-operation with PsaK, a similar 'nuts and bolt' function as PsaI and PsaJ.
5. **PsaH, PsaG, and PsaN :** PsaH, PsaG, and PsaN are the three eukaryote-specific subunits of PS I. The conding genes for these three polypeptide *psa*H, *psa*G, and *psa*N are all present in the nucleus.
   a. **PsaG** is an integral-membrane protein with about 100 amino acids, which has been only found in eukaryotic PS I complex.
   b. **PsaH** is another PS I subunit that has only been detected in algae and higher plants. The mature PsaH protein contains an N-terminal acidic domain, a central hydrophobic domain and a C-terminal basic region. PsaH is an extrinsic subunit on the stromal side of PS I.
   c. **PsaN:** The mature PsaN is 85 amino acids in length, with no putative transmembrane domain. It is a 9 kDa extrinsic polypeptide of plant PS I located on the *p*-side of thylakoids. It does not bind any cofactors and is not required for $NADP^+$ photoreduction. The cDNA clones encoding the PsaN subunits have been isolated from barley and maize.

The Cytochrome b_6f Complex: The Q Cycle

The cyt b_6f complex transfers electrons from PSII to PSI during linear electron flow from water to NADP. It can be considered as a plastoquinol-plastocyanin oxidoreductase. Recent work by J. Anderson (in Canberra) and others suggests that the cyt b_6f complexes are distributed in three domains of the thylakoid membrane: the appressed granal stacks (along with PSII), the non-appressed stromal thylakoids (along with PSI), and the non-appressed grana margins.

The b_6f complex can be isolated from chloroplasts by detergent fractionation and then further resolved into individual protein components by treatment with denaturing agents. The isolated complex contains four major polypeptides: cyt b_6 (molecular mass 23 kDa, *pet* B gene product), cyt *f* (31 kDa, *pet* A), the Rieske (discoverer) iron-sulphur protein FeS_R (20 kDa, *pet* C), and subunit IV (17 kDa, function unknown). The electron transfer sequence within the complex is: $PQH_2 \rightarrow FeS_R \rightarrow cytf \rightarrow PC \rightarrow PSI$. Cyt*f* is held in the thylakoid by a single transmembrane span located near its hydrophobic C-terminus, with a globular hydrophilic segment protruding into the lumen. Concomitant with electron transfer, the complex also translocates protons from the stromal to the lumenal side of the membrane.

Other functions of the b_6f complex are electron transfer from reduced Fd to PC via b_6 during cyclic photophosphorylation, and possibly in the regulation of light energy distribution between the two photosystems (in response to ATP demand) via LHC II phosphorylation-dephosphorylation reactions. It has been postulated that the lateral movement of phosphorylated LHC II from the grana ($PSII_\alpha$) membranes is accompanied by a simultaneous movement of the cyt b_6f from the grana to the stromal thylakoids ($PSII_\beta$ and PSI).

The cytochrome b_6f complex is present in both grana and stroma lamellae. It has generally been assumed that the cyt b_6f complex is mobile and can free move between the grana and the stroma lamellae i.e. it is the same complex in the two membrane regions. It has, however, also been suggested that the complex in the grana, cyt $(b_6f)_\alpha$, differs in structure and function from the complex in the storms lamellae cyt $(b_6f)_\beta$. The two pools of cyt b_6f, cyt $(b_6f)_\alpha$ and cyt $(b_6f)_\beta$, may have different functions. It has been suggested that the cyt bf of the appressed grana membrane is reduced by PSII light while the stroma lamellae cyt bf is oxidized by PSIβ.

This conclusion has gained experimental support by the work of Joliot and Joliot (1992) where they showed only 2/3 of cyt f (very likely located in the grana regions) is accessible to electrons arising from PS II reaction. This also means that plastocyanin can not be a shuttle for rapid equilibrium between cyt $(b_6f)_\alpha$ and cyt $(b_6f)_\beta$. Romanowska and Albertsson (1994) also showed that plastocyanin is bound to both Cyt b_6f complex populations.

It has been suggested that PSI_β of the stroma lamellae carries out cyclic photosynthetic electron transport while the $PSII_\alpha$ and PSI_α in the grana carry out non-cyclic oxygenic electron transport (Albertsson et al., 1990). The grana cyt b_6f complex participates in non-cyclic electron transport while the cyt b_6f complex of the stroma lamellae is restricted to function in cyclic electron transport. These functional differences between the two complexes also involve some differences in structure. The 15 KDa protein present in the grana cyt b_6f complex anchors the grana cyt b_6f complex to the grana partitions (by biding to $PSII_{\propto}$) and also regulates the activity of the complex for non-cyclic photosynthetic electron transport.

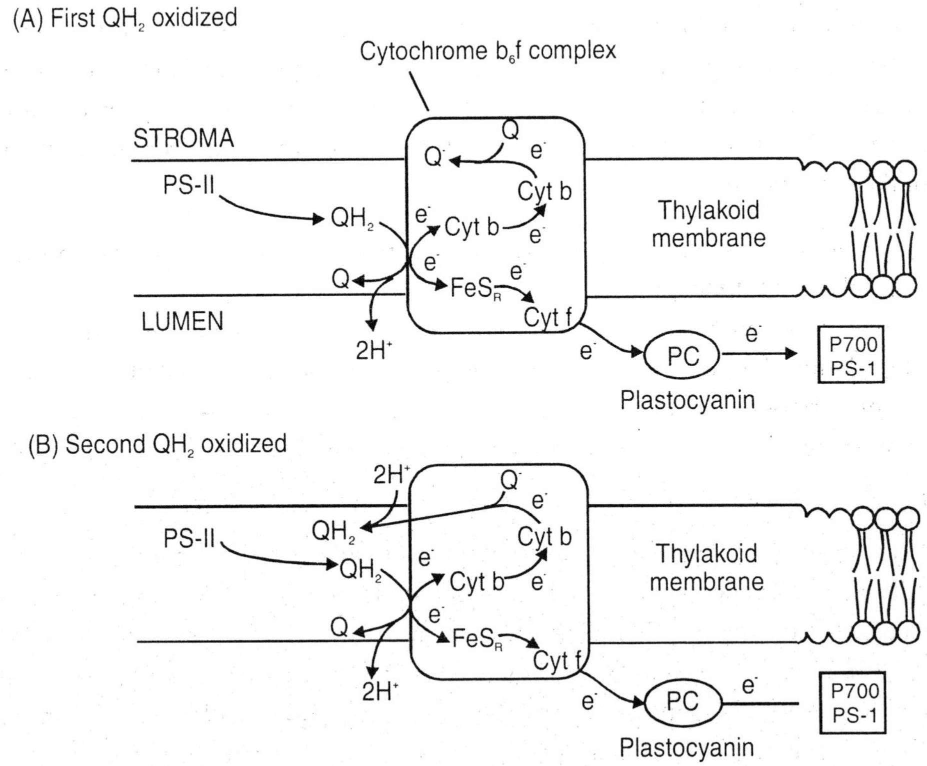

FIG. 4.15: *Mechanism of electron and proton transfer in the cytochrome* b_6f *complex. This complex contains two* b*-type cytochromes (Cyt* b*), a* c*-type cytochrome (Cyt* c*, historically called cytochrome* f*) a Rieske Fe-protein (FeS$_R$), and two quinone oxidation-reduction sites. (A) A plastohydroquinone (QH_2) molecule produced by the action of P II is oxidized near the lumenal side of the complex, transferring its two electrons to the Rieske Fe-S protein and one of the* b*-type cytochromes and simultaneously expelling two protons to the lumen. The electron transferred to FeS$_R$ is passed to cytochrome* f *(Cyt* f*) and then to plastocyanin (PC), which reduces P700 of PS-I. The reduced* b*-type cytochrome transfers an electron to the other* b*-type cytochrome, which reduces a quinone (Q) to the semiquinone (Q^-) state. (B) A second QH_2 is oxidized, with one electron going from FeS$_R$ to PC and finally to P700. The second electron goes through the two* b*-type cytochromes and reduces the semiquinone to the plastohydroquinoen, at the same time picking up two protons from the stroma. Over a four protons are transported across the membrane for every two electrons delivered to P700.*

The mechanism of proton translocation across the thylakoid membrane mediated by the b_6f complex is still not understood. There are many similarities (and differences) in the structure, composition and function of the chloroplast b_6f complex and the *bc* complex in the mitochondrial respiratory chain of mammals and bacteria. Cytochrome *f*, so named because it was detected by its unique absorption at 555 nm in leaves, is truly a *c*-type cytochrome. Mitchell had proposed a *Q cycle* to explain the mechanism of proton translocation via the mitochondrial ubiquinone-bc_1 complex, the key step being an electrogenic electron transfer through the site of quinol oxidation. Does a similar Q cycle operate in the chloroplast cyt b_6f complex? The electron and protein transfer across the membrane can be measured from light-induced changes in the absorption at 320 nm due to proton uptake. The participation of the b-type hemes in the reaction sequences can be detected from the electrochromic absorption changes at 515/524 nm that occur when thylakoids are illuminated by single turnover flashes.

Proponents of the Q cycle operation propose that during electron transport FeS_R accepts one electron from PQH_2 (Q_B doubly reduced by PSII picks up two stromal protons and forms PQH_2) and donates it to cyt*f*. Simultaneously, PQH_2 transfers its second electron to one of the b hemes (low potential) and releases the two protons to the lumen. The electron from the low potential b heme is transferred to PQ via the high potential b heme of the cyt b_6. The PQ gets doubly reduced after two cycles of electron transfer; doubly reduced PQ picks up two more stromal protons and enters the quinone pool.

Although some authors believe that the b_6f complex functions electrogenically during electron transport, there are others who consider the evidence as equivocal. Furthermore, if the Q cycle is truly operative, $2H^+$ should be translocated per electron transferred to cyt f; experimentally determined values of H^+/e^- range from 2 to 3. The mechanism of proton translocation across the thylakoid membrane is still an open question; maybe a modified catalytic Q cycle or a proton pump is involved in the translocation.

Plastocyanin

Plastocyanin (PC) is a small (97-104 amino acids) copper-binding protein whose function in oxygenic photosynthesis is the catalysis of electron transfer from Cyt f in the Cyt b_6f complex to $P700^+$ in the photosystem I. It is referred to as a "blue" copper protein on the basis of its spectroscopic properties. The protein is an eight-standard, antiparallel β-barrel with a single copper atom liganded by the side chains of two histidines, a cysteine and a methionine. In the green algae and cyanobacteria, plastocyanin function is replaced by C-type cytochrome.

Plastocyanin performs three main functions:

- First, it must maintain its copper binding site so that the redox potential of the protein lies between that of its physiological reductant (cytf) and oxidant ($P700^+$).

- Second, it must provide specific binding site for its physiological reaction partners.
- Third, it must provide the proper structural scaffolding for controlled electron transfer into and out of the copper center.

CFo-CF1 Complex: ATP Synthase

The CFo-CF1 complex catalyzes the synthesis of ATP by creating an electrochemical potential gradient of protons. Thus, the rate of cyclic and non-cyclic electron transport controls the rate of proton build-up which in turn determines the rate of ATP formation. The enzyme consists of an extramembraneous F_1 catalytic domain linked by means of a central stalk to an intrinsic membrane domain called F_0. The globular F_1 domain is an assembly of five different subunit with the stoichiometry of $\propto_3\beta_3\gamma_1\delta_1\varepsilon_1$. In the atomic structure of bovine F_1, the $\propto$ and β subunits are arranged alternately around a coiled coil of two anti parallel $\propto$ helices in the γ subunit. The catalytic sites are in the β subunits at the $\propto/\beta$ subunit interface. The remainder of γ subunit protrudes from the $\propto_3\beta_3$ assembly and can be cross-linked to the polar loop region of the c subunits in F_0. In mitochondria, the δ and ε subunits are associated with the γ subunit in the central stalk assembly as are the bacterial and chloroplast ε subunits, the counterparts of mitochondrial δ. ATP-dependent rotation of γ and ε within an immobilized $\propto_3\beta_3$ complex from the thermophilic bacterium Bacillus PS3 has been observed directly.

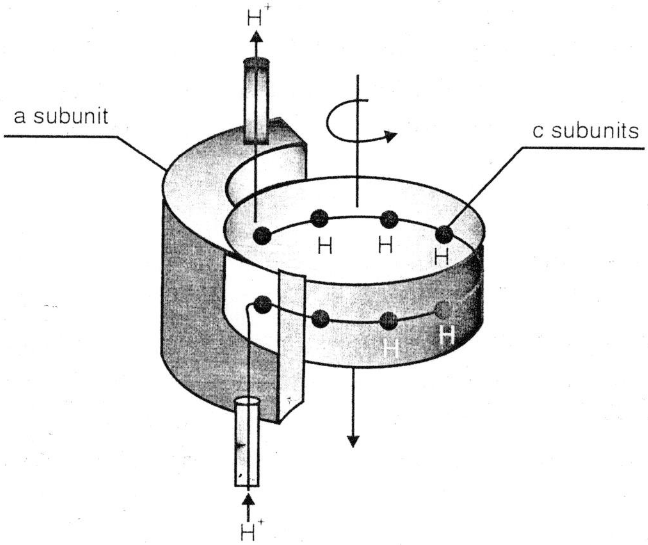

FIG. 4.16: *Hypothetical model for the generation of rotation by portion transport through the F_0 domain of ATP synthase. The central cylinder consists of c subunits; the external part corresponds to a single a subunit. The line indicates the proton path.*

The rotation of the γ subunit in ATP synthase is thought to be generated by the passage of proton through F_0. Because there is only one intrinsically asymmetric γ subunit, it interacts differently with each of the three catalytic β subunits in F_1, endowing them with different nucleotide affinities. Rotation of the central stalk is accompanied by conformational changes in the β-subunits, which cycle sequentially through structural states corresponding to low, medium and high nucleotide affinities. These three states are probably associated with release of product (ATP), binding of substrates (ADP and inorganic phosphate) and ATP formation respectively. The cycle is known as the "binding exchange mechanism" of ATP synthesis. In eubacteria, the procedure is reversible and hydrolysis of ATP in F_1 generates rotation of γ resulting in the pumping of proton back across the membrane through F_0.

HETEROGENEITY OF PHOTOSYSTEM

The studies have shown that the thylakoid has a very complex membrane structures, both with respect to its domain organization and the heterogeneity of the two photo systems. A major fraction of photosystem II, in the form of PSII∝, is located in the appressed, central core of the grana while the remaining photosystem II, $PSII_\beta$, is in the stroma lamellae and the grana margins. $PSII_\propto$ has a larger antennae than $PSII_\beta$ and can reduce plastoquinone while $PSII_\beta$ can not. About 36 per cent of the PSI is localized in the periphery of the grana (margins) in the form of $PSI_\propto$ which has a greater antenna size than the PSI_β of the stroma lamellae.

CHAPTER 5

Photoionization and Photophosphorylation

In fact, photosynthesis involves two quite different types of reactions: **physical reaction** (it is closely tied to the absorption of radiation energy) and **chemical reaction(s)** (related with assimilation of CO_2 and its conversion to carbohydrates- produced after reduction of CO_2 - and related cellular constituents). During these reactions oxygen is also evolved. The reduction of CO_2 in green plants is not directly related with light energy but with a product of the light reaction.

The presence of water molecule is essential, during light reaction bonds of water molecules are broken and electrons are captured; and oxygen and hydrogen ions are released. The electrons from water replace those lost electron from the reaction centre of PS II (after capturing of light by PS II). The energy is coupled to a movement of electrons to an acceptor Q, which looses electron to electron transfer chain with each sequential acceptor representing lower redox potential. In each step absorbed radiation energy is transferred to enzymatically copuled ATP synthesis (photophosphorylation). The electrons are then migrated into PS I where light reaction results in transfer to its acceptor Ao. NADP coenzyme would be available as an acceptor for electrons. ATP and NADP (NADPH) are used during Calvin cycle, where CO_2 is reduced.

LIGHT REACTION

The chemistry of photosynthesis can be summarized as a process by which water and carbon dioxide (CO_2) react to form carbohydrates. The photochemical reactions of photosynthesis starts with the capture of light, which requires intimate association of many chlorophyll molecules with compounds that catalyze chemical reactions. The details of the process are still the matter for active research because they are partly understood. The reaction starts, however, by the packet of light energy being absorbed by chlorophyll. When this happens, the chlorophyll molecule becomes excited. This excited chlorophyll molecule is capable of doing work, in the chemical sense that it can pass on its energy to other compounds. In the chloroplast, chlorophyll molecules are arranged in linked networks, wherein some 200 molecules feed energy into a reactive centre. Two such reactive centres which work sequentially (under normal conditions) are involved in the process of capturing light energy and storing it in energy-rich molecules that can be used in synthetic reactions.

EVIDENCES FOR THE EXISTENCE OF LIGHT AND DARK REACTIONS

It is well established that efficient photosynthesis process includes two reactions- light and dark. The evidences are:

(a) If radioactive CO_2 is supplied to the *Chlorella* suspension after the light is turned off, photosynthetic products are formed as usual.

(b) If the suspension of *Chlorella* is irradiated with intermittent light, it performs more efficient photosynthesis as compared to the suspension exposed with continuous light. This shows that amount of light is more efficiently used by *Chlorella* if a dark period follows light period. This occurs because in intermittent light all or most of the products of photochemical reaction are removed by a dark reaction during the intervening dark period, and the photosynthetic output per unit of light is more. Under continuous light, the products of light phase accumulate because the pace of light reaction is far greater than that of dark.

Hill Reaction or Photochemical Reaction or Photoexcitation or Splitting of Water

This is the first reaction in photosynthesis which occurs in presence of light. Absorption of photon of light of wavelength less than 680 nm by PS II triggers the separation of electron from P_{680} chlorophyll molecule. The result is an oxidized form of P_{680}. It is the direct pathway for utilization of electron energy in which chlorophyll on absorbing a light quantum ($h\upsilon$) becomes *excited* (photoexcitation). The electrons are raised to a high energy level and this energy is directly utilized by grana for separation of water into H and OH (not ions). H atoms are ionized into H^+ and e^- and OH radicles are rearranged to form water and molecular oxygen. One quantum is sufficient for the production of one high-energy electron and one such electron suffices for the liberation of 1H atom from water.

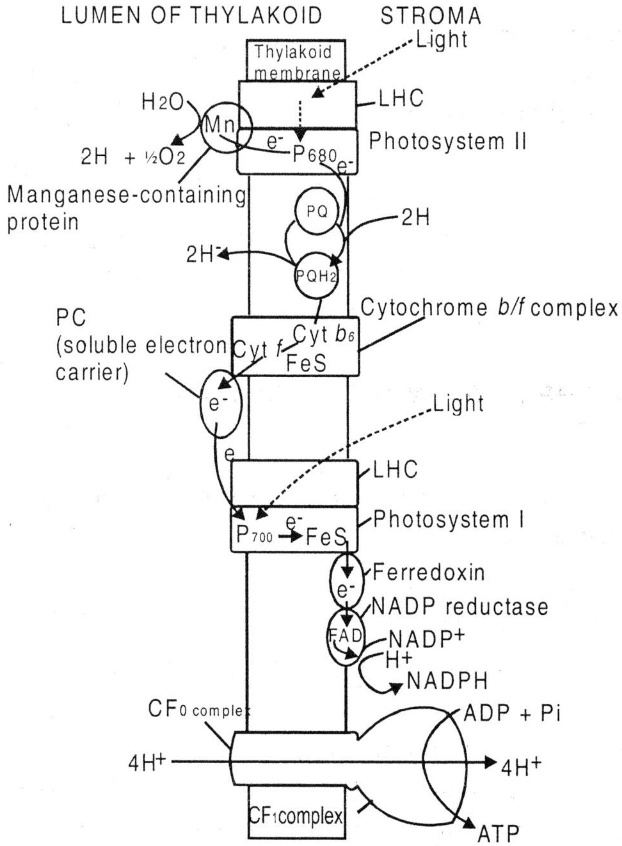

FIG. 5.1: *The orientation of the principal membrane components involved in photosynthesis.*

In photosynthesis 4H atoms ($2H_2$) are needed for later combination with one CO_2 molecule, therefore, 4 quanta are required. Since chlorophyll molecule can absorb only one quanta *at a time*, it means accumulation of energy up to 4 quanta packets must exist in the grana.

The reduction of P_{680} requires an *input* of electrons which are pulled from water molecules tightly bound to *manganese-containing protein*, present on the luminal surface of the P_{680} complex. A glance at the stoichiometry indicates that the oxidation of water is chemically not an easy task.

$$2H_2O \longrightarrow O_2 + 4H+ + 4e^-$$

Since four electrons must be withdrawn from two molecules of water before molecular oxygen can be evolved. PSII can be involved. PSII can only withdraw one electron at a time. There are then two possibilities, either four PSII reaction centers co-operate and pool their oxidizing equivalents to oxidize two molecules of water or each PSII reaction center works independently and has a means of storing oxidizing equivalents until four are collected and can be used together. Each PSII reaction center is imagined to have a charge storage mechanism, which can assume five states: S_0, S_1, S_2, S_3 and S_4 where the subscript denotes the oxidizing equivalents (power to extract electrons) stored. Each absorbed photon transforms one state to the next with equal efficiency S_4 exists only transiently and reverts to S_0 after oxidizing water.

The reaction centre of each PS II contain two manganese ions (Mn^{2+}) in the form of Mn^{2+} associated protein and help in splitting two water molecules into 4 protons, 4 electrons, and one oxygen molecule (O_2). The fait of these photolysis products is as follows:

(i) Protons released from water remain in the lumen of the thylakoid.

(ii) Electron split from P_{680} are immediately transported to outer surface of the thylakoid membrane. The nature of the initial acceptor of the electrons released from PS II is not known.

(iii) The reduction potential of the electron is increased, after excitation of electron by light to +0.8 V- that of H_2O-O_2 couple - to about 0 V. This is equivalent to a change in free energy of about 18 kcal/mol of electrons.

Visualize that light has been absorbed by PS II, an electron has been activated, and O_2 has been evolved. As the activated electron moves along a series of electron carriers between the PS II and PS I networks, it loses some of its chemical potential energy. In that process however, as the electron **flow downhill** on the chemical potential energy scale, one of the excited proteins catalyzes the reaction of ADP with an inorganic phosphate ion to form ATP. Thus, part of energy derived from the photon is transferred to a high-energy phosphate bond in ATP. This is called photophosphorylation. The second alternative for utilization of electron energy is indirect one which comes into play primarily after photoionization- that is complete **removal** of high-energy electron from excited chlorophyll molecules. It is based on oxidation and reduction of electron carriers. Oxidation is the addition of oxygen or the withdrawal of hydrogen or the withdrawal of electrons with or without protons. Reduction is the withdrawal of oxygen or the addition of hydrogen or the addition of electrons with or without the addition of protons.

There are currently three major views about how the energy of the transported electrons is transferred to form ATP:

- **Chemical model:** The phosphorylation of ADP to ATP is accomplished enzymatically (much like glycolytic phosphorylation) and is coupled to the oxidation of the electron carriers.

- **Chemiosmotic model:** The electron flow along the electron transport chain pumps protons across the membrane, against a proton gradient. As the protons *fall* back to their original position, the energy released is used by a membrane-bound protein to phosphorylated ADP to ATP.
- **Conformational model:** The energy released by the flow of electrons changes the conformation of the large molecules residing in the chloroplast/mitochondrion . The return of these molecules to their original shape releases sufficient energy for phosphorylation to occur.

MECHANISM OF ELECTRON TRANSPORT

There exit two separate and alternate pathways of electron transfer away from chlorophyll molecule. PS I is used for both cyclic and non-cyclic photophosphorylation.

Cyclic Photophosphorylation

In this pathway, electrons of the chlorophyll molecules become electrically neutral and ready to be ionized once more. The net result is that electrons have travelled a **complete cycle**, from chlorophyll to chlorophyll. The activated electrons can flow through a series of cytochromes back across the membrane, until they reach PS I on the interior surface.

A series of ferredoxins and cytochromes mediates this electron transport. Concomitant with this movement of electrons is the pumping of additional protons from stromatal space to the thylakoid leumn. Additional ATP is synthesized, via the proton gradient generated during electron transport, but no NADPH is produced. PS II is not involved, and no oxygen is evolved. PS I is used for generation of proton gradient.

Non-cyclic Photophosphorylation

Non-cyclic electron transport involves the flow of electrons from water through PSII and PSI in series to the terminal electron acceptor NADP. Two fates are possible for electrons excited by PS I. Together with a proton, they can be transferred, via the electron carrier FAD, to $NADPH^+$, thus forming the reduced molecule of NADP.

A second possibility is that electrons might be transferred from ferredoxin back to oxygen, reducing it to H_2O. It is possible that this process might also involve an electron transport chain and make ATP. The experimental basis for pseudocyclic photophosphorylation, as it is called, is not nearly so firm as that for cyclic or noncyclic phosphorylation.

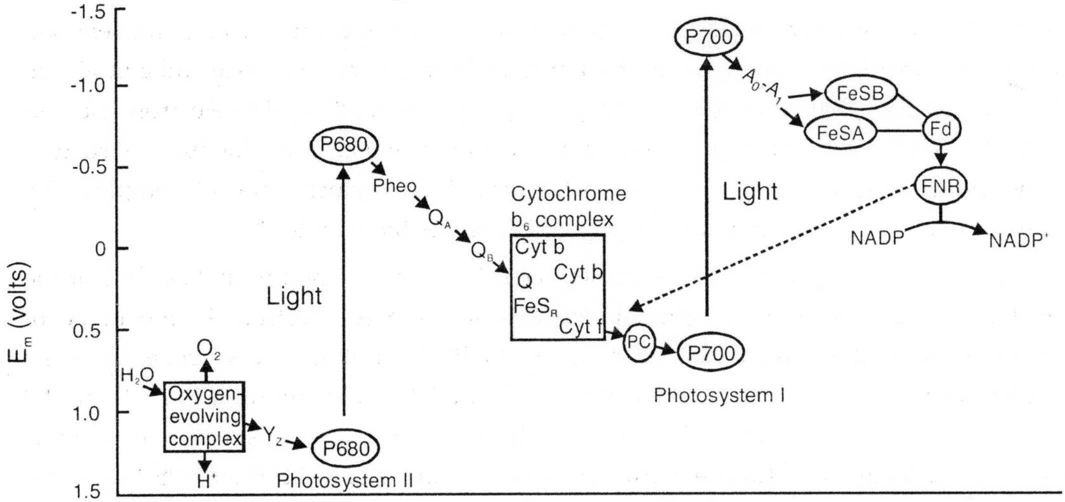

FIG. 5. 2: *Detailed Z scheme for O_2-evolving photosynthetic organisms. The redox carriers are placed at their accepted midpoint redox potentials (at pH 7). The vertical arrows signify photon absorption by the reaction center chlorophylls : P680 for photosystem II and P700 for photosystem I. The excited PS-II reaction center chlorophyll, P680*, transfers an electron to pheophytin (Pheo). On the oxidizing side of PS-II (to the left of the arrow joining P680 with P680*), P680 oxidized by light is rereduced by Y_Z, the immediate donor to PS-II. Y_Z is a tyrosine on the reaction center protein D1. (The electron donor, Z, has been shown to be a tyrosine in the reaction center protein D1. The tyrosine amino acid is designated by Y. Hence what has been designated Z in chemical schemes of electron transport, is now called Y_Z). Electrons are extracted from water by the oxygen-evolving complex and rereduce Y_{Z+} (oxidized Y_Z). On the reducing side of PS-II (to the right of the arrow joining P680 with P680*), pheophytin transfers electrosn to the plastoquinone (PQ) acceptors Q_A and Q_B. The cytochrome b_6f complex transfers electrons to plastocyanin (PC), a soluble protein, which in turn reduces $P700^+$ (oxidized P700). The b_6f complex contains a Rieske iron-sulfur protein (FeS_R), two b-type cytochromes (Cyt b) and cytochrome f (Cyt f). Electron flow within this complex is shown. The acceptor of electrons from P700* (A_0) is thought to be a chlorophyll, and the next acceptor (A_1) is a quinone. A series of membrane-bound iron-sulfur proteins (FeS_X, FeS_B, and FeS_A) transfers electrons to soluble ferredoxin (Fd) reduces $NADP^+$ to NADPH, which is used in the Calvin cycle to reduce CO_2. The dashed line indicates cyclic electron flow around PS-I. Pseudocyclic phosphorylation.*

TABLE 5.1: Differences between cyclic and non-cyclic photophosphorylation

Cyclic photophosphorylation	Non-cyclic photophosphorylation
Only PS I is functional	Both PS I and PS II are functional
It functions in close circle	It requires independent electron donor.
Electron freed from chlorophyll to acceptor and return to chlorophyll	Water is the ultimate source of electrons and NADP+is final electron acceptor
$NADPH_2$ is not formed	$NADPH_2$ is the reduced compound
O_2 is not evolved	Oxygen is evolved.
The system is common in bacteria	Very common in green plants.
Dichlorophenyldimethyl urea(DCMU) has no effect on the system.	DCMU stops the reaction.

The mechanism of electron transport from PS II to PS I is complex. Electron released from PS II immediately combine with two protons from the stroma space and a molecule of plastoquinione (PQ), to generate reduded hydroquinone, PQH_2. The electron released by PQH_2 are transferred to the cytochrome *b/f* complex, which is analogous in structure to the cytochrome *b/c* complex of mitochondria. From cytochrome *b/f* complex, the electrons move to the carrier plastocyanin, which passes them on to PS I.

Plastocyanin is a peripheral protein present closely bound to the luminal surface of the thylakoid membrane. It may immediately diffuse, with its electron, in the plane of membrane from the cytochrome *b/f* complex to PS I, or it may dissociate from the membrane and diffuse in the luminal space of the thylakoid. In chloroplast, PS II and PS I complexes are not uniformly distributed between stacked and unstacked membrane, apparently a **single structural electron system** containing both PS II and PS I does not exist. Rather, **mobile** carriers such as PC and PQ are used to transfer electrons between these systems.

Z SCHEME

In the "Z" scheme, light striking the aggregation of molecules designated PSII results in the transfer of an electron from a donor called Z (now known to be a tyrosine residue) to an acceptor called Q_A, a bound plastoquinone molecule. The PSII where this reaction occurs, contains several polypeptides including a 32-KDa D1 protein, a 34-KDa D_2 protein, a cytochrome b_{559} and three extrinsic polypeptides having molecular masses of 17, 23and 33 KDa. Z is located on D_1 and Q_A on D_2. The electron transport from Z to Q_A involves several steps:

- Light energy produces the excited reaction center chlorophyll a molecule (P680*) from its unexcited state (P_{680}) located between D1 and D2.
- Within 3 picosecods, an electron from P_{680^*} is transferred to a pheophytin molecule producing an oxidized P_{680} and a reduced pheophytin.
- Within 200 picoseconds, the reduced pheophytin transfers the electron to Q_A.
- Within 20-300 manoseconds, Z is oxidized and P680* is reduced.

From reduced Q_A the electrons are transferred down the electrochemical gradient to PSI. Several kinds of molecules serve as intermediaries including plastoquinone molecule Q_B located on D1, mobile plastoquinone molecule, a Rieske iron sulphur protein, a cytochrome f and a copper-containing protein, plastocyanin. The cytochrome b_6f complex contains two type of cytochrome b and a 17 KDa protein in addition to Rieske iron-sulphur protein and cytochrome f.

On absorbing a quantum of light, PSI promotes the transfer of an electron from plastocyanin to another iron-sulphur protein (FeS). The electron transport from plastocyanin to FeS involves following steps:

- Light energy produces the excited reaction centre chlorophyll a molecule (P_{700}) from its unexcited state.
- Within about 10 picoseconds, an electron from excited P_{700} is transferred to A_0 producing an oxidised P_{700} and a reduced A_0^-.
- Within 40 to 200 picoseconds, the reduced A_0 transfers its electron to A1 (most likely, a phylloquinone).
- The reduced A1 transfers its electron to the FeS centre within nanosecond.
- Within few microseconds oxidized P700 recovers an electron from plastocyanin. The oxidized plastocyanin is then reduced by an electron that flows from reduced QA via intermediates mentioned above. The reduced FeS transfers electron to ferredoxin. The ferredoxin, with the help of the enzyme ferredoxin-NADP reductase (FNR), ultimately reduces NADP, a molecule with many roles in metabolism.

Energetically the transfer of electrons from reduced Q_A to the oxidized $P700^+$ is a downhill process and is coupled, through the proton and membrane potential gradient, with the production of ATP from ADP and inorganic phosphate. According to the chemiosmotic hypothesis of P. Mitchell ATP is synthesized by the membrane-bound ATP synthase using the energy of proton motive force includes the proton gradient and the membrane potential created by the electron transport process. The charge separations in PSI and PSII lead to victorial transport of electrons across the membrane from P_{700} to Fx and from P_{680} to Q_A creating a membrane potential. Since reduction of PQ pool by Q_A occurs near the outside of the membrane and since plastoquinol formation formation requires proton, protons are picked up from the outside and plastoqninol reduces the electron carrier Rieske iron-sulphur center in the cyf b_6f complex on the innerside releasing protons into the lumen. This, then, creates the proton gradient across the membrane. Furthermore, the oxidation of water by the oxidized "M" also leads to the release of protons into the lumen.

MEHLER REACTION

In isolated chloroplasts, electron transfer to O_2 has been experimentally demonstrated with methylviologen (MV) as the electron acceptor, this reaction is called Mehler reaction in which two molecules of oxygen are used per photosynthetically produced molecule O_2. MV oxidizes various membrane components e.g., pigments and lipids and is used in agriculture (as named paraquat) as a herbicide. This reaction represents a mirror-image of the photosynthetic production of O_2 and is a Hill reaction including whole of electron transport chain.

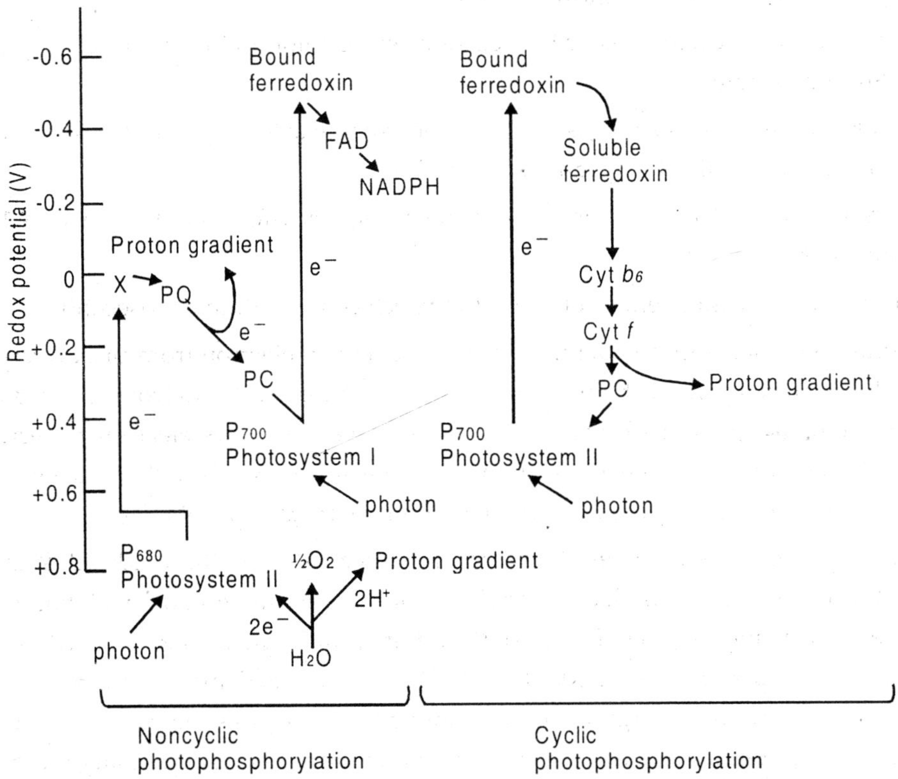

FIG. 5. 3: *Presentation of cyclic and non-cyclic photophosphorylation along with redox-potential (V).*

ENERGETICS OF ELECTRONS AND REDOX POTENTIAL

Photons absorbed by PS II excite electrons from a potential of +0.8 V, that of the H_2O-O_2 couple, to a potential of about 0 V. This a change in potential equal to free energy of 18 kcal/mol of electrons. The nature of the primary electrons move from plastoquinone (PQ) through the cytochrome *b/f* complex to plastocyanin (PC), some of this acquired energy is used to transport protons into the thylakoid vesicles.

Absorption of photons by PS I causes an additional increase of 1.0 V in the potential of electrons. Electrons excited by PS I can either be transferred directly to $NADP^+$, to form NADPH (non-cyclic photophosphorylation) or be transferred via a series of ferredoxins and cytochromes back to PS I (cyclic photophosphorylation) coupled to the transport of protons into the thylakoid lumen.

OXYGEN EVOLUTION DURING PHOTOSYNTHESIS

For each electron moved through the complete system, two photons are required, one each at PS I and one at PS II. The formation of a molecule of O_2 involves the removal of four electrons from two water molecules. A number of intermediate are involved, but little is known for their identity or mechanisms of action. Manganese and chloride ions are involved at the same point in the oxygen evolution pathway.

Two models have been presented for oxygen evolution.

Cooperative Chains

In this chain, either positive charge is stored on P_{680} and can migrate to another chain and once four such charges are stored, oxygen could be evolved from reaction with water or chemical species equivalent to 0 atoms may be formed before oxygen is evolved.

Independent Chains

In this chain, four successive reactions on one and the same centre are needed. Oxygen will not be evolved unless 4 oxidizing equivalents are stored on the same reaction chain and this could be accomplished with a series of *four short saturating light flashes* separated by appropriate dark time between the flashes. This is essential for the recovery of the reactions centres. This model is well accepted. This is also called Kok's oxygen Clock (refer to Chapter 3 for details on Oxygen evolution).

PLANT FERREDOXIN

Plant ferredoxins (Fds) are first recognized as component of the photosynthetic electron transport chain, Fds have been shown to play a central function in many light-dependent metabolic processes. They are localized in the stroma of chloroplasts and serve to mediate electron flow from PS I to a variety of Fd-linked enzymes. Recently it has been reported that non-photosynthetic organs, such as maize roots and mesocotyls, and radish white roots also contain Fd. These Fds are most probably involved in the electron transfers from pyridine dinucleotides to some Fd-linked enzymes in a light-independent manner, although subcellular location of the Fds and detailed metabolic processes are still unknown. The non-photosynthetic Fd is known to be distinct from the photosynthetic Fd by immunological and chemical basis. In addition to the structural difference the two types of Fd differ in the manner of expression. It has been reported that young maize seedlings contain at least four Fd isoproteins (Fd I to Fd IV) and Fd I and Fd II are only found in leaves, whereas Fd III and Fd IV are distributed in all parts of seedlings. The leaf-specific Fds are *light-inducible,* where as others are not affected by light.

ARTIFICIAL ELECTRON TRANSPORT AND INHIBITORS

Use of inhibitors of electron transport chain helps to isolate functional parts of chain and to analyze partial reactions spectroscopically. Furthermore, it is possible to follow the electron flux over some redox systems by using artificial electron donors and acceptors. Hill (1939) successfully reduced an artificial electron acceptor by green cell homogenate. Iron oxalate and later potassium ferricyanide (Fe^{3+}) which reduced to Fe^{2+} by electron uptake·Currently, intensely coloured dyes are used for Hill reaction, e.g., dichlorophenolindophenol (DCPIP, deep blue in colour) which is available in oxidized state and colourless in reduced state. By irradiation of a dilute chloroplast suspension, the bleaching can be followed by the naked eyes.

Presently we know number of artificial electron donors and acceptors. Herbicides inhibit the electron transport chain at well-defined positions and block electron transport at well-defined positions.

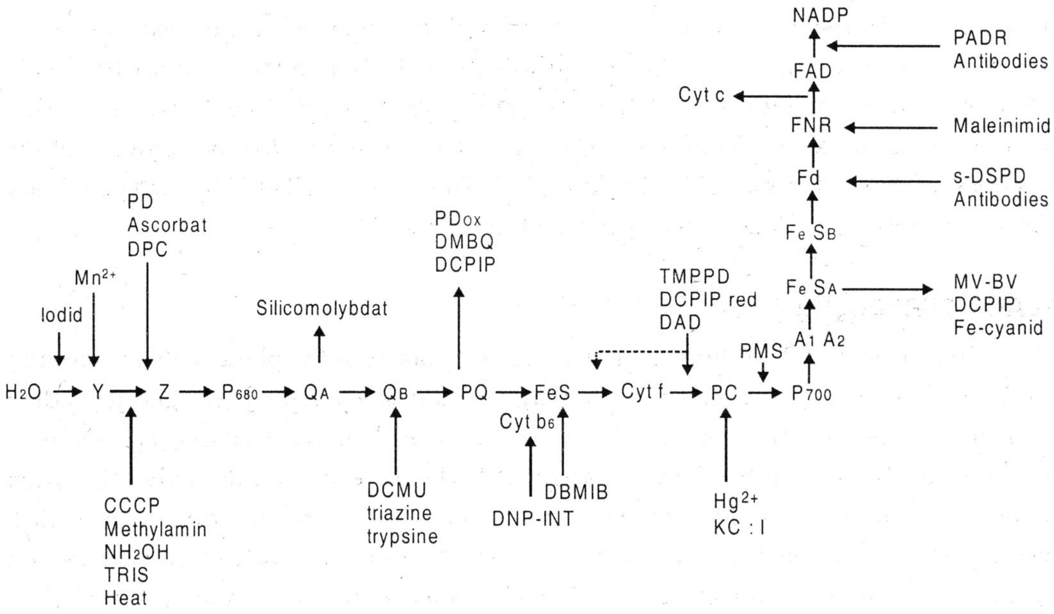

FIG. 5. 4: *Inhibitors and artificial electron donors and acceptors (straight lines) in the photosynthetic electron transport. BV = benzylviologen, CCCP = carbonyl cyanide chlorophenlhydrazone, DAD = diaminodurole, DBMIB = dibromomethyl isopropyl benzoquinone, DCPIP = dichlorphenolindophenole, DCMU = diuron = dichlorophenldimethylurea, DPC = diphenylcarbazid, s-DSPD = sulfodisalicylidin propandiamine, DNP-INT = dinitrophenoxyl nitrobenzene, DMBQ = dimethylbenzoquinone, MV = methylviologen, PADR = phosphoadenosine-P-P-ribose, PD = p=phenylendiamine, PMS = phenazinemethosulfate, TMPD = tetramethyl-p-phenylenediamine.*

TRANSFER OF ATP AND ELECTRONS FROM CHLOROPLAST TO CYTOSOL

During photosynthesis, NADPH and ATP are formed in the chloroplast but are not directly available for reactions outside the chloroplast since neither nucleotide can pass through the inner membrane of the chloroplast. A shuttle system has been proposed that involves dihydroxyacetone phosphate (DHAP) and 3-phosphoglyceric acid (3-PGA), both of which are freely permeable. Thus, 3-PGA synthesized in the chloroplast stroma phase by the carboxylation of ribulose bisphosphate is phosphorylated in the presence of ATP and is reduced to glyceraldehyde-3-phosphate by NADPH. After isomerization to dihydroxyacetone phosphate, the latter moves out of the chloroplast on the phosphate transporter into the cytosol. In the cytosol, 3-phosphoglycerate is regenerated by a reversal of the sequence of reactions just described, producing NADH and ATP. These compounds are then available in the cytosol for biosynthesis while the 3-PGA enters the chloroplast and completes the shuttle.

REGULATION OF THE ELECTRON TRANSPORT CHAIN

Balancing the rate of photosynthetic electron transport with the rate at which reductant is consumed by metabolism is one of the major challenges facing a plant. The rate of excitation of PSI and PSII is determined by the light intensity incident upon the plant but is modulated by leaf and chloroplast movement and by thermal dissipation of absorbed energy. The rate at which the photosynthetically derived reductant is used is determined by factors such as CO_2 availability, sink capacity and temperature. In natural environments, all of these factors will vary markedly over time-scales ranging from seconds to days or longer. Over longer time periods plants have the ability to alter the relative amounts of different proteins involved in photosynthesis. On short time-scales, a number of regulatory processes are known to occur modulating the efficiency of light capture and metabolism (Genty and Harbinson, 1996).

If the rate of photosynthetic electron transport exceeds the capacity of metabolism, excess reductant may drive the reduction of molecular oxygen-the Mehler reaction. The Mehler reaction can be regarded as a simple chemical reaction, the rate of which is proportional to the concentration of its reductants-molecular oxygen and reduced ferredoxin. The rate of the Mehler reaction can therefore, only be regulated by regulating by redox state of the ferredoxin pool. In recent years, two contrasting views concerning the Mehler reaction have been expressed (Wiese *et al.*, 1998). Reactions of superoxide lead to the formation of harmful radical species which may damage the cell. Plants invest heavily in detoxifying active oxygen species. Hence, it is argued that minimizing superoxide formation will be beneficial for the plant. The alternative point of view is that the Mehler reaction is to be encouraged. Oxygen acts as an alternative pathway for the consumption of electrons so limiting the accumulation of reduced intermediates in the electron transport chain. This will reduce the excitation pressure on PSI and so minimize the risk of PSII photoinhibition. Electron transport to O_2 will also contribute to the formation of a pH gradient across the thylakoid membrane without dissipating that ΔpH through ATP consumption. This will enhance other protective processes particularly high-energy-state quenching.

Charge seperation in the PSI reaction center effectively represents the committed step in ferredoxin reduction. Thus, the rate of charge seperation needs to be regulated. This might be achieved by regulating the efficiency of energy capture by PSI, through increasing thermal dissipation of absorbed energy or by decreasing the rate at which the primary donor of PSI is reduced following charge seperation. The latter requires regulation of the electron transport chain at some point between PSII and PSI.

The rate-limiting reaction in photosynthetic electron transport is thought to be the oxidation of plastoquinol by the Cyf b_6f complex. It is known that this reaction can be regulated *in vitro* by the pH in the thylakoid lumen. This provides a potential route for regulating PSI turnover and hence ferredoxin redox state. Various studies have shown that regulation of electron transport is possible in leaves. Harbinson (1994) examined the effects of limitations in electron transport induced by a reduction of CO_2 below ambient levels, in the presence of 2% O_2. Under such conditions, a significant decrease can be measured in the kinetics of electron transport, measured as a change in the rate constant for P_{700} reduction following a light-to-dark transition. Such conditions are not, however, relevant to any physiological conditions. At ambient CO_2 and O_2, Harbinson and Hedley (1989) were unable to detect any variation in the reduction kinetics of P_{700} in pea leaves in response to changing light intensity, suggesting that regulation of electron transport is not physiologically important under non-stressful conditions in this species. They did, however, find evidence for the regulation of electron transport during photosynthetic induction in pea but not under steady-state conditions. They suggested that, under most conditions, other regulatory processes were sufficient to deal with excess light.

Laisk and Oja (1994) examined the effects of both gas composition and low temperatures on electron transport kinetics. They found that electron transport is inhibited at low temperatures, however, this effect may be due to a direct effect of low temperatures on diffusion processes in the electron transport chain rather than on regulation. Using the plant, *Silene dioica*, Ott et al. (1999) showed first of all that down-regulation of electron transport can occur in a leaf in a normal atmosphere and at normal temperatures as well as under conditions where the leaf is exposed to drought stress or at different temperatures. The mechanism of this regulation remains an open question.

PHOTOSYSTEM DAMAGE AND REPAIR CYCLE IN CHLOROPLAST

Life on earth is sustained by oxygenic photosynthesis, a process that begins with the utilization of sunlight for oxidation of water molecules. The chemical energy stored in this endergonic oxidation is processed through the electron-transport chain of the chloroplast thylakoids and eventually, is delivered in the form of reductant. The absorption of light and the conversion of excitation energy to chemical energy take place in PSII and PSI in the thylakoid membrane. From a biochemical point of view, PSII is a specialized water to plastoquinone oxidoreductase. This specialized enzyme features a sizable holocomplex consisting of 25-35 transmembrane and peripheral proteins. The functional center of this holocomplex contains the so-called D1/D2 32/34 KDa heterodimer proteins, which perform the light utilization, water oxidation and electron transfer reactions to plastoquinone. These highly specialized functions of PSII take place in an oxygen-saturated microenvironment, where photons, in the form of excitation energy, arrive at a rate up to 10000 per second. The transient formation of strong oxidants, the abundance of oxygen and the arrival of excitation energy at high rates can lead to photo-oxidative damage (Anderson et al., 1998). Such photodamage occurs frequently within the reaction of PSII. It causes an irreversible inhibition in the function of the reaction center chlorophyll (P_{680}) in the D1 protein and stop photosynthesis.

The PSII damage and repair cycle is important for the function and productivity of photosynthesis. It has been estimated that in the absence of repair mechanism, photodamage would lower the yield of photosynthesis to > 5 per cent of the yield achieved now.

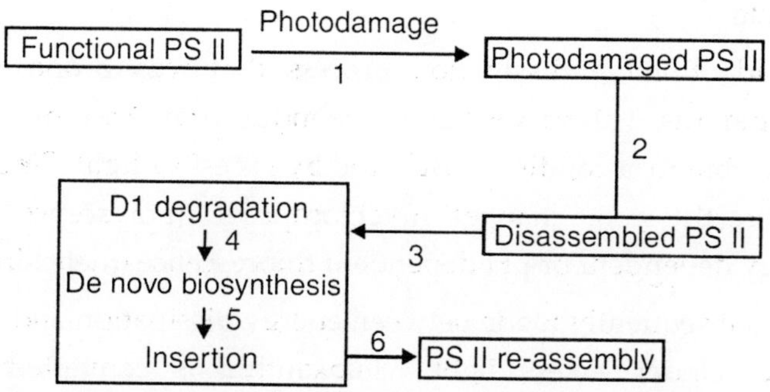

FIG. 5. 5: *Temporal sequence of events in PSII damage, holocomplex disassembly and degradation and replacement of the D1/32 KD reaction center protein.*

PSII Damage and Repair Cycle

In broad terms, photodamage to D1 followed by:

- Prompt partial disassembly of the PSII holocomplex.
- Exposure of the photodamaged PSII core to the stroma of the chloroplast
- Degradation of photodamaged D1
- De novo biosynthesis and insertion in the thylakoid membrane.
- Re-assembly of the PSII holocomplex followed by the activation of the electron transport process through the reconstituted D1/D2 heterodimer.

There is a dynamic relationship between photodamage and repair. The interplay between these two processes will define whether there is an adverse effect on photosynthesis. When the chloroplast repair process can not keep up with the rate of photodamage, the productivity of the photosynthetic apparatus declines and plant growth diminishes. This condition is known as photoinhibition of photosynthesis; it occurs whenever the rate of photodamage exceeds the capacity for repair.

Results in the literature support the theory that the probability for PSII photodamage depends on the redox state of QA. Photodamage will occur with a low probability, when QA is oxidized and excitation energy is utilized in electron transport. When QA is reduced in course of steady-state photosynthesis, excitation energy is dissipated by non-assimilatory *charge recombination* processes. The latter might lead to a generation of long-lived excited states of chlorophyll which, in the presence of oxygen can cause irreversible photodamage to D1. The picture emerging therefore, is that physiological and environmental parameters modulate the redox state of QA, which in turn defines the photochemical or non-assimilatory dissipation of excitation energy and thus low or high probability of photodamage in the PSII reaction center complex.

Xanthophyll Cycle

Early researches identified a dissipation process that was proposed to protect the photosynthetic apparatus. This process could be induced by lowering the pH within the photosynthetic membrane, a condition triggered by excess of light. Since the process was characterized indirectly from changes in chlorophyll fluorescence emission, it was termed high-energy dependent or pH dependent fluorescence quenching.

A connection was subsequently made between energy dissipation and carotenoids of the xanthophyll cycle. Under excess light, violoxanthin is converted rapidly via the intermediate antheraxanthin to zeaxanthin and this reaction is reversed under low light levels. This reversible sequence of two independent reactions is termed the xanthophyll cycle and is present throughout the plant kingdom.

Much recent debate has centered on whether energy dissipation can be induced by low pH alone or whether the presence of zeaxanthin (and antheraxanthin) is obligatory for dissipation to occur. A series of important experiments has now clearly established that under physiological conditions, the key energy dissipation process is obligatorily and stoichiometrically dependent on the presence of zeaxanthin and antheraxanthin. It is thus concluded that zeaxanthin and pH act synergistically and are both required for energy dissipation in vivo.

Major advances have been made recently in the photophysics of carotenoids, opening up the possibility of a simple downhill transfer of energy from excited singlet chlorophyll to zeaxanthin (Frank et al., 1994). In essence, zeaxanthin could act much like a lightning rod, receiving the excess energy from chlorophyll and then dissipating it harmlessly. Carotenoids are known to be able to lose excitation energy relatively rapidly in the form of heat. In contrast, chlorophyll has high a probability of passing energy on to the ever-present O_2 leading to the potential for photo-oxidative damage. Energy transfer to zeaxanthin would thus prevent this dangerous transfer of energy to O_2.

Frank and Owens (1994; 1992) have shown that the properties of most higher plant carotenoids are different from those of lower photosynthetic organization. In bacteria and certain algal groups, carotenoids that are structurally different from those of higher plants have been shown to act as accessory light-harvesting pigments that pass excitation energy down to chloroplast. In contrast, most of the higher plant carotenoids have now been shown to possess a low-lying energy level that can accept energy from the excited singlet state of chlorophyll. This analysis has led to the proposal of a novel model (Frank et al. 1994) in which the xanthophyll cycle conversions constitute a *molecular gear shift* from violaxanthin (with nine conjugated carbon-carbon double bonds) to antheraxanin and zeaxanthin (with ten and eleven carbon-carbon double bonds respectively) which could potentially accept energy from chlorophyll and then dissipate this energy as heat.

The antenna system is also the key site of energy dissipation under physiological conditions. Dissipation of excess energy in the antenna protects the photochemical reaction centers from over-excitation. The antenna are layered around the PSII centers with Chl a - binding core antennae being most closely associated with the centers followed by the inner, minor components and the outer, major components of the Chl a and Chl b - binding, light harvesting antennae. Whereas the major peripheral complexes bind majority of the chlorophyll molecules, the minor, proximal antennae bind only a small fraction of the total chlorophyll but are enriched with the xanthophyll cycle components. The proximal antenna are emerging as key sites of dissipation.

With a high pH gradient across the thylakoid (e.g., at high intensity that is saturating for photosynthesis), the epoxide groups of violaxanthin are removed via the action of a deepoxidising enzyme, forming first antheraxanthin and then zeatin. Violaxanthin is regenerated by light-dependent epoxidation. Violaxanthin is found in the LHCs of PSI and PSII and the violaxanthin cycle takes place in both systems. The de-epoxidase is a 43 KD water-soluble protein that is present in small amounts (approximately one enzyme per 20-100 electron transport chain) in the thylakoid lumen at neutral pH and becomes tightly bound to the lumenal surface of the thylakoid vesicles upon light-induced acidification of the lumen (Eskling et al., 1997). The zeatin-epoxidase enzyme is an FAD-containing multicomponent enzyme system active at the stromal side of the thylakoids, with optimum pH for activity around 7.5 and NAD(P)H and O_2 as cosubstrate.

The transmembrane organization of the cycle with the de-epoxidase located on one side of the membrane and epoxidase located on the other side implies a certain freedom of movement of the xanthophyll cycle pigments in the membrane lipid phase to allow contact between the de-epoxidized/epoxidized head group of the xanthophylls. The purified de-epoxidase of violaxanthin is active in vitro, in the presence of monogalactosyldiacylglycerol (MGDG). It is likely that in vivo de-epoxidation occurs at a similar lipid interface through a rapid light-dependent equilibrium between violaxanthin in the pigment-protein complexes and a pool of MGDG.

There are two pools of convertible violaxanthin; violaxanthin that can be de-epoxidized consists of molecules loosely bound to pigment-protein complexes and of molecules that can diffuse freely in the lipid phase where both fractions can equilibrate. As a corollary, zeaxanthin formed by violaxanthin de-epoxidation in the xanthophyll cycle must initially be located in the thylakoid membrane lipid phase and can subsequently re-attach to the light-harvesting complexes leading to a dynamic equilibrium between bound and unbound zeaxanthin.

Zeaxanthin has been suggested to re-bind preferentially to some minor LHCII and/or after prolonged exposure to high light irradiance, to light stress-induced proteins (ELIPs; early light inducible proteins). Zeaxanthin-biding ELIPs constitute light-harvesting complexes with enhanced energy dissipation capacities (Krol et al., 1995).

Recently, Li et al. (2000) showed that PsbS, an intrinsic chlorophyll-biding protein of PSII, contributes to photoprotective energy dissipation rather than the photosynthetic light harvesting. Upon protonation of PsbS, a conformational change allows for quenching of single excited chlorophyll either by direct energy transfer from chlorophyll to zeaxanthin bound to PsbS or by a zeaxanthin-dependent chlorophyll-chlorophyll or chlorophyll-protein interaction occurring within PsbS. It is also possible that PsbS is simply necessary to maintain LHC proteins in a proper supramolecular organization that allows quenching to occur.

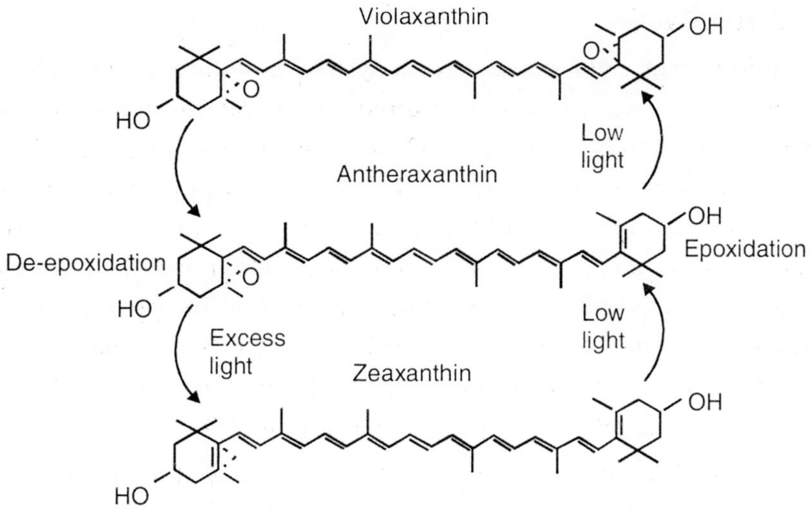

FIG. 5. 6: *Scheme of the xanthophylls cycle and its regulation by excess or limiting light. Stepwise removal (de-epoxidation) of two oxygen functions (the epoxy groups) in violaxanthin results in a lengthening of the conjugated system of double bonds from nine in violaxanthin to ten in antheraxanthin to 11 in zeaxanthin. De-epoxidation occurs within minutes. Epoxidation occurs within minutes to hours, but can take days under additional stresses.*

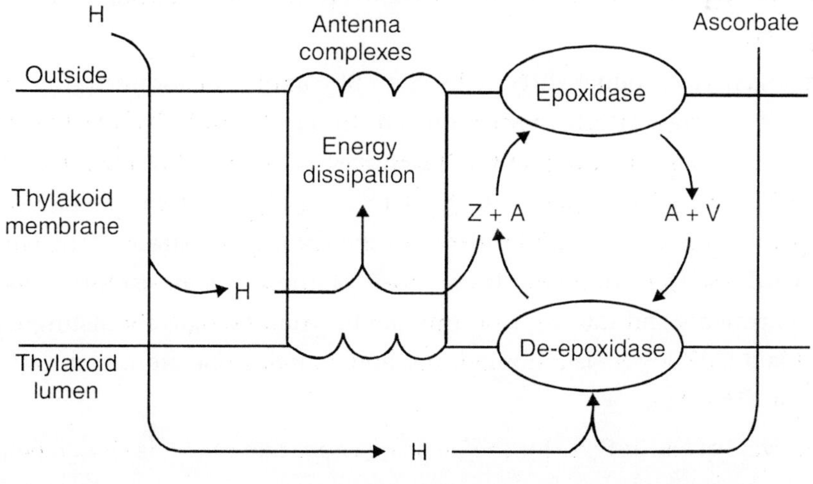

FIG. 5. 7: *Presentation of the regulation of the biochemistry of the xanthophylls cycle as well as the induction of xanthophylls cycle-dependent energy dissipation by pH. The de-epoxidation to anthraxanthin (A) and zeaxanthin (Z) (from violaxanthin V). requires a low pH in the lumen of the thylakoid, as well as reduced ascorbate. In addition, a low pH of certain domains within the membrane, together with the presence of zeaxanthin or antheraxanthin, is required to induce the actual energy dissipation. This dissipation takes place within the light-collecting antenna complexes, particularly by minor, proximal antennae.*

The Water-Water Cycle

The results indicate that chloroplasts have a system to reduce O_2 to water using the electrons derived from water: This is referred to as water-water cycle because the electrons from water in PSII are donated to O_2 for reduction to water in PSI without a net change of O_2. The thylakoid-bound [4Fe-4S] clusters X on psaA and psaB or A/B on psaC of the PSI complex is the most likely donor to O_2. PSII membranes photoproduce O_2 under specified conditions via autooxidation of the electron acceptor of the reaction center complex but not in an intact state of the membranes. The photoproduction of O_2 as observed in intact thylakoids is inhibited by either 3- (3,4-dichlorophenyl)-1,1-dimethylurea (DCMU) or dibromothymo-quinone (DBMIB), indicating little contribution of either PSII or the plastoquinones to the photoreduction of O_2.

The stromal components are thought to participate in the photo-reduction of O_2 in chloroplasts and a possible candidate for the mediator is flavodehydrogenase. The flavodehydrogenases including the chloroplastic enzyme FNR (thylakoid-bound Fd-$NADP^+$ redcutase), glutathione reductase and MDA (monodehydroascrobate) reductase stimulate the photoproduction of O_2 in thylakoids. The flavodehydrogenase-stimulated photoproduction of $O_2$is inhibited by DCMU and DBMIB and the addition of FD does not affect the rate. Thus, these flavodehydrogenasie are directly reduced on the reducing side of PSI, and the reduced enzymes successively donate electrons to O_2 producing molecules of O_2.

The O_2^- photogenerated either directly by the PSI complex or indirectly by the stromal factor-mediated reaction is disproportionated to H_2O_2 and O_2 catalyzed by SOD (superoxide dismutase). In all plants, chloroplasts contain CuZn-SOD as the major isoform of SOD and several plants such as tobacco also contain Fe-SOD. In Fe-SOD containing plants, it is exclusively localized in the chloroplast stroma. The chloroplastic isoform of CuZn-SOD is different from those of the cytosolic isoforms localized in cellular compartments and the two isoforms can be immunologically distinguished from each other. Plant CuZn-SOD is a homodimer and contains one atom each of Cu and Zn in each subunit (16KDa).

The catalytic cycle of CuZn-SOD for the disproportionation of O_2^- is shown below:

$$\text{SOD} - \text{Cu (II)} + O_2^- \longrightarrow \text{SOD} - \text{Cu (I)} + O_2$$

$$\text{SOD} - \text{Cu (I)} + O_2^- + 2\text{H} \longrightarrow \text{SOD} - \text{Cu (II)} + H_2O_2$$

(where SOD – Cu (II) and SOD – Cu (I) represent oxidized and reduced enzymes respectively).

Fe-SOD also is localized in the chloroplast stroma and the tobacco transformants overproducing Fe-SOD in chloroplasts acquire tolerance against photooxidative stress (van Camp et al., 1996). In cyanobacteria, Fe – SOD is located in the cytosol and Mn – SOD is bound to thylakoids. An Fe – SOD lacking mutant of *Synechoccus* is sensitive to photooxidative stress and its PSI complex is inactivated, similar to the chloroplastic CuZn – SOD lacking tobacco. These observations indicate the attachment of Fe – SOD on the thylakoid membranes as well and scavenging of O_2^- at the site where it is generated prior to its diffusion to the stroma or cytosol.

The H_2O_2 produced via the disproportionation of O_2 catalyzed with the thylakoid membrane-attached SOD is reduced to water by ascorbate catalyzed with ascorbate-specific peroxidase (APX). APX is a heme peroxidase and its amino acid sequence indicates that APX belongs to the some superfamily of heme-peroxidase (class 1) as cytochrome c peroxidase (CPX). Chloroplasts contain APX in two isoforms, thylakoid-bound (tAPX) and soluble stromal enzymes (sAPX). tAPX is about 5.5 KDa larger than sAPX (32.2 KDa) but both the isoforms are monomers and share similar enzymatic properties. Chloroplastic APXs are highly specific to ascorbate as the electron donor.

The water-water cycle is composed of the following reactions:

- $2H_2O \rightarrow 4H^+ + O_2$ [Photooxidation of H_2O in PSII]
- $2O_2 + 2\,[e^-] \rightarrow 2O_2^-$ [Photoreduction of O_2 mediated by stroma factor in PSI]
- $2O_2^- + 2H^+ \rightarrow H_2O_2 + O_2$ [SOD – catalyzed disproportionation of O_2^-]
- H_2O_2 + 2 ascorbate $\rightarrow$ $2H_2O$ + 2MDA [APX–catalyzed reduction of H_2O_2 by ascorbate]
- 2MDA + 2 $[e^-]$ + $2H^+$ $\rightarrow$ 2 ascorbate [reduction of oxidized ascorbates]

$$\Sigma\ 2H_2O + O_2^* \longrightarrow O_2 + 2H_2O^*$$

The enzymes participating in the thylakoid system are microcompartmented in the vicinity of the PSI complex where O_2^- is photogenerated. CuZn – SOD or Fe – SOD is attached onto and APX (tAPX) binds to the membranes of the stroma thylakoids. Fd associates with the PSI complex. When O_2^- and H_2O_2 fail to be scavenged in the thylakoid system, O_2^- is disproportionated with the CuZn – SOD located in the stroma and H_2O_2 is then reduced to water with the stromal APX (sAPX).

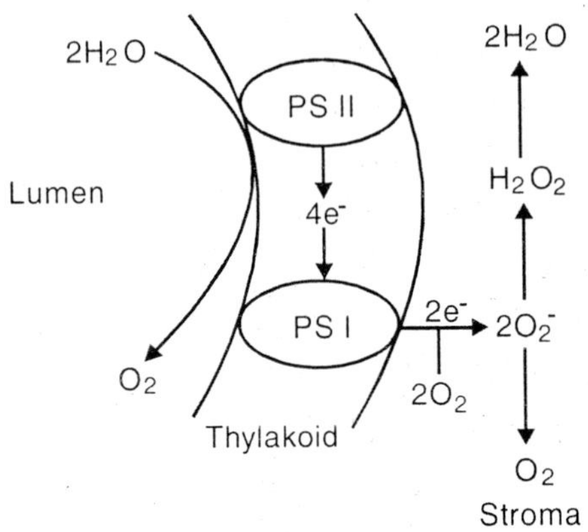

FIG. 5. 8 : *Outline of the water-water cycle in chloroplast.*

ATP SYNTHESIS: PHOTOPHOSPHORYLATION

Transthylakoid proton motive force (pmf) is one of the most important parameters of bioenergetics. It is the key chemiosmotic intermediate between light-driven electron transfer and the synthesis of ATP. It is generally accepted that the two components of pmf, $\Delta\Psi$ and ΔpH, are energetically equivalent but the chloroplast of higher plants store pmf almost exclusively as ΔpH in the steady state and electric field component ($\Delta\Psi$) is negligible. At present, the bulk of experimental evidence indicates that only a fraction of pmf in the chloroplasts is in the form of a $\Delta\Psi$ but this fraction is sufficient to support optimal turnover of the ATP synthase. On the other hand, there is great deal of confusion in the literature over the pH of the lumen during steady-state photosynthesis. The available data generally fall into two categories, supporting either a moderately (pH between ~ 5.8 and 6.5) or strongly (pH <5) acidic lumen during normal photosynthesis.

A transthylakoid pmf is required, not only to drive ATP synthesis, but also to activate the ATP synthase itself. Overall estimates of pmf based on the activation of ATP synthase indicate that the lower limit of ΔpH is between 1.7 and 2.6 units depending upon which estimate is used and whether n = 4 or 3 (n = the stoichiometry of protons pumped through the ATP synthase per ATP synthesized). If all of pmf is in the form of a ΔpH, these estimates translate into upper bounds (at low light levels where photosynthesis is just beginning to engaged) for lumen pH values between 5.2 and 6.1 (assuming a stromal pH of 7.8) on the other hand, even a relatively small contribution of $\Delta\Psi$ to pmf would allow activation of the ATP synthase while maintaining the lumen at moderate acidity.

The relationship between ΔpH and the free energy stored in the ATP (ΔG_{ATP}) is determined by the value of n. In a sense, the ATP synthase acts as a *transmission* connecting pmf with ΔG_{ATP} and n denotes the *gear ratio* at which transmission is set. The higher the value of n the larger the value of ΔG_{ATP} for given ΔpH. The value of n also sets the level of ΔG_{ATP} at which feedback regulation via ΔpH is initiated. The accepted values of n for CF1CFO ATP synthase have changed over the past years with estimates ranging from 2 to 3 and more recently to 4 (Kobayashi et al., 1995; Strotman and Lohse, 1988; Rumberg et al., 1990). Additional support for n=4 has come from mechanistic studies of ATP synthases. It is generally accepted, as postulated by the binding exchange mechanism, that protons are not direct participants in the ATP synthetic reactions but rather drive the conformational changes within the ATP synthase required for release of tightly bound ATP. Most models propose that these conformational changes are associated with rotation of the smaller subunits (ε and γ) with respect to the $\propto_3\beta_3$ ring. (Boyer, 1993) and compelling evidence has been presented in favour of such models (Hasler et al. 1998; Kato-Yamada et al. 1998). As there are three nucleotide binding sites, a complete 360° rotation of F_1 will result in the synthesis of 3 molecules of ATP. In the most recent models, rotation is powered by protonation and deprotonation of acidic groups (on aspartate or glutamate residues) in the c subunits of F_0, which form a transmembrane ring. The torque generated by this rotation is then transmitted to F1 by a combination of smaller subunits of both F0 and F1. Because each protonation/deprotonation causes rotation by an increment of one c subunit around the F_0 ring, the value of n depends upon the stoichiometry of c subunits per complex, i.e. n should correspond to 1/3 the number of c subunit in the c multimer.

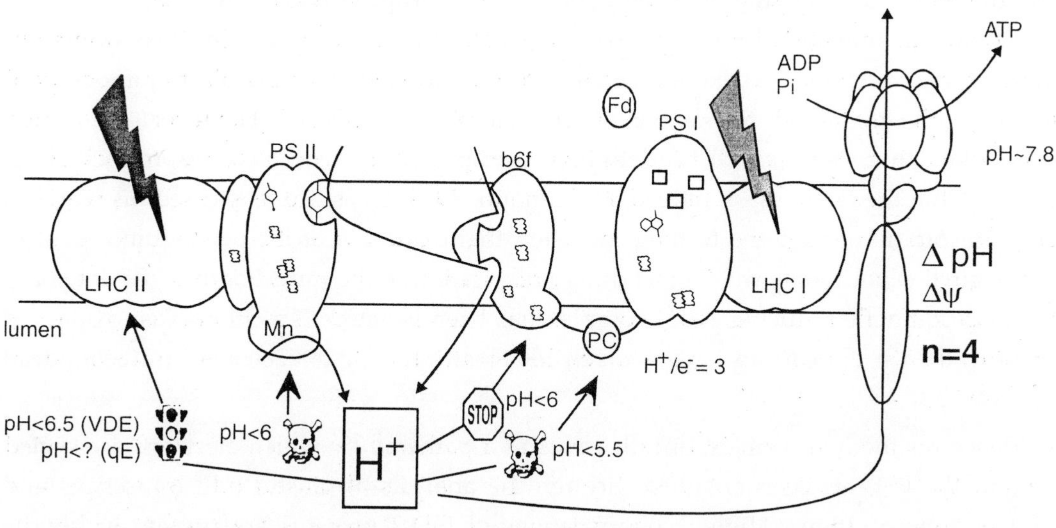

FIG. 5. 9: *Photophosphorylation Abbreviations: QE-quenching-dissipation of excitation energy in PSII antenna bed by processes initiated by thylakoid membrane energization; VDE-Violaxanthin deepoxidase.*

Recent estimates have indicated that a stoichiometry of 12 c subunits per F_0. This would lead to result in translocation of 4H+ per one ATP molecule synthesized (Saraste, 1999).

In summary, light energy drives electron transfer, generating reduced Fd and a transthylakoid pmf. In the steady state most of the pmf is in the form of a delocalized ΔpH but a fraction may remain in the form of delocalized $\Delta\Psi$. A value of n=4 allows the moderate lumen pH model to adequately power the chloroplast, especially with a stromal pH near 8 and if $\Delta\Psi$ contributes 10-20% of total pmf. Electron transfer is coupled to an obligatory Q cycle yielding a H+/e ratio of 3 (Sacksteder et al. 1998) and providing an ATP/NADPH ratio of 3/2 which matches the requiremental for CO_2 fixation by the Calvin cycle (Kramer et al. 1999).

CHLORORESPIRATION

The cyclic electron flow through PSI may be dependent on a particular redox state or "poising" of electron carriers involved. In some cases, this redox poising may be set by electrons derived from carbohydrate metabolism, a phenomenon known as chlororespiration. A variation of chlororespiration, light-dependent electron flow from carbohydrate oxidation through PSI may also occur especially in nutrient-deprived cells.

Bennoun (1982, 1983) studied Chlorella and *Chlamydomonas* and coined the term chlororespiration to describe the respiratory-like O_2 uptake in chloroplast in the dark. Analysis of tobacco ndh mutants by Pal Maliga group (1998) has led to the conclusion that the NDH complex might be component of a chloroplast respiratory chain similar to that found in mitochondrion. This so-called chlororespiratory "Chain" was originally invoked by P. Bennoun to explain changes in the redox state of the plastoquinone pool found in the thylakoid membrane of *Chlamydomonas reinhadtii*. However, it is often overlooked that there is still no definitive evidence that a complete respiratory chain exists within the chloroplast. Indeed, P. Bennoun (1994) has since reassessed his work on chlororespiration and come to the conclusion that oxidation of the plastoquinol pool in his original experiments was in fact due to respiration in the mitochondria. To date only the plastoquinone-reducing NDH complex has been isolated. If a chloroplast oxidase is present, then it appears to have a much lower affinity for oxygen than mitochondrial oxidase.

Evidence for the involvement of ndh genes in a pathway by which electrons are cycled around the PSI has been obtained through the analysis of plastid ndh transcripts and protein concentrations. Higher concentrations of NDH proteins are present in bundle sheath plastids than in mesophyll plastids of the C4 plant *Sorghum bicolor*. PSII activity is much lower in bundle sheath chloroplast and so is the capacity for linear electron transport.

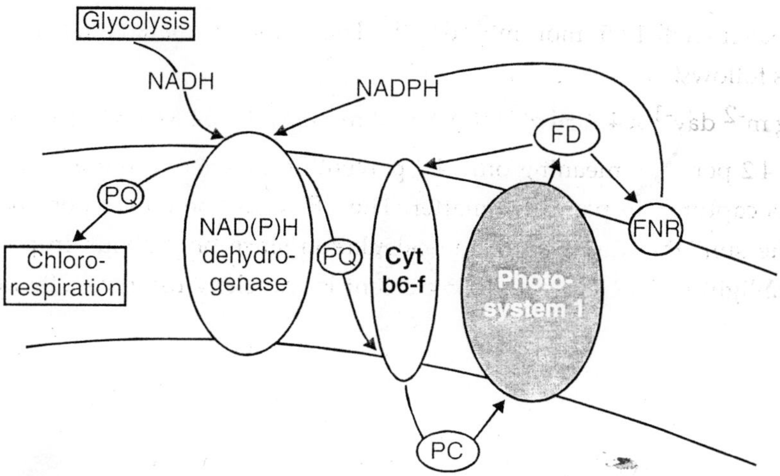

FIG. 5. 10: *A schematic depicting the possible paths of non-oxygenic electron flow through PSI. Arrows represent known or possible paths of electron flow. PQ – plastoquinone; PC – plastocyanin (cytochrome c_{553} in cyanobacteria); FD – ferredoxin; FNR – ferredoxin-NADP oxidoreductase.*

Thus, the NDH complex appears to play a role in cyclic electron transport pathway in bundle sheath chloroplasts. Evidence has also been obtained of higher levels of ndh B transcripts in senescing leaves than in young leaves of barley and the product of the ndhA gene has been identified in thylakoids of barley seedlings exposed to photooxidative conditions. Together, these findings suggest that chlororespiration plays a role in protecting against excess production of active oxygen species in the chloroplast.

PHOTOSYNTHETIC EFFICIENCY

To determine photosynthetic efficiency the light energy incident on the crop and the energy stored by the plant in dry matter must be known. Such measurements are done with closely spaced plants where all PAR (photosynthetically active region, usually taken to be 400-700 nm) is intercepted by the leaf canopy. The energy in the PAR incident on the crop can be calculated from PAR level and the light source or measured with a radiometer. The energy in the plant dry matter can be measured in a bomb calorimeter or estimated. Crop growth rate (CGR) is expressed as g plant dry mass produced per m^2 per day (g m^{-2} day^{-1}) and is measured by periodic harvests to determine the net increase in dry matter over time. Maximum CGR is about 50 g m^{-2} day^{-1} but is typically much lower.

Hershey and Merritt (1987) calculated photosynthetic efficiency for a green-house crop of wax begonia. They assumed a plant tissue energy content of 4 Kcal g^{-1} and a PAR energy content of 52 Kcal mol^{-1}, both typical values. The CGR was 10.1 g m^{-2} day^{-1} for a period when PAR averaged 18.5 mol m^{-2} day^{-1}. The photosynthetic efficiency (PE) was calculated as follows:

$$PE = [(10.1 \text{ g m}^{-2} \text{ day}^{-1} \times 4 \text{ Kcal g}^{-1})/(18.5 \text{ mol m}^{-2} \text{ day}^{-1} \times 52 \text{ Kcal g}^{-1})] \times 100$$

The PE was 4.2 per cent, meaning only 4.2 percent of the energy in the PAR incident on the crop was captured in plant dry matter. The PE would be only 2 percent if the total energy in the sunlight was used in the calculation since only about 45 percent of the energy in sunlight is PAR. The other 55 percent is wavelengths above 700 nm or below 400 nm.

CHAPTER 6

Bacterial Photosynthesis

Microorganisms are classified on the basis of light/organic/inorganic molecules requirements as phototrophs (use light as their energy source), and chemotrophs (obtain from the oxidation of chemical compounds, either organic or inorganic).

- **Lithotrophs (*rock-eaters*):** They use reduced inorganic substances as their electron source.
- **Organotrophs:** They extract electrons or hydrogen from organic compounds.
- **Photolithotrophic autotrophs** (often called photoautotrophs): They use light energy and CO_2 as carbon source. Eukaryotic algae and blue-green bacteria (cyanobacteria) employ water as the electron donor and release oxygen. Purple and green sulfur bacteria cannot oxidize water but extract electrons from inorganic donors like hydrogen, hydrogen sulfide, and elemental sulfur.
- **Chemoorganotrophic heterotrophs** (often called chemohetertophs, or even heterotrophs): They are organic compounds as source of energy, hydrogen, electron, and carbon for biosynthesis. Essentially all pathogenic microorganisms are chemoheterotrophs.
- **Photoorganotrophic heterotrophs**: Some purple and green bacteria are photosynthetic and use organic matter as their electron donor and carbon source. These are common inhabitants of polluted lakes and streams. Some of these bacteria also can grow as photoautotrophs with molecular hydrogen as electron donor.
- **Chemolithotrophic autotrophs**: They oxidizes reduced inorganic compounds such as iron, nitrogen, or sulfur molecules to derive both energy and electrons for biosynthesis. These organisms contribute greatly to the chemical transformations of elements (e.g., the conversion of ammonia to nitrate or sulfur to sulfate) that continually occur in the ecosystem.

NUTRIENT REQUIREMENTS FOR MICROBIAL GROWTH

Ninety fiver per cent of cell dry weight of an organism is made of a few major elements; carbon, oxygen, hydrogen, nitrogen, sulphur, phosphorus, potassium, calcium, magnesium, and iron. These are called **macroelements** or macronutrients because they are required by microorganisms in relatively large amounts. The first six (C, O, H, N, S, and P) are component of carbohydrates, lipids, proteins, and nucleic acids. The remaining four macroelements exists in the cell as cations and play a variety of roles.

The microorganisms require several **trace elements** or microelements or micronutrients besides macroelements. The trace elements- manganese, zinc, cobalt, molybdenum, nickel, and copper- are needed by most cells. Trace elements are normally a part of enzymes and cofactors, and they aid in the catalysis of reactions and maintenance of protein structure. Vitamins are needed for the promotion of physical growth (as growth factor).

The major elements (C, O, H, N, S, P) are needed in gram quantities in a liter of culture medium. The minor elements (K, Ca, Mg, Fe) of ten are required in milligram quantities. Trace elements (Mn, Zn, Co, Mo, Ni, Cu) must be available in microgram amounts.

Carbon is required for the skeleton or backbone of all organic molecules, and molecules serving as carbon sources usually also contribute both oxygen and hydrogen atoms. One carbon source for which this is not true is carbon dioxide (CO_2) because it is oxidized and lack hydrogen. Probably all macroorganisms can fix CO_2- that is, reduce it and incorporate it into organic molecules. However, by definition, only **autotrophs** can use CO_2 as their sole or principal source of carbon. Many microorganisms are autotrophic; many of these are photosynthetic, but some autotrophs oxidize inorganic molecules to obtain energy.

TABLE 6.1: Source of Carbon, Energy, and Hydrogen/Electrons

Carbon Source	Autotrophs	CO_2 sole or principle biosynthetic carbon source
	Hetertophs	Reduced, preformed, organic molecules from other organisms
Energy sources	Phototrophs	Light
	Chemotrophs	Oxidation of organic or inorganic compounds
Hydrogen or Electron Sources	Lithotrophs	Reduced inorganic molecules
	Organotrophs	Organic molecules

The reduction of CO_2 is a very energy expensive process. Thus, many microorganisms cannot use CO_2 as their sole carbon source but must rely on the presence of more reduced, complex molecules for a supply of carbon. Organism that reduce preformed organic molecules as carbon sources are called **heterotrophs.** These performed molecules normally come from other organisms. Most heterotrophs use organic nutrients as a source of both carbon and energy. For example, the glycolytic pathway traps energy as ATP and NADH and also produce carbon skeleton for use in biosynthesis.

Nutritional characteristic of microorganisms is extraordinary flexible, specially with respect to carbon sources. They utilize different carbon sources

Actinomycetes	It can degrade amyl alcohol, papraffin, and even rubber
Pseudomonas cepacia	It can use over 100 different carbon compounds.
Methylotrophic bacteria	They can metabolize only methane, methanol, carbon monoxide, formic acid, and a few related one-carbon molecules.
Leptospira (parasitic members)	They use only long-chain fatty acids as their major source of carbon and energy.

Photosynthetic or Photoautotrophic Bacteria

There are few species of bacteria which contain two photosynthetic pigments for manufacturing organic food from inorganic sources. They are principally of two kinds viz., *purple sulphur* or red bacteria and *sulphur bacteria.* These are often found in sulphur springs where hydrogen sulphide is often present. Purple or red bacteria contain only one type of chlorophyll pigment called bacteriochlorophyll along with carotenoids. Green bacteria contain two types of chlorophyll. Therefore, bacteria are only plants which manufacture their food without chlorophyll a, in presence of light. Moreover, the process of photosynthesis in photosynthetic bacteria, for example of *Rhodosprillum* spp. and *Rhodoseudomonas* spp. is different to those of photosynthetic plants. Free oxygen is not liberated during photosynthesis. In them hydrogen sulphide serves as a source of hydrogen e.g., *chlorobium limicola.* The involvement of chemical changes in bacterial photosynthesis may be represented as follows:

$$6CO_2 + 12H_2S \xrightarrow[\text{BacterialChlorophyl}]{\text{Light}} C_6H_{12}O_6 + 6H_2O + 12S$$

The photosynthetic bacteria may be separated on the basis of nature of H_2A which they are capable of using.

1. **First:** Those which can utilize organic materials as H_2A. A simple example is a reaction carried out by *Rhodopseudomonas* spp. which in light can dehydrogenate isopropyl alcohol but can not use the resulting acetone.
2. **Second:** Those capable of using an inorganic material as H_2A. This material is characteristically H_2S, and in light the bacteria generally form S (compare $2H_2S \rightarrow S_2$ with $2H_2O \rightarrow O_2$). There are two kinds of such bacteria: (i) Purple bacteria and (ii) Green bacteria.

Purple Sulphur Bacteria

Purple bacteria are the largest, most diverse and most thoroughly studied group of the anoxygenic photosynthetic bacteria. Physiologically and ecologically this group can be subdivided into the purple sulphur and purple non-sulphur bacteria. The purple sulphur bacteria are mostly anaerobic photoautotrophs using hydrogen or reduced sulphur to fix CO_2 via the reductive pentose phosphate cycle. The purple non-sulphur bacteries grow best as photoheterotrophs although many are capable of autotrophic CO_2 fixation via the reductive pentose phosphate cycle using hydrogen or reduced sulphur and most can also fix nitrogen.

Green Bacteria

The cells appear green because the colour of the chlorophyll is not masked by other pigments. Their photosynthetic metabolism involves the reduction of CO2 at the ultimate expense of H2S and other reduced sulphur compounds. No growth factors are required, growth occurring in mineral medium. They are obligately photosynthetic and strictly anaerobic. Most physiological and biochemical studies have been done on species of Chlorobium and Prosthecochloris. Carbon dioxide is fixed via the reductive tricarboxylic acid cycle rather than the reductive pentose phosphate cycle.

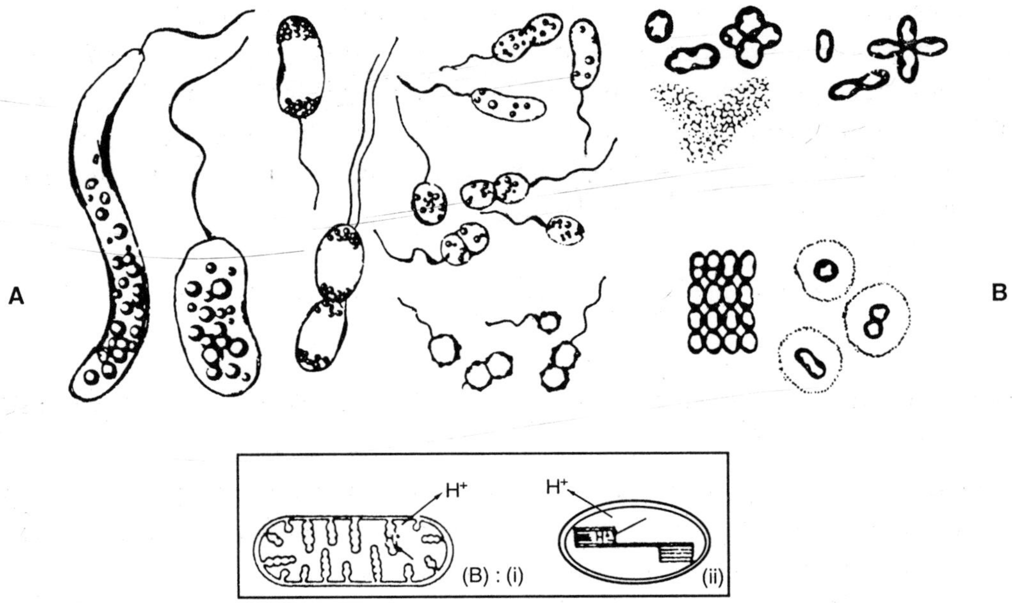

FIG. 6.1: *(A) Some sulphur purple bacteria (Chromatiaceae) and non-sulphur bacteria (Rhodosprillaceae). (B) Proton transport in illuminated cell and organelles (i)* Rhodopseudomonas sphaeroides *(ii) chloroplast of higher plant.*

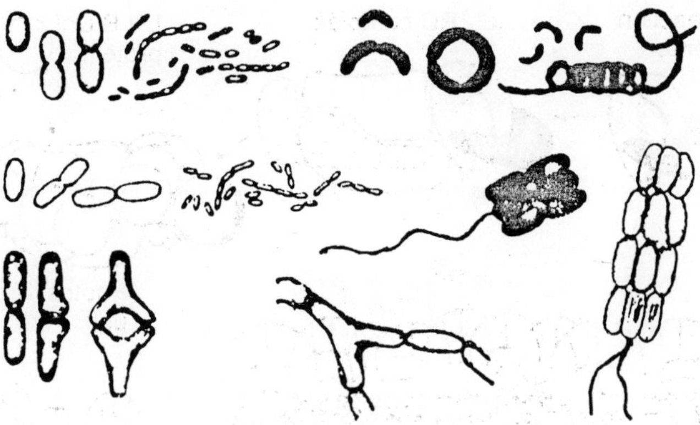

FIG. 6.2: *Phototrophic green sulphur bacteria (Chlorobiaceae).*

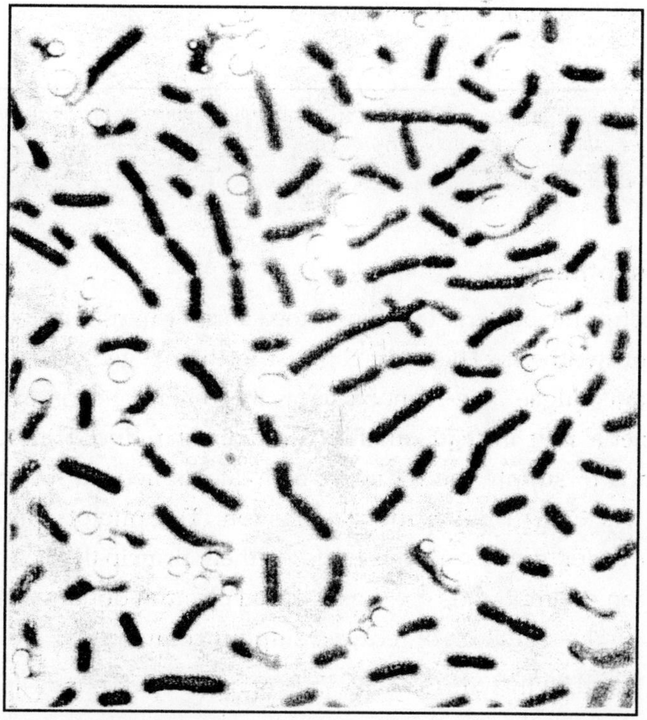

FIG. 6.3: *The microphotograph of green sulfur bacteria Chlorobium limicola deposites sulfur granules in the environment outside of the cell.*

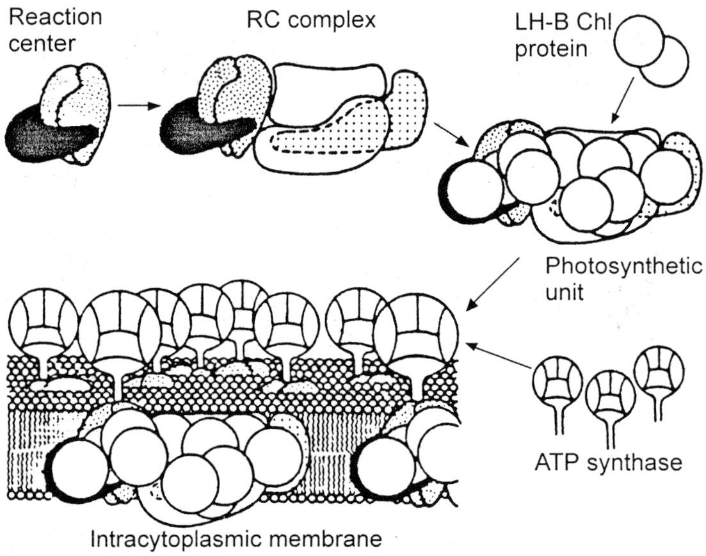

FIG. 6.4: *Intracytoplasmic membrane (photosynthetic part) of photosynthetic purple bacteria and the individual particles shown separately (from left to right). LH-B-Chl = Light harvesting bacteriochlorophyll protein.*

Bacterial Photosynthesis

Anoxygenic phototrophic bacteria (anoxyphotobacteria) can not use water as an electron donor during photosynthesis. This is in contrast to the oxygenic photosynthesis carried out by higher plants, algae and cyanobacteria, in which water is oxidized to O_2. The phototrophic bacteria rely instead on the oxidation of reduced inorganic and simple organic compounds to supply the reducing equivalents needed for the biosynthesis of cellular material via CO_2 fixation and N_2 fixation. The phototrophic bacteria can be divided into two major groups on the basis of their principal electron donors. In the organotrophs, organic compounds are the favoured electron donors while the lithotrophs are characterized by the oxidation of reduced sulphur compounds.

Photoorganotrophic growth with a variety of organic compounds is common for the purple non-sulphur bacteria and the green gliding bacteria. In these bacteria, oxidation of substrates that are more reduced than cell materials requires the presence of a terminal electro acceptor that is more oxidized than the cell material synthesized. Thus, for example, butyrate oxidation occurs only in the presence of CO_2. Growth with aromatic compounds serving as both electron donors and carbon sources has also been observed in purple non-sulphur bacteria. The oxidation of organic compounds such as benzoate poses considerable problems for phototrophs since phototrophic growth under anaerobic conditions precludes the use of O_2 to initiate oxidation of aromatic substrates. Some purple non-sulphur bacteria require the presence of reduced sulphur compounds, in addition to organic compounds for growth but in such cases the reduced sulphur compounds are needed for the biosynthesis of sulphur-containing compounds rather than as electron donors.

Photolithotrophic growth with reduced inorganic compounds (e.g. H_2 and reduced sulphur compounds) characterizes the purple sulphur bacteria (Ectothiorhodospiraceae and Chromaticeae) and the green sulphur bacteria (Chlorobiaceae). Sulphide and elemental sulphur can be utilized as electron donors by all species of these three families of bacteria while thiosulphate oxidation is confined to a limited number of Chlorobiaceae species, to the Ectolhiorhosperaceae and to small cell forms of the Chromatiaceae. During the process of sulfide oxidation, elemental sulphur is accumulated either intracellularly or extracellularly. In the former case, sulphide is completely oxidized to sulphate while in the latter sulfide oxidation is often incomplete.

Facultatively chemoautotrophic bacteria display an enormous metabolic versatility which allow them to grow on a wide range of substrates. In contrast to obligate autotrophs, catabolic and anabolic pathways of facultatively autotrophic becteria are inducible. A characteristic of these bacteria is the ability to grow mixotrophically, i.e. substrates supporting autotrophic and heterotrophic growth are used simultaneously.

All photosynthetic bacteria contain photochemical reaction centers (Rcs) containing one or more chlorophyll molecules. Each reaction center consists of a primary electron donor (bacteriochlorophyll), an initial electron acceptor (bacteriochlorophyll or bacteriopheophytin) and one or more secondary acceptors (Fe-S centers, quinines). Sometimes a secondary electron donor (cytc) is tightly bound to the RC. When a quantum of excitation reaches the RC, the primary donor is excited to a new state in which it is a powerful reducing agent. It then transfers an electron to the initial acceptor. To prevent the electron from falling back to primary donor, the secondary acceptor takes the electron from initial electron acceptor and stabilizes the charge separation. To further stabilize this separation the secondary donor gives an electron to primary electron donor. These reaction taking place in the RC are the primary chemical reactions of photosynthesis.

The absorption cross-section of a single RC is so small that it can not trap light fast enough to drive the organisms electron transport system up to capacity. LHCs exist to enlarge the effective absorption cross-section for each RC. The LHCs range in size from 50 Bchla molecules per RC in some purple bacteria to 2000 Bchlc molecules per RC in some green sulphur bacteria. Some LHCs are integral components of a membrane (purple bacteria) while others are housed in extramembranous bodies (e.g. green bacterial chlorosomes) which are attached to the membrane. In all cases the function of an LHC is to absorb light and to transfer the resulting excited state to the RC.

Bacterial Photosynthetic Reaction Centers

Photosynthetic reaction centres (RCs) are membrane spanning complexes of polypeptide chains and co-factors that catalyze the first steps in the conversion of light energy to chemical energy during photosynthesis. Absorption of the energy of a photon in the RC is followed by rapid, efficient electron transfer from a primary donor along a chain of acceptors leading across the photosynthetic membrane. In RCs from photosynthetic bacteria or from PSII of green plants and cyanobacteria, the end of this chain is a quinone; during two functional cycles of the RC this quinone accepts two electrons and two protons and dissociates as a quinol from the RC. The end of the electron-transfer chain in PSI is an iron-sulphur center. This center reduces the small, water-soluble, electron-transport protein ferredoxin.

Photosynthetic purple bacteria were the first organisms in which RCs were discovered. Their relative simplicity made them logical targets for studying photosynthetic light reactions, RCs of photosynthetic purple bacteria consists of at least three protein subunits called L (light), M (medium) and H (heavy), for the apparent molecular weights determined with SDS gel electrophoresis. Subunits L and M have very hydrophobic amino acid sequences, in which five membranes spanning segments can be distinguished. Subunit H has a single membrane spanning segment near its amino terminus. The subunits L and M bind bacteriochlorophylls (Bchl), bacteropheophytin (BPh), quinones, a ferrous iron ion and a carotenoid as co-factor. These co-factors come in a variety of slightly different forms, e.g. Bchl and BPh of type a or b, in different species.

Amino acid sequence homologies between subunits from four different species range from 60 to 78% for subunit L and from 50 to 70% for subunit M. The sequences of three H-subunits show homologies between 38 to 64%. A major difference between RCs from different species of photosynthetic purple bacteria is the presence or absence of a four-heme C-type cytochrome as the fourth protein subunit. Such a subunit part of the RC from e.g. *Rhodopseudomonas virdis* and *Thiocapsa pfennigii*, RCs of other bacteria, e.g. *Rhodobacter sphaeroides, Rhodobacter capsulatus* and *Rhodospirillum rubrum,* consist of only the three protein subunits L, M and H.

Reaction Centre of Purple Bacteria

The RC from the purple bacterium *RPs.viridis* is a complex of 4 protein subunits and 14 co-factors. The protein subunits are cytochrome, subunit M, subunit L and subunit H.

The complete amino acid sequences of these subunits were derived from the sequences of genes coding for them. The co-factors are four heme groups, covalently linked to the cytochrome subunit, four BChl-b, two BPh-b, a menaquinone-9 and a ubiquinone-9, a ferrous, iron ion and the carotenoid dihydroneurosporene. The RC from *Rb.sphaeroids* consists of protein subunits M, L and H of ten cofactors (4 BChl-1, 2 BPh-a, 2 ubiquinone-10, 1 non-heme iron, 1 carotenoid). In contrast to the RC from *RPs.viridis,* it lacks a bound cytochrome subunit. It has BChl-a and BPh-a instead of the by-type chromophores. It has two identical quinones (ubiquinone-10) and its carotenoid is spheroidene instead of dihydroneurosporene. The amino acid sequences of the subunits M, L and H in *Rb.sphaeroides* and *RPs.viridis* show 59, 49 and 30% homology respectively.

Similarity to PSII Reaction Centre

The elucidation of the three-dimensional structure of RCs from purple bacteria significantly impacted the identification of PSII reaction center of green plants and cyanobacteria. Weak sequence homologies between subunits L and M from purple bacteria and proteins D1 and D2 of PSII had been found. The three-dimensional RC structures revealed that the sequence homologies include key amino acid residues e.g. histidine residues that bind DL, DM and the nonheme iron and tryptophane and phenylalanine residues in contact with the quinones. These and other findings strengthen the hypothesis that the RCs of purple bacteria and of PSII are structurally related and proteins D1 and D2 are equivalent to the subunits L and M respectively. Interestingly, the histidines binding B_A and B_B in the bacterial RC are not conserved in the sequences of D1 and D2 which indicates either that the accessory chlorophylls are absent or they utilize a different mode of binding. Additional evidence comes from the study of herbicide-resistant PSII mutants in which sequence changes are found in a region analogous to the QB-binding site of the bacterial RC. Structural information from the bacterial RC was a valuable guideline during the identification of two intermediate electron carriers as tyrosine residues in proteins D1 and D2.

Pathway of Electron Transfer

The current understanding of the function of RC was developed by combining structural information with information from other experimental techniques. The special pair, D, is the starting point for a light-driven electron transfer reaction across the membrane. Absorption of a photon or energy transfer from light-harvesting complexes in the membrane raises D to its first excited singlet state, D*. From D* an electron is transferred to ϕA with a time constant of 3 picoseconds. The cofactor ϕA can be identified as the electron acceptor because ϕA and ϕB have absorption maxima at slightly different wavelengths. From ϕA, the electron is transferred to Q_A with a time constant of 200 picoseconds at room temperature. In moving from D to Q_A, the electron has crossed most of the membrane. From Q_A, the electron moves on to Q_B within about 10 μs. the nonheme iron does not play an essential role in this step. Q_B can pick up two electrons and two protons. Q_BH_2 then dissociaties from the R_C and Q_B site is refilled from a pool of quinones dissolved in the membrane. Electrons and protons on Q_BH_2 are transferred back through the membranes by the cytochrome bc1 complex. Cytochrome C_2 molecules deliver the electrons back to the RC. The whole process can be described as light-driven cyclic electron flow, the net effect of which is the generation of a proton gradient across the membrane that is used to synthesize ATP in accord with Mitchell's chemiosmotic theory.

In RCs without a bound cytochrome subunit, cytochrome C_2 directly reduces D^+. In RCs with bound cytochrome subunit, D^+ first receives an electron from the closest heme group. Cytochrome C_2 in turn reduces the bound cytochrome. It is unknown which of the four hemes of the bound cytochrome of *RPs.viridis* accepts the electron from cytochrome C_2.

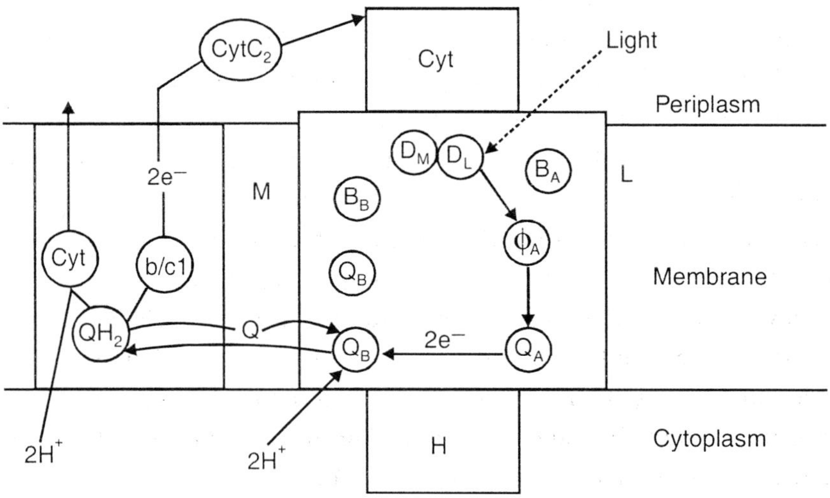

FIG. 6.5: *Purple Bacteria.*

Reaction Centre of Green Sulphur Bacteria

All species contain two different chlorophylls, Bchla and an additional light-harvesting pigments either Bchl-c, d or e. These light-harvesting pigments are housed in chlorosomes located adjacent to the cytoplasmic membrane. In *Chlorobium,* Bchl-c is associated with a 7.5 KDa polypeptide. A small amount of energy-transferring Bchl-a is also found in the chlorosomes but most of the Bchl-a is found in a water-soluble protein associated with the cytoplasmic membrane. The RC is similar to that of PSI in cyanobacteria and chloroplasts. The primary electron donor P-840 is a Bchl-a dimer and is associated with a 65 KDa polypeptide. The initial electron acceptor is Bchl-663, a special lipophilic form of Bchl-c and the secondary acceptor is a –0.54 V Fe-S center capable of reducing ferredoxin directly.

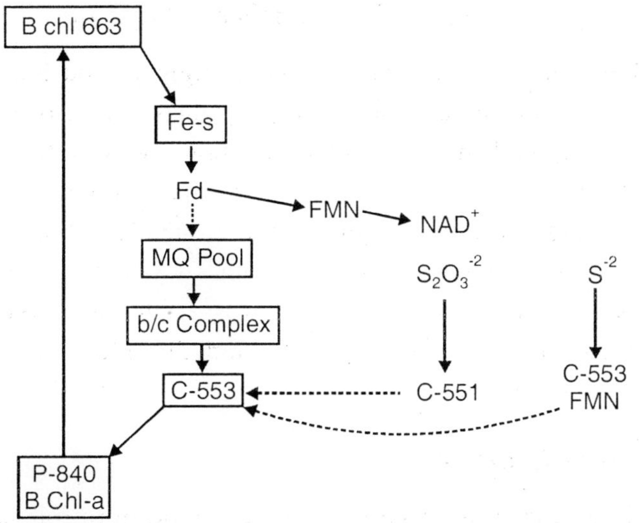

FIG. 6.6: *Green Sulfur Bacteria.*

Pathway of Electron Transfer

The electron transfer system contains cytochrome of the b and c types, a Rieske-type Fe-S protein, menaquinone. A membrane-bound Cyt c-553 is the immediate-electron donor to P-680 in both *Chlorobium* and *Prosthecochloris.* Another membrane-bound Cyt c is Cyt c-550.5 which may be part of a Cyt b/c complex along with a Rieske-Type S center. There are three b-tyep cytochromes in the membrane: Cyt b-563, Cyt b-564 and Cyt b-562. Cyt b-562 is thought to be part of a b/c complex. There is indirect evidence for cyclic electron flow in green sulphur bacteria and in analogy to the path of cyclic electron flow in PSI. It is supposed that electrons from the –0.54 V Fe-S center in the RC flow to soluble ferredoxin and from there to the presumed MQ pool. Electrons from the MQ pool are thought to enter the b/c complex and to return to the RC via Cyt c-553.

It is well established that green sulphur bacteria carry out linear electron flow from external donors such as S^{2-}, S^0, $S_2O_3^{-2}$ and SO_3^{-2} to ferredoxin and NAD^+. The use of O_2 –uptake as a valid assay of linear flow in membrane preparations has also been demonstrated. Sulphide oxidation is mediated by a soluble flavocytochrome c-553 and thiosulphate oxidation seems to require the soluble Cyt c-551. Soluble Cyt c-555 is reduced by both S^{-2} and $S_2O_3^{-2}$ but its exact role is not yet known. It is supposed that linear electron flow via the soluble c-type cytochromes enters the RC via the membrane-bound Cyt c-553 and leaves the RC via ferredoxin. Soluble ferredoxin and flavoprotein ferredoxin-NAD (P) reductase have been found in several green sulphur bacteria.

Reaction Centre of Filamentous Photosynthetic Bacteria

The BChl-c functions entirely as a light-harvesting pigment and is associated with 5.6 KDa polypeptide in small sub-cellular structure called chlorosomes adjacent to the cytoplasmic membrane. The Bchl-a serves two functions. Most of the pigment is located in the cytplasmic membrane in the B808-866 antenna complex and functions as an energy-transfer and light-harvesting pigment. A very small fraction of the Bchl-a (Bchl-790) is associated with the chlorosomes and functiosn as an energy-transfer pigment. Some of the B Chl-a is found in the RC where it is associated with two protein subunits. The primary electron donor is a BChl-a dimer and the initial acceptor is Bpha. The secondary acceptor is menaquinone.

Pathway of Electron Transfer

The pathway of electron transfer is similar to those found in purple bacteria. The cyclic pathways involve b and c-type cytochromes, Fe-S centres and Quinones. *Chloroflexus* contains only MQ. On the electron acceptor side of the RC in *chloroflexus* there are two MQ molecules in series (MQ_A and MQ_B) which are assumed to feed electrons in to the MQ pool. Electrons enter a Cyt b/c complex from the MQ pool and leaves the complex via Cyt c and Cyt-554. The membrane-bound Cyt c-554 is the direct electron donor to P-865 in the RC. It contains two hemes with redox potentials of +0.14 and +0.26 V and is absent from aerobically grown cells. During electron flow the b/c complex functions as an energy transducer that converts a substantial fraction of the energy difference between the quinone pool and Cyt c-554 to potential energy in the from of a transmembrane proton gradient.

Reaction Centre of Heliobacterium chlorum

H. chlorum is a recently described anoxygenic N_2-fixing photosynthetic bacterium. It is a relatively large unicellular organism with gliding mobility. Phylogenetic analysis using sequence data for 16 rRNA has revealed that this organism is more closely related to the gram-positive bacteria. Physiologically it is an obligate photoheterotroph, since no respiratory metabolism has been observed. Autotrophic CO_2 fixation is not sustained by either hydrogen or sulphide.

H. chlorum has one major chlorophyll, a unique pigment called Bchl-g which has a major absorption band at 788 nm in vivo. The light-harvesting Bchl-g, which exists in three spectroscopically different forms, is housed in the cytoplasmic membrane, as there are no intracytoplasmic membranes or chlorosomes. The cytoplasmic membrane is unusually rich in protein. *H. chlorum* also has a pigment absorbin gat 670 nm and an unusually low content of a single carotenoid, neurosporene. Both the 670-nm pigment and the cerotenoid transfer excitation energy to Bch-g with 70% efficiency.

Pathway of Electron Transfer

The primary electron donor is P-798. It has properties similar to those of P-840 in the RC of green sulphur bacteria but it is not clear whether P-798 is a dimer or monomer of Bchl-g. The initial acceptor is a molecule absorbing at 670 nm and it might be related to Bchl-c. The secondary electron acceptor is an Fe – S center as in green sulphur bacteria but quinone-type secondary acceptor has also been suggested. This new RC is unique in having both an Fe-S center and a quinone-like molecule as secondary acceptors. Nothing is yet known about the polypeptide (s) associated with RC but a 50-KDa protein may be associated with membrane-bound Cyt c-553 which is electron donor to P-798. The other components of the electron transport system are as yet unknown.

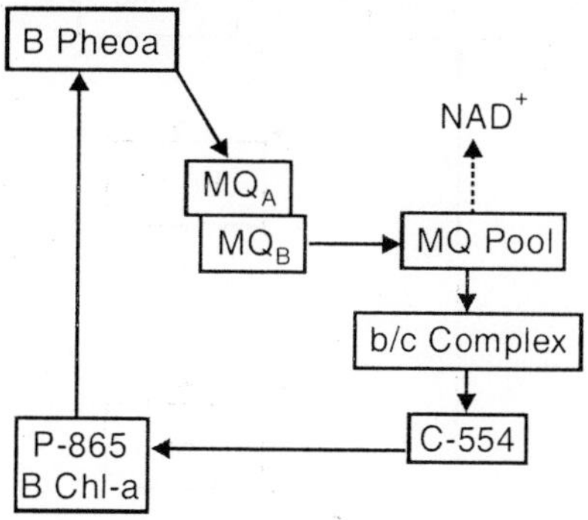

FIG. 6.7: *Filamentous bacteria.*

The Calvin Cycle

CO_2 fixation by the Calvin cycle is dependent on 13 enzymatic reactions. The enzyme responsible for the actual fixation of CO_2, Rubisco, catalyzes the carboxylation of RuBP to form two molecules of 3-phosphoglycerate. The other enzymes of the Calvin cycle are dedicated to the regeneration of RuBP. Three glycolytic enzymes-phosphoglycerate kinase, glyceraldehyde-3-phosphate dehydrogenase and triosephosphate isomerase-convert two molecules of 3-phosphoglycerate to glyceraldehyde-3-phosphate and dihydroxyacetone phosphate at the expense of two molecules of ATP and two of NADH. This is followed by a series of rearrangement reactions that result in the production of ribulose-5-phosphate.

The rearrangement reactions are divided into two similar metabolic units, made up of an aldolase, a phosphatase and a transketolase (APT). The first APT unit converts glyceraldehyde-3-phosphate and dihydroxyacetone phosphate into xylulose-5-phosphate and erythrose-4-phosphate which is followed by the formation of xylulose-5-phosphate and ribulose-5-phosphate from dihydroxyacetone phosphate and erythrose 4-phosphate by the second unit. The phosphatase used in the second APT unit (sedoheptulose bisphosphatase, SBPase) catalyzes the dephosphorylation of SBP which is, in contrast to the gluconeogenic fructose bisphosphatase (FBPase), an activity unique to the Calvin cycle. In principle, the pentose phosphate cycle would provide an alternative for these arrangement reactions. In this scenario, the aldolase and phosphatase of the second APT unit are replaced by transaldolase. The main difference is that the use of phosphate in the aldolase/SBPase variant renders the second APT unit irreversible whereas it is reversible in the transaldolase variant. All autotrphic organisms studied to date employ the aldolase/SBPase variant of the Calvin cycle.

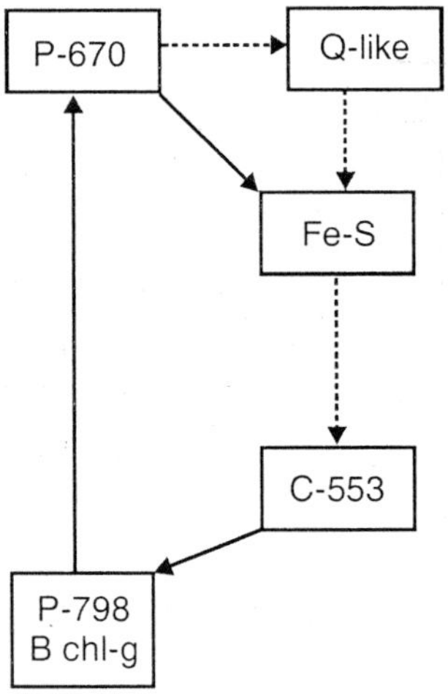

FIG. 6.8: H. chlorum.

The sequential reactions of the two APT units result in the formation of xylulose-5-phosphate and ribose-5-phosphate. These are converted to ribulose-5-phosphate by pentose phosphate epimerase and pentose phosphate isomerase respectively. The final step in the regeneration of RuBP is catalyzed by another unique Calvin cycle enzyme, phosphoribulokinase (PRK), which phosphorylates ribulose-5-phosphate at the expense of an ATP. The net result of the activity of the Calvin cycle is the formation of one molecule of triose phosphate from three molecules of CO_2 at the expense of nine molecules of ATP and six of NADPH.

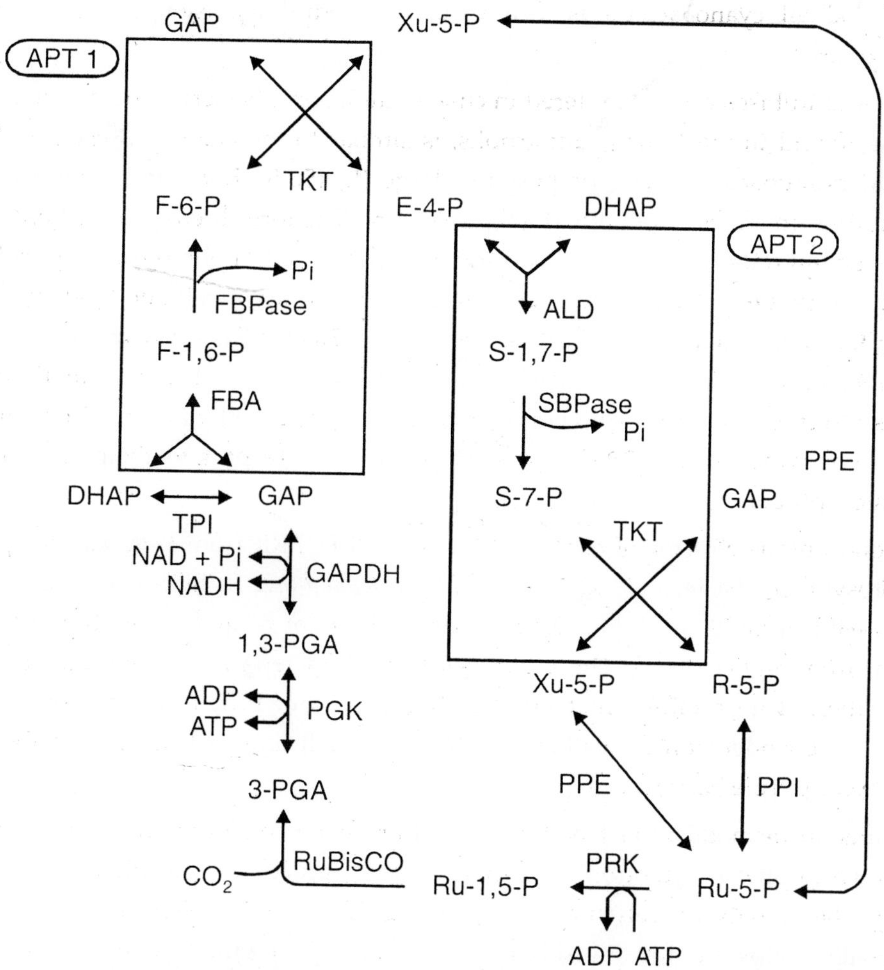

FIG. 6.9: *The Calvin cycle. Metabolities: DHAP, dihydroxyacetonephosphate; E-4-P, erythrose-4-phosphate; F-6-P, fructose-6-phosphate; F-1.6-P, fructose-1,6-bisphosphate; GAP, glyceraldehyde-3-phosphate; 3-PGA, 3-phosphoglycerate; 1,3-PGA, 1,3-diphosphoglycerate; R-5-P, ribose-5-phosphate; Ru-5-P, ribulose-5-phosphate; Ru-1, 5-P, ribulose-1,5-bisphosphate; S-7-P, sedoheptulose-7-phosphate; S-1,7-P, sedoheptulose-1, 7-bisphosphate; Xu-5-P, xylulose-5-phosphate; Pi, inorganic phosphate. Enzymes; ALD, aldolase; FBA, fructose-1,6-bisphosphate aldolase; FBPase, fructose-1,6-bisphosphatase; GAPDH, glyceraldehydes-3-phosphate dehydrogenase; PGK, phosphoglycerate kinase; PPE, pentose-5-phosphate epimerase; PPI, pentose-5-phosphate isomerase; PRK, phosphoribulokinase; RuBisCO, ribulose-1,5-bisphosphate carboxylase/oxygenase; SBPase, sedoheptulose bisphosphatase; TKT, transketolase; TPI, triosephosphate isomerase.*

Calvin Cycle Enzymes in Bacteria

The enzyme of the Calvin Cycle are located in the cytoplasm in most autotrophic bacteria. An notable exception is Rubisco, which in a number of chemoautotrophic bacteria and all cyanobacteria is located in polyhedral inclusion bodies known as carboxysomes.

Two types of Rubisco are encountered in chemoautotrophic bacteria. The dominant form I enzyme, found in most chemoautotrophs, is similar to the enzyme present in plants, algae and cyanobacteria. It is composed of large (L) (50-55 KDa) and small (S) (12-18 KDa) subunits in a hexadecameric (L_8 S_8) structure. The form II enzyme, which consists of large subunit only (L_2, L_4 or L_8) was initially discovered in phototrophic purple non-sulfur bacteria but it has since been found in certain symbiotic and free-living chemoautotrophic bacteria. *Thiobacillus denitrificans, Thiobacillus intermedius, Thiobacillus neapolitanus* and *Hydrogenovibrio marinus* possess both a form I and form II Rubisco, which is also the case for purple non-sulfur bacteria. The bacterial form I enzymes are similar to enzymes encountered in higher plants with respect to their structure and biochemical properties.

The second unique enzyme of the Calvin cycle in PRK. PRK from *Ralstonia eutropha* and the photosynthetic bacteria, *Rhodopseudomonas acidophila* and *R. sphaeroides* are octameric proteins with subunits of 32-36 KDa. Bacterial PRK is not related to the dimeric enzyme (45 KDa) from higher plants. The activity of PRK in bacteria is tightly regulated by the concentration of intracellular metabolities. The activity of PRK from chemoautotrophic bacteria is dependent on activation by NADH, which is also true for the PRK from phototrophic purple bacteria.

The change in metabolism of facultatively autotrophic bacteria following the transition from heterotrophic to autotrophic growth is the induction of Rubisco and PRK. In addition, the activity of Calvin cycle enzymes others than Rubisco and PRK increase dramatically following the transition to autotrophic growth. Mutants with strongly reduced activities in phosphoglycerate kinase and triphosphate isomerase have been isolated and only had marginally reduced growth rates while growing on heterotrophic growth substrates such as succinate and gluconate. In contrast; neither mutant was able to grow autotrophically, despite the fact that the other Calvin cycle enzymes are fully induced. The sharp increase in activity of enzymes other than Rubisco and PRK has, therefore, been interpreted as a reflection of the increased carbon flux required to assimilate sufficient CO_2 to allow autotrophic growth to proceed.

Efficiency of Rubisco

The efficiency of Rubisco is commonly evaluated by determining its specificity factor, tau (τ) i.e. the enzyme's ability to discriminate between carboxylation and oxygenation. The form II enzymes have a characteristically low τ value between 10 and 20. The form I enzymes of higher plants generally have τ values above 80. Interestingly, very high τ values have been recently reported in several eukaryotic algae. The form I enzymes of bacteria including chemo-and photoautotrophic as well as green algae are intermediate commonly falling between 30 and 60. A number of factors have been shown to alter specificity including mutations resulting in alterations of both the large and small subunits, the creation of hybrid Rubisco and the interaction of the enzymes makes it understandable why *H. marinus* predominantly synthesizes form II Rubisco when grown in an atmosphere containing 10% CO_2, whereas form I Rubisco is the dominant enzyme during growth in the presence of 2% CO_2. A similar differential regulation of form I and form II Rubisco was observed in the phototroph *Rhodobacter sphaeroides*.

CHAPTER 7

Path of Carbon

The rate of photosynthetic carbon assimilation is determined by both supply and demand of CO_2. The supply of CO_2 to the chloroplast is governed by diffusion in the gas and liquid phases. It can be limited at several points in the pathway from the surrounding the leaf to the site of carboxylation in-side. The demand for CO_2, is determined by the rate of processing the CO_2 in the chloroplast which is governed by the structure and biochemistry of the chloroplast, by environmental factors such as irradiance and plant demand for carbohydrates. Limitations imposed by either supply or demand can regulate overall rate of carbon assimilation.

Dark reaction is the pathway by which CO_2 is reduced to sugar. Its components are found in the stroma of the chloroplast. This reaction does not actually require darkness; it is merely light-independent. Since CO_2 is an energy poor compound, its conversion to an energy-rich carbohydrate involves a sizable jump up the energy ladder. This is accomplished through a series of complex steps involving small bits of energy. The CO_2 assimilation takes place both in light and darkness when the substrates NADPH and ATP are available. Because of the need for NADPH as a reductant and ATP as energy equivalent, CO_2 fixation is closely linked to the light reactions. During evolution three different ecological varients have evolved with different CO_2 incorporation mechanism: C_3, C_4 and CAM plants. In C_3 plants the first detectable CO_2 fixation product is a C_3 compound - phosphoglyceric acid, in C_4 plants we find C_4 compounds such as malate or oxalate. In CAM plants, most of which are of crassulaceae, COx is incorporated into C_4 acids during the night: these are used as CO_2 reservoir for CO_2 fixation during day-time, which follows the same pattern as C_3 plants. Plants such as crown, sugar cane, crabgrass, and other that can grow in a hot, dry environment must keep their stomata (the gas exchange pores in the leaves) closed much of the time to prevent excessive loss of moisture. This causes the CO_2 level within the leaf to fall below the K_m of ribulose 1,5-bisphosphate carboxylase; under these conditions, photorespiration is greatly favoured over photosynthesis. To avoid this problem, C_4 plants have evolved two steps pathway of CO_2 fixation that involves two types of cells: mesophyll cells - adjacent to surface of the leaf, and interior bundle sheath cells.

CO_2 Response Cruve

The response of photosynthetic rate to CO_2 concentration is the basic tool to analyze the demand for CO_2. The graph giving CO_2 assimilation (A) as a function of intercellular CO_2 partial pressure (p_i) is generally referrd to as the A-pi curve, or A-c_i when CO_2 concentration is given as mole fraction in air. There is no net CO_2 assimilation until the production of CO_2 in respiration (mainly photorespiration, but also some dark respiration that occurs in the light) is fully compensated by the fixation of CO_2 in photosynthesis. The CO_2 pressure at which this is reached is the CO_2-compensation point (τ). In C_3 plants this is largely determined by kinetic properties of Rubsico, with values for τ in the range of 4 to 5 Pa.

CO_2 – Stomatal and Boundary Layer Conductances

The rate of supply of CO_2 to the chloroplast can be described as a diffusion process (**supply function**). When considering fluxes, however, it is more convenient to use conductance, which is the reciprocal of resistance, because the flux varies in proportion to the conductance. Leaf conductance for CO_2 transport can be derived from measurements on leaf transpiration, which can also be described by Fick's law. The total leaf resistance for water vapor transfer, is largely composed of two components that are in series: the **boundary layer resistance** and the **stomatal resistance**. For the transport of CO_2 from the substomatal cavity to the chloroplast, an internal conductance also plays a role. As a consequence, the internal resistance and partial CO_2 pressure in the chloroplast is less than p_i. The evidence for this aspect largely comes from carbon isotope ratio of the products of photosynthesis.

Carbon: Principles of Methodology

Several techniques have been largely responsible for the successful analysis of the thermochemical dark reaction in photosynthesis. One of these is the use of radioactive carbon making it possible to tag the CO_2 ($C^{14}O_2$) which is incorporated into organic compounds during photosynthesis. The basic experimental approach was as follows:

In a plant that is carrying out photosynthesis at a steady rate, CO_2 is being converted to carbohydrates and other plant compounds through a series of intermediates.

$$CO_2 \longrightarrow \text{Compound A} \longrightarrow \text{Compound B} \longrightarrow \underset{\text{(Acetyl CoA)}}{\overset{\text{(Amino acid)}}{\text{Compound C}}} \longrightarrow \underset{\text{(Lipids)}}{\overset{\text{(Proteins)}}{\text{Carbohydrate}}}$$

If, at time zero, radioactive CO_2 ($^{14}CO_2$) is introduced into the system, some of the labelled carbon atoms are be converted to these different products, and during the time it takes for this to occur all the intermediates are become labelled. After a relatively shorter period of time, the photosynthesizing plant is plunged into hot alcohol to inactivate its enzymes and stop all reactions. If the reaction is stopped in a short period (just after the formation of first stable intermediate), only compound A will be labelled and will be known as **first product** of CO_2 fixation.

Method

Using these tools Calvin's group was able to identify the early stable intermediates in the path of carbon from CO_2 to plant metabolites. They used suspension of *Chlorella pyrenoidosa* or *Scenesdesmus obliques* (because they could be handled with ease) grown in culture medium, at a constant rate in presence of light and CO_2. Radioactive CO_2 was introduced into the reaction mixture at zero time, and a period of time was allowed to elapse. The cells were killed and extracted in boiling alcohol. The soluble constituents of alcohol solution were analysed by 2-dimentional paper chromatography and autoradiography. The compounds were identified in the chromatographs both their radioactive emissions and by chemical reactions. For confirmation, the spots were cut, extracted and rechromatographed with authentic samples.

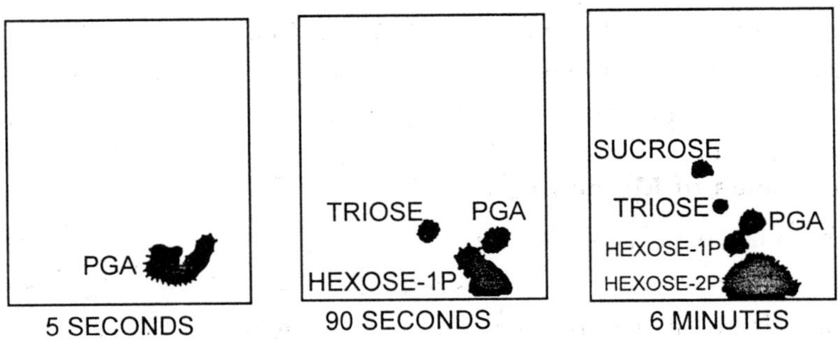

FIG. 7.1: *$^{14}CO_2$ supplied to the algal suspension. after an interval of 5, 90 seconds and 5 minutes different chemicals were synthesized. Black spots indicate C^{14} incorporation in different compounds.*

When algae were exposed to $^{14}CO_2$ for 30 seconds, hexose phosphate, triose phosphate and phosphoglyceric acid were labelled. With long periods, these compounds as well as amino acids, organic acids, lipids, and proteins were labelled. With 5 second exposure, however, most of the radioactive carbon was located in 3-phosphoglyceric acid and within this compound, the carboxyl group contained the majority of the radioactivity. The results reveal that phosphoglyceric acid was formed by the carboxylation of some unknown compounds containing 2-carbon atoms. Further closer analysis of photosynthetic products revealed that sedodoheptulose-7-phosphate and several phosphorylated pentose are also produced during the assimilation of $^{14}CO_2$. Since PGA is a 3-carbon compound, it was thought that CO_2 was accepted by a 2-Carbon molecule to form the 3-carbon PGA. But no such compound was detectable in plant.

TABLE 7.1: Exposure time (second) or $^{14}CO_2$ product in which ^{14}C appeared as metabolite of Calvin cycle

Product	Time of exposure (Second)
3-phosphoglyceric acid (3-PGA)	5
Triose phosphates	30
8-hexose phosphate	60

Later on, A. A. Benson working in the same laboratory found that a 5-carbon sugar ribulose bisphosphate (RuBP) was the acceptor molecule of CO_2. Benson, further conducted many experiments in this regard. In one experiment in which light was switched off after a short period of $^{14}CO_2$ feeding, PGA accumulated and RuBP disappeared completely. In another complementary experiment in which the light was kept on but $^{14}CO_2$ was withdrawn RuBP accumulated and PGA disappeared. After accepting CO_2, plant synthesize many unstable carbon intermediates which cleave into 2 molecules of PGA. In vitro studies further suggested that unstable 6-carbon compound is formed when RuBP accepted CO_2.

The end products of photosynthesis, namely sugars or starch are synthesized from PGA by reversed sequence of EMP pathway of glycolysis. Some PGA is used to regenerate RuBP for accepting some more CO_2 molecules. This is a regular cyclic-reaction that ultimately produce the end-products of photosynthesis and this cyclic metabolic pathway has been termed **Calvin cycle.** Calvin received Nobel Prize for his work in 1961.

The Reductive Pentose Phosphate (RPP) or Calvin Cycle

Horecker, Racker, and others established the relationships between trioses, tetroses, pentoses, hexoses, and heptoses. Studies on ^{14}C labelling of the sugars produced during short periods of photosynthesis permitted Calvin's laboratory to postulate the operation of a reductive pentose phosphate (RPP) cycle. This cycle involves essentially only one new reaction, the **carboxylation** of ribulose- 1,5-bisphosphate. The remainder of the reactions are identical or similar to reactions encountered previously in glycolysis and pentose phosphate metabolism. The enzymes required to catalyze the reactions in the RPP cycle occur in the stromal phase of chloroplasts.

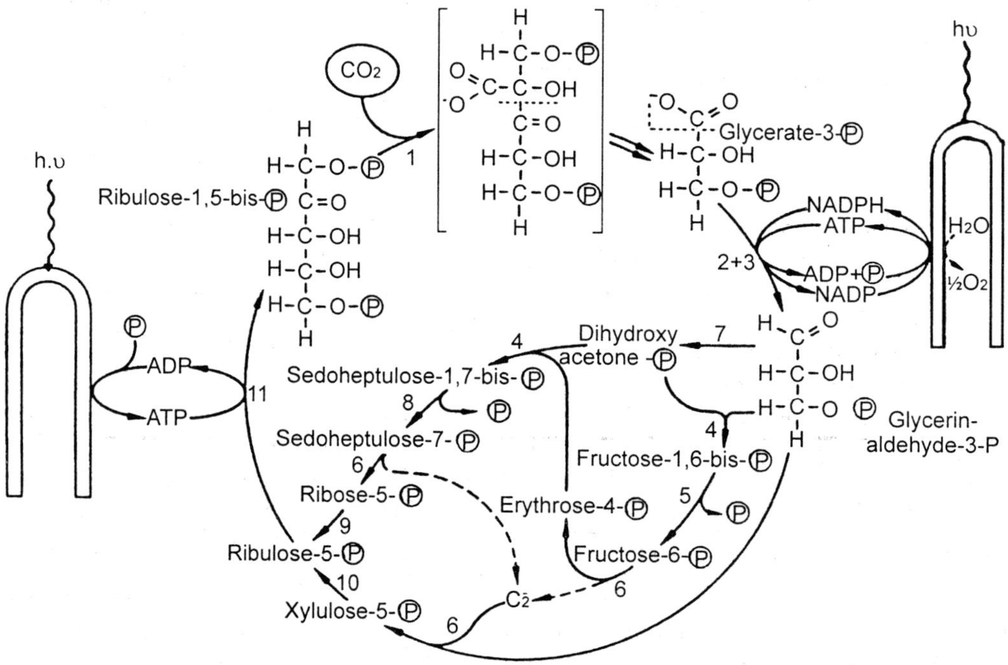

FIG. 7.2: *Calvin cycle.*

Carboxylation Phase

This scheme has been grouped into three phases. The **first** phase is the carboxylation phase, involves the single reaction catalyzed by ribulose-1,5-bisphosphate carboxylase-oxygenase (also called **Rubisco**). The reaction involves the carboxylation of five-carbon compound, ribulose-1,5-bisphosphate, and yield two molecules of 3-phosphoglyceric acid. In the presence of the enzyme Rubisco, CO_2 adds to the enediol form of ribulose-1,5-bisphosphate to form an unstable beta-keto acid that undergoes hydrolytic cleavage to form two molecules of 3-phosphologlyceric acid. The equilibrium of the reaction is far to the right.

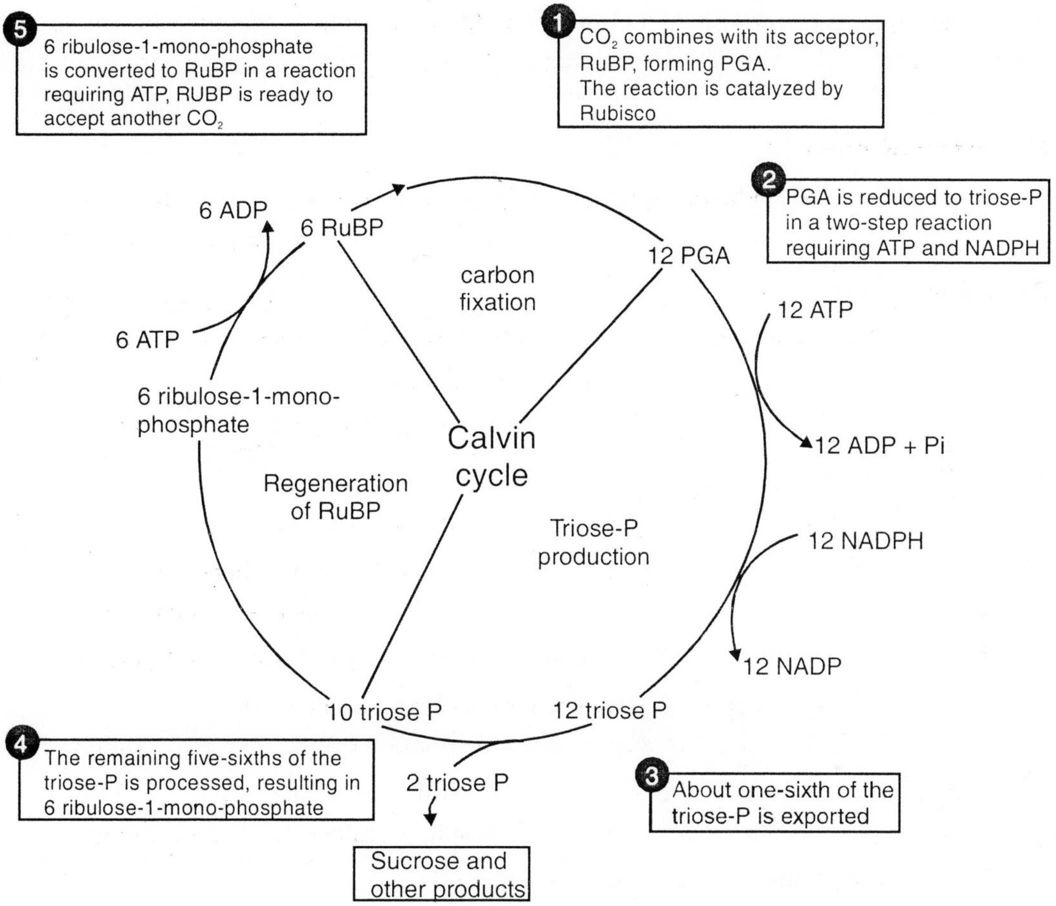

Fig 7.3: Schematic representation of the photosynthetic carbon reduction cycle (Calvin cycle) that shows major steps : carbon fixation, triose-P production, and regeneration of RuBP. 1 : CO_2 combines with its substrate, ribulose-1,5-bisphosphate (RuBP), catalyzed by ribulose bisphosphate carboxylase/oxygenase (Rubisco), producing phosphoglyceric acid (PGA). 2 : PGA is reduced to triose-phosphate (triose-P), in a two-step reaction; the reaction for which ATP is required is the conversion of PGA to 1,3-bisphosphoglycerate, catalyzed by phosphoglycerate kinase. 3 and 4 : part of the triose-P is exported to the cytosol, in exchange for P_i; the remainder is used to regenerate ribulose-1-monophosphate. 5 : ribulose-1-monophosphate is phosphorylated, catalyzed by ribulose-5-phosphate kinase, producing RuBP.

Reduction Phase

A **second** phase of the RPP cycle is known as **reduction phase**. It consists of two reactions encountered in glycolysis. ATP and a reduced nicontinamide nucleotide are utilized. The first reaction involves the phosphorylation of 3-phosphoglycerate by ATP to form 1,3-bisphosphoglycerate. While in the second reaction 1,3-bisphosphoglyceric acid is reduced by NADPH in the presence of a NADP-specific glyceraldehyde-3-phosphate dehydrogenase. ATP and NADPH activate the chloroplast enzyme. In these two reactions all of the NADPH and two-thirds of the ATP are required to drive the RPP cycle.

Carboxylation phase

1. 3-Ribulose- 1,5-bisphosphate + 3 CO_2 + 3 $H_2O \longrightarrow$
6-3-Phosphoglycerate

Reduction phase

2. 6 3-Phosphoglycerate + 6ATP $\longrightarrow$ 6 1,3-bisphosphoglycerate + 6 ADP
3. 6 1,3-bisphosphoglycerate + 6 NADPH + 6 $H^+ \longrightarrow$
6-Glyceraldehyde-3- phosphate+ 6 $NADP^+$ + H_3PO_4

Regeneration phase

4. 2 Glyceraldehyde-3-phosphate $\longrightarrow$ 2 Dihydroxy-acetone phosphate
5. Glyceraldehyde-3-phosphate + dihydroxy-acetone phosphate $\longrightarrow$
Fructose- 1,6-bisphosphate
6. Fructose-1,6-bisphosphate + $H_2O \longrightarrow$ Fructose-6-phosphate + H_3PO_4
7. Fructose-6-phpsphate + glyceraldehyde-3-phosphate $\longrightarrow$
Xylulose-5- phosphate + Erythrose-4-phosphate
8. Erythrose-4-phosphate + dihydroxy-acetone phosphate $\longrightarrow$
Sedoheptulose-1,7- bisphosphate
9. Sedoheptulose-1,7-bisphosphate + $H_2O \longrightarrow$
Sedoheptulose-7-phosphate + H_3PO_4
10. Sedoheptulose-7-phosphate + Glyceraldehyde-3-phosphate $\longrightarrow$
Ribose-5- phosphate +xylulose-5-phosphate
11. Ribose-5-phosphate $\longrightarrow$ Ribulose-5-phosphate
12. 2 Xylulose-5-phosphate $\longrightarrow$ 2 Ribulose-5-phosphate
13. 3 Ribulose-5-phosphate + 3 ATP $\longrightarrow$ 3 Ribulose-1,5-bisphosphate + 3 ADP

Sum

$3CO_2$ + 9 ATP + 6 NADPH = 6 H^+ + 5 $H_2O \longrightarrow$
Glyceraldehyde-3-phosphate + 9 ADP + 6 $NADP^+$ + 8 H_3PO_4.

Regeneration Phase

The **third** reaction of the RPP cycle is the **regeneration phase** of ribulose-1,5-bisphosphate which is necessary to keep the cycle in normal operation. At the end of the reduction phase, a total of 18 carbon atoms will be present in six molecules of glyceraldehyde-3-phosphate. One of these can be considered as the photosynthetic product. The 15 carbon atoms in the other five molecules of glyceraldehyde-3-phosphate are, in turn, rearranged by the reactions of the regeneration phase into three molecules of ribulose-5-phosphate ($5C_3 \rightarrow 3C_5$). The phosphorylation of these pentose phosphates into ribulose-1,5-bisphosphate completes the RPP cycle and produces the substrates for assimilation of another three moles of CO_2. Examination of steps 2,3, and 13 show that a total of 2 NADPH and 3 ATP are required for each CO_2 assimilated by this process.

Rubisco: Structure and Function

Ribulose bisphosphate carboxylase-oxygenase (Rubisco) is probably the most abundant protein on earth; it constitutes about half of the leaf proteins. Rubisco catalyzes the biological fixation of 10^{11} tonnes of CO_2 annually from the atmosphere. In the last two decades considerable research effort has been expended towards the elucidation of the mechanism of carboxylation by the enzyme and in the determination of its three-dimensional structure. Besides fundamental understanding, a main objective behind this research is to design, if at all possible, a Rubisco molecule with an increased carboxylation to oxygenation activity ratio than is found in the naturally occurring enzyme.

Rubisco isolated from oxygenic photosynthetic organisms is an octamer consisting of eight large subunits (LSU, molecular mass 50-55 kDa) and eight small subunits (SSU, molecular mass-12-15 kDa).

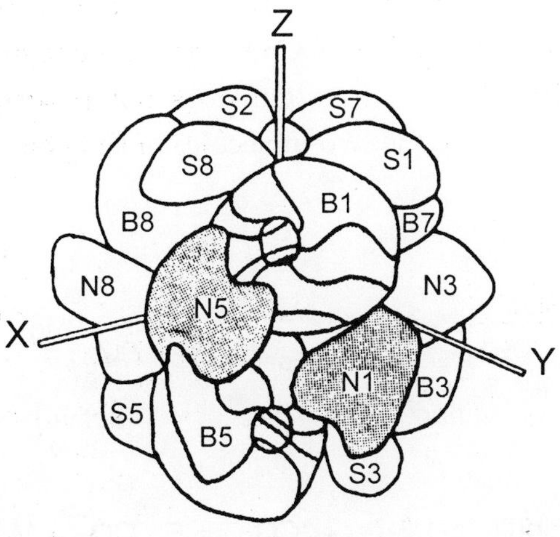

FIG. 7.4: *The three dimensional relationship between two L-subunits and associated S subunits of Ribisco (Molecular configuration).*

The LSU, which is coded by the chloroplast DNA, is synthesized within the chloroplasts; the catalytically active site(s) is located in the LSU. The SSU is coded by the nuclear genome and is assembled in the cytoplasm as a pre-protein that contains a domain of 50 extra amino acids at the aminoterminal end; this domain is cleaved off when the preprotein enters the chloroplast through the envelope. Assembly of the matured SSU and LSU within the chloroplast is mediated by a *chaperone*, the Rubisco assembling protein.

Rubisco is one of the proteins subjected to extensive gene sequence determination in recent years. The genetic dissection of the enzyme in the green alga *Chlamydomonas reinhardtii* had identified three large-subunit (L) regions that should provide loci for future investigations via genetic engineering.

In contrast to the Rubisco enzymes of chloroplast stroma, the functionally active Rubisco of *Rhodospirillum rubrum* is a dimmer of 50 kDa subunits which lack any SSU. The X-ray structure of *R. rubrum* enzyme has also been determined, and many residues in the enzyme molecule have been altered by site-directed mutagenesis. These studies have helped to identify the activator and substrate-binding sites on the Rubisco molecule.

Activation of Rubisco

The inactive form of the enzyme first combines with a molecule of CO_2 in a slow, reversible reaction. When this enzyme – CO_2 complex subsequently reacts with Mg^{2+} in a second reversible step, the complex enzyme – CO_2 – Mg^{2+} is rapidly formed and it is only this form of enzyme that is fully active. The Mg^{2+} is necessary for the activation of the enzyme but not for the carboxylase reaction. The actual carboxylation of ribulose –1-5-P_2 occurs when a second molecule of CO_2 reacts with enzyme. The molecule of CO_2 that binds to the enzyme during activation does not participate in the carboxylase reaction and it is most likely that the two molecules of CO_2 bind at different sites on the enzyme.

$$\underset{\text{(inactive)}}{E + CO_2} \xleftrightarrow{\text{Slow}} \underset{\text{(inactive)}}{E + CO_2} \xleftrightarrow[\text{FasL}]{Mg^{2+}} \underset{\text{(active)}}{E + CO_2 + Mg^{2+}} \text{ (activation)}$$

$$E + CO_2 + Mg + \text{Ribulose-1,5-}P_2 + {}^{*}CO_2 \Leftrightarrow E - CO_2 + Mg^{2+} + PGA + PGA$$

(carboxylation reaction)

The carbamylation (condensation between CO_2 and enzyme – NH_2 group) occurs at a specific Lys residue at the activator site of the enzyme which is distinct from the catalytic site. Activation of Rubisco also requires light. Recent work with a mutant of *Arabidopsis* (a plant amenable to mutations and genetic engineering) has led to the discovery of a protein named *Rubisco activase* which is involved in the light-regulated activation of Rubisco.

RUBISCO ACTIVASE

Rubisco activase is a nuclear-encoded chloroplast protein and in most plants examined to date, there are two isoforms (46- and 43-KDa) of Rubisco activase. Both isoforms differ only at the carboxyl terminus. Zhang and Portis (1999) suggested that in plants containing both isoforms, Rubisco activase regulates the activity of Rubisco in response to light-induced changes in both the ADP / ATP ratio and the redox potential via thioredoxin-f.

The action of Rubisco activase is dependent on the availability of ATP and a membrane pH gradient, which are generated during light-induced electron transport. Many plant species produce 2-carboxy arabinitol-phosphate, an analogue of a six-carbon intermediate formed during Rubisco catalysis, which inhibits the enzyme by occupying the catalytic site. It has been proposed that Rubisco activase functions by removing this inhibitor from the catalytic domain of Rubisco during transition from darkness to light. The activity of Rubisco is regulated by the ADP/ATP ratio in the stroma because ADP inhibits the ATP hydrolysis reaction of the activase which is required to remove the inhibitors. Rubisco activase operates in the form of a large complex. The A/a substitution results indicate that Cys-392 and Cys-411 probably form the critical disulphide bond in the larger isoform that is reduced by thioredoxin-f. The alteration of the activity of the larger isoform is sufficient to regulate the activity of both isoforms when they are present together. Rubisco is one of the light-regulated chloroplast enzymes not controlled by the Fd-thioredoxin system.

Catalysis by Rubisco

The catalysis takes place in following three steps:

1. Translocation of a proton from the C_3 position of the substrate RuBP to the C2 position and formation of an enediol.

2. The enediol reacts with a molecule of CO_2, forming the intermediate adduct 3-keto-2-carboxy arabinitol bisphosphate (CABP). This intermediate can be isolated in a stable form. Crystal structure analysis of the quaternary complex of spinach Rubisco

co-ordinated to CABP has confirmed the nature of the substrate-binding sites on the enzyme. These had been previously deduced from comparative sequence analysis, chemical modifications, and site-directed mutagenesis. The residues involved are thought to be a specific Lys and Glu.

3. The intermemdiate is hydrolytically cleaved into one molecule of PGA and a C_3 carbanion; the carbanion is immediately protonated to yield the second molecule of PGA.

Although, many derivatives of naturally occurring Rubisco have been prepared by chemical or genetic modification of the molecule, none of these derivatives has manifested an enhanced ratio of carboxylation to oxygenation activity compared to the native enzyme.

Fallover of Rubisco

In vitro, the activity of purified Rubisco from higher plants decreases with a half-time of several minutes after contact with substrate, ribulose-1,5-P_2. This loss in activity is termed as fallover. Three following reasons have been given for this fallover :

- Rubisco becomes catalytically competent only after carbamylation by CO_2 at the catalytic site.
- Readily reversible binding of ribulose-1,5-P_2 at a regulatory site.
- Slow, tight-binding inhibition by isomerization or degradation products which accumulate in ribulose-1,5-P_2 preparations during storage.

Carboxyarabinitol 1-Phosphate

Carboxyarabinitol 1-phosphate (**CA1-P**) a low molecular weight carbohydrate accumulates in the chloroplasts of numerous plant species Because of structural similarities to the carboxylation reaction intermediate, 2-carboxy-3-ketoarabinitol-1,5-bisphosphate, CA1-P binds tightly to the active site of Rubisco. CA1-P accumulation is readily detected in leaf extracts by measuring the inhibition of Mg^{2+}/CO_2^- activated Rubisco activity, since the decrease in maximal catalytic rate is a function of the proportion of active sites occupied by the inhibitor. Among species in which CA1-P accumulates, maximal synthesis of the inhibitor in the dark can cause a decrease in carboxylation activity ranging from 25 to 95 per cent . When leaves are subject to high irradiance CA1-P is metabolized, causing inhibition of Rubisco activity to be fully reversed within 10 min. The metabolic reactions that synthesize CA1-P in the chloroplast of some plant species have not been elucidated. Metabolism of CA1-P by the stromal protein exhibited an apparent requirement for NADPH.

Carbonic Anhydrase

Carbonic anhydrase (EC 4.2.1.1) catalyzes the freely reversible hydration of CO_2 to bicarbonate and it is present in both animals and plants. Since bicarbonate is a substrate for phosphoenolpyruvate carboxylase (the primary carboxylation enzyme in C_4 species), carbonic anhydrase is regarded as the first enzyme of the C_4 photosynthetic pathway. In C_4 plants this enzyme is localized in cytosol of mesophyll cells while it is essentially absent in the bundle sheath cells. In C_3 species it is found to be localized in chloroplasts. Recent immunological studies of this enzyme from C_3 and C_4 plants have indicated that there are isozymic forms of the enzyme in leaf tissue in both C_3 and C_4 plants which are similar in size but differ antigenically. Further studies have revealed that both C_3 and C_4 species show considerably high activity of this enzyme. It has been established that CO_2 and not HCO_3^- is the species responsible for both the activation process and the actual carboxylase reaction. Although the alkaline environment in stroma of an illuminated chloroplast would keep most of the carbon as the bicarbonate ion, the equilibrium between CO_2 and HCO_3^- is maintained by the enzyme carbonic anhydrase that catalyzes the reaction $H + HCO_3^- \rightarrow CO_2 + H_2O$. This would ensure that CO_2 removed by the carboxylasae reaction is rapidly replaced from the pool of bicarbonate.

Regulation of the RPP Cycle

The RPP cycle (Benson-Calvin cycle) is the main biochemical pathway for the conversion of CO_2 to organic compounds. Two unique system that link light-triggered events in thylakoid membranes with enzyme regulation are located in the soluble portion of chloroplast (stroma): the ferredoxin-thioredoxin system and rebulose 1.5 – bisphosphate carboxylase/oxygenase-Activase (Rubisco Activase). The ferredoxin-thioredoxin system transforms inactive to catalytically components forms. On the other hand, Rubisco-Activase facilitates the combination of CO_2 with Rubsico and the subsequent stabilization of the carbamylated enzyme by Mg^{2+}. A convincing body of evidence supports the view that the inability of chloroplast to fix CO_2 in the dark is not only due to lack of metabolic precursors but also to inactive enzymes. Upon illumination, the concentration of intermediates of the cycle increases in the chloroplast stroma and most notably regulatory enzymes are converted to active forms.

The assimilation of CO_2 by the enzymes of the Reductive Pentose Phosphate (RPP) cycle is regulated primarily by the activity of the carboxylation enzyme, Rubisco which exists in an active from, that is a complex of CO_2 and Mg^{2+} formed sequentially as follows.

(i) The preliminary binding of CO_2 is a relatively slow process, but the reaction with Mg^{2+} is more rapid and is accelerated at higher pH (8.5).

(ii) Light strongly affects the above process, since the stromal compartment becomes more alkaline in the light as protons move into the thylakoids.

(iii) The intermembrane space of the thylakoid becomes acidic by movement of Mg^{2+} ions out into the stroma where they can complex to form the active form of Rubisco.

(iv) Since formation of the active Rubisco is reversible, the enzyme is converted to its inactive form in the dark when protons move out of the thylakoids and Mg^{2+} leaves the stroma.

Regulation by Light

Light was considered as an energy source for the primary photochemical processes for the production of NADPH and ATP after irradiation. In contrast to the primary reactions, the biochemical secondary reactions in the Calvin cycle reach their optimal activity only after a **lag phase** of a few minutes. The delay in CO_2 fixation is due to light activation of enzymes. Two mechanisms have been observed:

- During initiation of photoreaction enzyme activation occurs. One of the examples for this process is the activation of triose phosphate dehydrogenase by reduced thioredoxin.
- Change in pH or ion concentration in the stroma activates enzymes. Examples for this mechanism are RuBP and PEP carboxylase which operate optimally at a pH of 8 and an increased Mg^+ concentration. Such conditions are generated in light by vectorial proton transport into the thylakoid interior and antiparallel Mg^+ transport into the stroma. This activity is also affected by a change in CO_2/O_2 partial pressures. Furthermore, RuBP carboxylase is thought to be induced by visible light and UV-A radiation according to the above-described activation by mediators.

REGULATION BY THIOREDOXIN

The major role of the photosynthetic electron transfer is to provide the chloroplast with NADPH and ATP for inorganic carbon assimilation. But it is now well established that a small fraction of the light-generated electron flow is diverted to serve a signaling purpose. The activities of several chlorophyll enzymes involved in anabolic process are enhanced or triggered by light and the activity of one catabolic enzyme is inhibited by light. The signaling system responsible for this effect is now well-characterized. It is called the ferredoxin-thioredoxin system.

In this system, the light signal sensed by chlorophyll is transferred via the photosynthetic electron flow to protein called thioredoxin. Thioredoxins are ubiquitous proteins which are very efficient in thiol-disulphide interchanges with various protein disulphides because of their very reactive active-site dithiol. The electron transfer between ferredoxin and thioredoxin is mediated by ferrodoxin-thioredoxin reductase (FTR). This heterotrimiric protein features an iron-sulphur cluster and disulphide bridge and converts the electronic signal to a thiol signal. Two thioredoxin isoforms are present in the chloroplasts which are named m and f according to their specificities toward their target enzymes. However, recent studies have shown that these specificities are much less strict.

Most of the chloroplast thioredoxin-regulated enzymes are related to carbon metabolism. Four of them, namely phsphoribulokinase (PRK), glyceraldehydes-3-phosphate dehydrogenase (G3PDH), FBPase and sedoheptulose-1,7-bisphosphate (SBPase) belong to the calvin cycle.

From an evolutionary point of view, it is clear that the appearance of the light-dependent redox regulation coincides with the appearance of oxygenic photosynthesis. Thus, this regulation might appear as defense mechanism against the potential oxidative stress that might be generated by the build-up of high oxygen concentration.

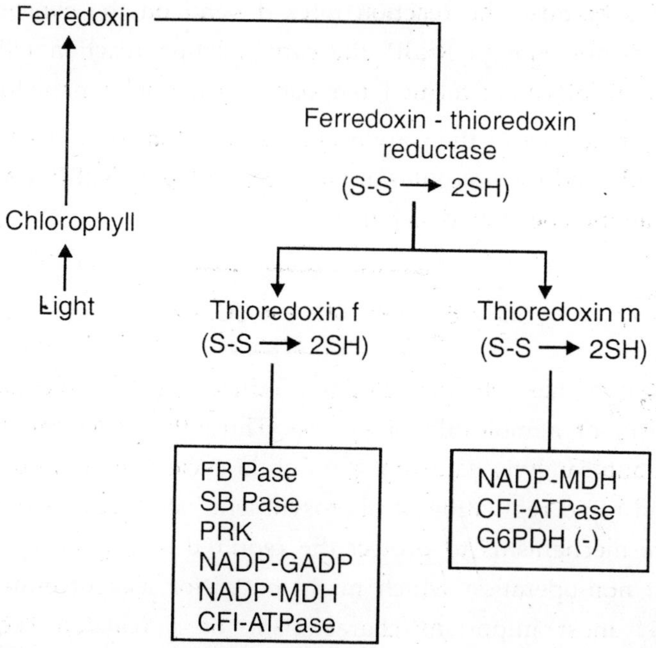

Fig 7.5: Enzyme regulated by the ferredoxin-thioredoxin system.

It has also been suggested that this regulation enables avoidance of the futile cycle that could occur in the chloroplast where anabolic and catabolic enzymes coexist in the same compartment. However, the evidence for the physiological importance of this mechanism in plant metabolism is rather scarce. The most cyanobacteria that are unable to grow autotrophically upon disruption of a gene encoding thioredoxin. This kind of evidence has never been obtained in higher plants.

Regulation by Nonphysiological Factors

Organic solvents miscible in water (cosolvents), neutral salts or high hydrostatic pressure enhance the specific activity of F-2,6-bisphosphatase, phosphoribulokinase and NADP-glyceraldehyde –3-phosphate dehydrogenase. Nonphysiological modulators yield enzyme conformations different from that elicited by theoredoxin. The underlying mechanism for their stimulatory effect should be the modification of enzyme noncovalent interactions.

Autocatalysis

RPP cycle is a very efficient autocatalytic device. The cycle works well so long as there is sufficient RuBP available. If there is not (for example, if RuBP leaves out of the chloroplast or was metabolized during a prolonged period of darkness), then the plant is in difficulty. This is because the reaction rates depend on the concentrations of the reactants. If there is not enough RuBP, the carboxylation reaction will proceed very slowly. Unless more RuBP can be formed, the reaction can not be made to go any faster.

Examination of the reactions of the Calvin cycle shows that it is possible to rearrange them so that the end product molecules are also converted to RuBP. Using a simplified outline, the cycle can then be altered as follos:

$$5C_1 + 5C_5 \longrightarrow 10C_3 \longrightarrow 5C_5 + C_5$$

In other words, the cycle has been modified to produce an extra molecule of RUBP as an end product instead of a molecule of hexose. Thus, the cycle can be described as autocatalytic. The autocatalytic nature of the Calvin cycle is most important because it permits simple and rapid regulation of photosynthetic rate. Because of this, there is no need for elaborate mechanisms to protect the required supply of cyclic intermediate during periods of non-operation which might occur for lack of either CO_2 or light. Autocatalysis is a most important characteristic of a reaction sequence such as photosynthesis that is intermittent but is required to get quickly to high speed when conditions are right.

C_4 CYCLE (HATCH AND SLACK CYCLE)

In addition to Calvin cycle, an alternative (supportive to RPP) pathway for CO_2 fixation in photosynthesis was discovered by Kortschak et al. (1954). They reported the formation of C_4 dicarboxylic acids as primary products of photosynthesis in sugarcane. When leaves of sugarcane were treated with $^{14}CO_2$ for a short period, 3 types of 4-carbon dicarboxylic acids (oxaloacetate, malate and aspartate) were produced at initial stage. When the period of exposure was extended, label appeared in 3-phosphoglycerate. M.D. Hatch and C.R. Slack (1966) further investigated the incorporation of $^{14}C_4$ in several plants and then proposed an alternative pathway, which was later found supportive to Calvin cycle, because CO_2 was converted to carbohydrates through Calvin cycle in bundle sheath cells of C_4 plants.

The C_4 pathway shows the cooperativity of metabolism in the two photosynthetic cell types, mesophyll cells and bundles sheath cells. Mesophyle cells carry out the initial steps of CO_2 fixation via the enzyme phosphoenol pyruvate (PEP) carboxylase to produce the four-carbon organic acid oxaloacetate. Oxaloacetate is converted to malate or aspartate which diffuse into the bundle sheath cells. In the bundle sheath cells, the C_4 acid is decarboxylated and the resulting CO_2 is refixed by Rubisco in the photosynthetic carbon reduction (PCR) cycle. A three carbon product (pyruvate, alanine or PEP) returns to the mesophyll cells to be recycled for another round of carboxylation by PEP carboxylase. C_4 plants have evolved three distinct options for decarboxylation of C_4 acids in the bundle sheath cells:

- A chloroplastic NADP-malic enzyme that decarboxylates malate to produce pyruvate (NADP-ME type). E.g., *Saccharum offinarum, Zea mays, Euphorbia hirta.*
- A mitochondrial NAD-malic enzyme (NAD-ME type) as the dicarboxylating enzyme. In NADP-ME type species, malate is the predo-dominant C_4 acid translocated. E.g., *Portulaca oleracea, Cynodon dactylon, Atriplex spongiosa.*
- A cytosolic PEP caroxylase that produces PEP from oxaloacetate (PCK – type). E.g., *Panicum maximum, Sphorobolus poirettii.*

In NADP-ME type species, malate is promdominant CA acid translocated. In other 2 and 3 types, it is mainly it is mainly aspartate that is translocated to the bundle sheath cells, where it is first converted to oxaloacetate by transamination.

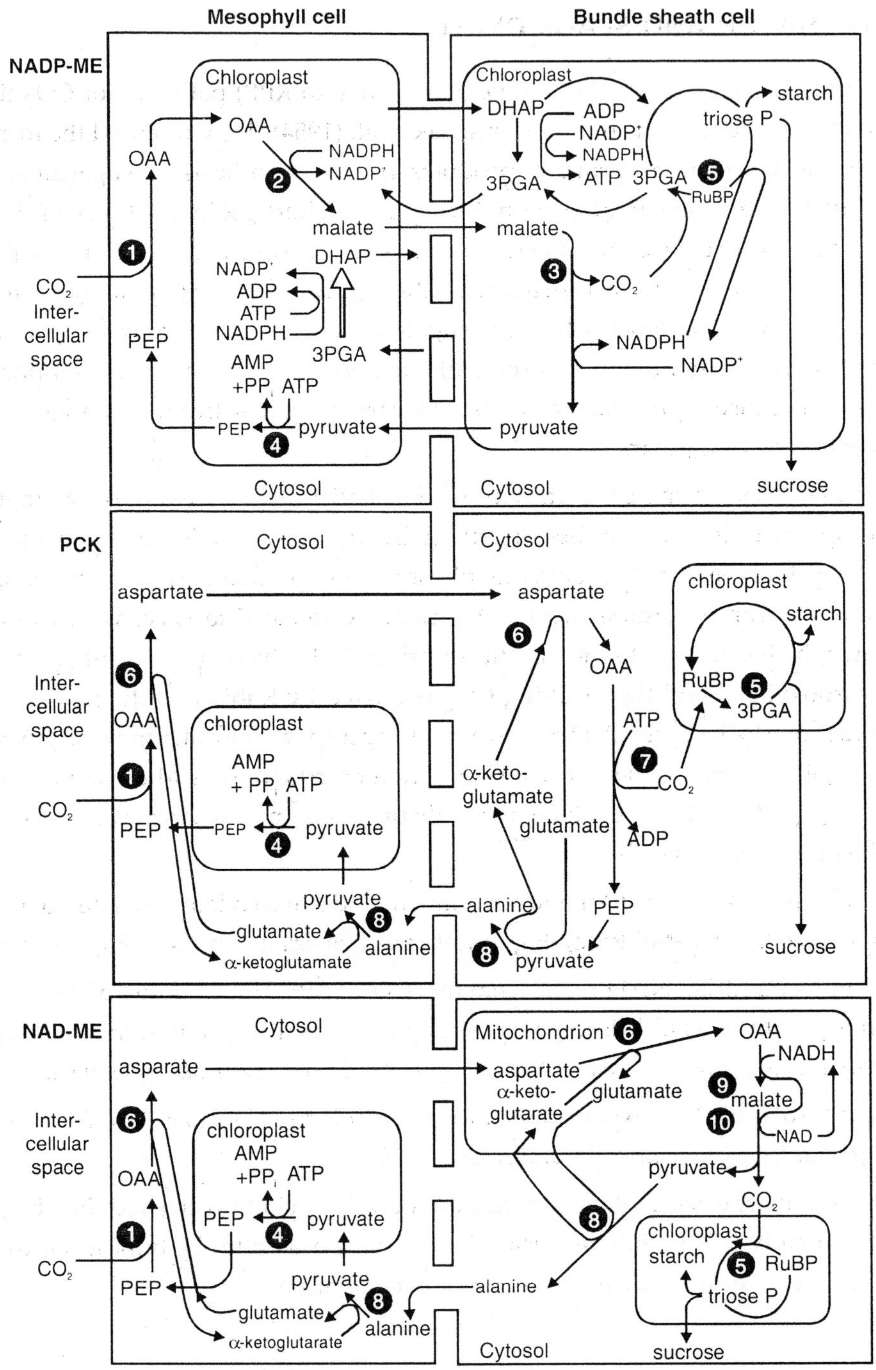

FIG. 7. 6: *A relation of mesophyll cells and bundle sheath cells during C_4 cycle. The enzyme involved are indicated by numbers. 1 = PEP carboxylase, 2 = NADP-MDH, 3 = NADP-ME, 4 = pyruvate-pi-dikinase, 5 = Rubisco, 6 = aspartate aminotransferase, 7 = PEP-CK, 8 = alanine aminotransferase, 9 = NAD-MDH, 10 = NAD-ME*

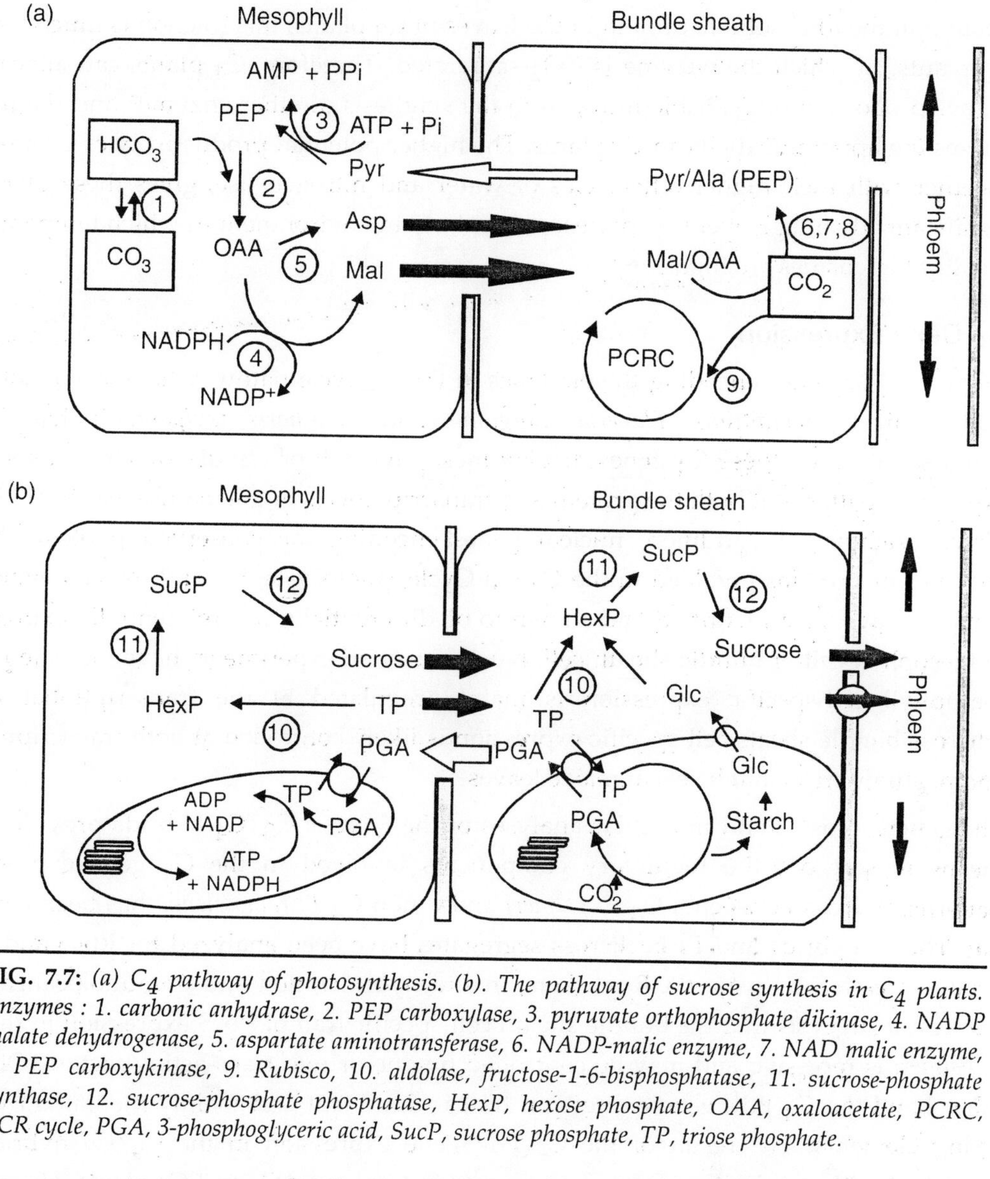

FIG. 7.7: *(a) C_4 pathway of photosynthesis. (b). The pathway of sucrose synthesis in C_4 plants. Enzymes : 1. carbonic anhydrase, 2. PEP carboxylase, 3. pyruvate orthophosphate dikinase, 4. NADP malate dehydrogenase, 5. aspartate aminotransferase, 6. NADP-malic enzyme, 7. NAD malic enzyme, 8. PEP carboxykinase, 9. Rubisco, 10. aldolase, fructose-1-6-bisphosphatase, 11. sucrose-phosphate synthase, 12. sucrose-phosphate phosphatase, HexP, hexose phosphate, OAA, oxaloacetate, PCRC, PCR cycle, PGA, 3-phosphoglyceric acid, SucP, sucrose phosphate, TP, triose phosphate.*

The C_4 pathway acts as a biochemical CO_2 "pump" that concentrates CO_2 in the bundle sheath cells. This biochemical pump can generate a CO_2 concentration up to 10-fold that in the atmosphere, effectively saturating Rubisco with CO_2 and inhibiting photorespiration. When grown under optimal conditions, the suppression of photorespiration in C_4 plants gives them significantly higher potential productivity than C_3 plants, despite the extra energetic cost of the C_4 pathway (two ATP per CO_2 fixed). The ability to concentrate CO_2 also allows C_4 plant to continue photosynthetic CO_2 fixation even with the stomata partly closed to decrease water loss. Rubisco accounts for about half the total soluble protein in the leaves of C_3 plants; this fraction is much less in C_4 plants, in which the enzyme is CO_2-satuarated. Therefore, C_4 plants can allocate a higher proportion of available nitrogen to the synthesis of other enzymes and the light-harvesting apparatus than can C_3 plants. The higher potential productively of C_4 species together with their higher efficiencies of water and nitrogen use, gives these plants a significant advantage over C_3 species, particularly in environment in which temperature and light intensities are high.

C_4 Gene Expression

Many nuclear genes encoding the enzymes in the C_4 cycle pathway have been isolated from maize, *Amaranthus, Flaveria, Sorghum* and *Eleocharis vivipara*. It has been demonstrated that these C_4 genes display mesophyll cell or bundle sheath cell-specific expression patterns at both the protein and transcript levels in the mature leaves of all C_4 plants examined. In addition, nuclear genes encoding the conserved photosynthetic enzymes or proteins involved in the Calvin Cycle, photosystem II and oxygen-evolving complex have been identified and shown to be differentially or preferentially expressed in mesophyll cell or bundle sheath cell. Nuclear run-on experiments have indicated that mesophyll cell-specific expression is mainly regulated at the transcriptional level whereas bundle sheats cell-specific expression is likely controlled at both transcriptional and posttranscriptional levels in maize leaves.

The genetic crosses and molecular analyses of the *Flaveria* C_3/C_4 hybrids provide now means to sort out the regulatory components involved in the C_4 gene expression patterns. A cross between a C_4 *F. palmeri* and a non-C_4 *F. ramossisima* has been carried out. The F1 hybrids and F1 backcross segregates have been analyzed for RbcS and Pdk gene expression by using PCR markers, in situ hybridization and immunolocalization. The result suggests that the bundle sheath cell-specific trait of RbcS expression in the C_4 *F. palmeri* is recessive and may be controlled by one or two trans-acting factors. On the other hand the C_4 Pdk gene expression trait is dominant and mainly mediated by cis-acting elements. The result of the C_4 Pdk gene expression in the C_4/C_3 hybrids is consistent with the similar analyses performed in transgenic C_4 and C_3 plants. However, the regulation of RbcS gene in bundle sheath cell seems to be complex in different C_4 plants.

C_4 PLANTS

Over 117 genera of plants belonging to 13 families possessing C_4 mechanism of CO_2 fixation are known and out of 13 families, 11 were dicotyledons while the remaining two were monocotyledons (Downton, 1975). At least 11 genera had both C_3 and C_4 species.

All these plants have some features in common: similar anatomical structure (Kranz type of leaf anatomy) and capability to grow successfully in environments with low levels of CO_2 in the air. C_4 plants possess several adaptive features such as low rates of photorespiration, higher water use efficiency, higher nitrogen use efficiency, temperature independent quantum yield and higher light saturation point which make them markedly superior to C_3 species with respect to harvest of solar energy.

In C_4 plants where modifications of the primary CO_2 trapping systems have been noticed. The secondary mechanisms are associated with distinctive anatomical characteristics and seem to have arisen independently several times in different angiosperm families during evolution.

TABLE 7.2: Some important crop plants and their characteristics.*

Common name	Scientific name	Photorespiration type*	Photoperiodic reaction*
Alfalfa	*Medicago sativa* L.	C_3	L
Barley	*Hordeum vulgare* L.	C_3	L;N
Bean, field	*Phaseolus vulgaris* L.	C_3	S;N
Bean, lima	*Phaseolus lunatus* Macf.	C_3	S
Bean, mung	*Phaseolus aureus* Roxb.	C_3	S
Beet, sugar	*Beta vulgaris saccharifera* L.	C_3	L
Bentgrass, creeping	*Agrostic palustris* Huds.	C_4	L
Bermudagrass	*Cynodon dactylon* (L.)Pers.	C_4	-
Birdsfoot trefoil	*Lotus corniculatus* L.	C_3	L(?)
Bluegrass, Kentucky	*Poa pratensis* L.	C_3	N
Bromegrass, smooth	*Bromus inermis* Leyss.	C_3	L
Broomcorn	*Sorghum vulgare technicum* Pers.	C_4	S
Buckwheat, common	*Fagopyrum esculentum* Gaertn.	C_3	L;N
Cassava	*Manihot utilissima esculenta* Pohl.	-	N
Castorbean	*Ricinus communis* L.	C_3	L
Clover alsike	*Trifolium hybridum* L.	C_3	L
crimson	*Trifolium incarnatum* L.	C_3	L
ladino	*Trifolium repens* L.	C_3	L
red	*Trifolium pratense* L.	C_3	L
sweet, white	*Melilotus alba* Med.	C_3	L
sweet, yellow	*Melilotus officinalis* Lam.	C_3	L
white	*Trifolium repens* L.	C_3	L
Corn or maize, dent	*Zea mays* L.	C_4	S
Corn or maize, sweet	*Zea mays saccharum* L.	C_4	L
Cotton, upland	*Gossypium hirsutum* L.	C_3	N
Cotton, Egyptian	*Gossypium barbadense* L.	C_3	N
Cowpea	*Vigna sinensis* Endl.	C_3	S
Crotalaria, showy	*Crotalaria spectabilis* Roth.	C_3	S

Common name	Scientific name	Photorespiration type*	Photoperiodic reaction*
Crownvetch	*Coronilla varia*	C_3	S
Flax	*Linum usitatissimum* L.	C_3	L
Guar	*Cyamopsis psoralides* DC.	-	-
Guayule	*Parthenium argentatum* Gray	-	W
Hemp	*Cannabis sativa* L.	C_3	S
Hop	*Humulus lupulus* L.	C_3	L
Kenaf	*Hibiscus cannabinua* L.	C_3	S
Kudzu	*Pueraria thunbergiana* Benth.	-	-
Lentil	*Lentilla lens*	C_3	L
Lespedeza, common	*Lespedeza striata* Hook and Arn	C_3	S
Lespedeza, Korean	*Lespedeza stipulacea* Maxim	C_3	S
Lespedeza, sericea	*Lespedeza cuneata* (Dum. de Cours)	C_3	S
Lupine, blue	*Lupinus angustifolius* L.	C_3	N;S
Lupine, white	*Lupinus albus* L.	C_3	N
Lupine, yellow	*Lupinus luteus* L.	C_3	N;L
Millet, foxtail	*Setaria italica* (L.) Beauv.	C_4	S
Millet, pearl	*Pennisetum glaucum* L.	C_4	S
Mint (peppermint)	*Mentha piperita* L.	C_3	-
Mint (spearmint)	*Mentha spicata* L.	C_3	-
Oats, common	*Avena sativa* L.	C_3.	L; N
Oats, red	*Avena byzantina* C. Koch	C_3	L; N
Orchardgrass	*Dactylis glomerata* L.	C_3	N; L
Pea, field	*Pisum sativum* L.	C_3	L
Peanut	*Arachis hypogaea* L.	C_3	S
Potato	*Solanum tuberosum* L.	C_3	L; N
Ramie	*Boehmeria nivea* Gaud.	C_3	-
Rape, oilseed	*Brassica napus annua* Koch	C_3.	L
Rape, winter	*Brassica napus biennis* (Schubl.and Mart.)	C_3	-
Redtop	*Agrostis alba* L.	C_3	-
Reed canarygrass	*Phalaris arundinacea* L.	C_3	L
Rice	*Oryza sativa* L.	C_3	S
Rye	*Secale cereale* L.	C_3	L; N
Ryegrass, Italian	*Lolium multiflorum* Lam.	C_3	L
Ryegrass, perennial	*Lolium perenne* L.	C_3	L
Safflower	*Carthamus tinctorius* L.	C_3	L; N
Sesame	*Sesamum indicum* L.	C_3	S
Sorghum (feterita)	*Sorghum bicolor* (L.) Moench	C_4	S
Sorghum (hegari)	*Sorghum bicolor* (L.) Moench	C_4	S
Sorghum (kafir)	*Sorghum bicolor* (L.) Moench	C_4	S
Sorghum (milo)	*Sorghum bicolor* (L.) Moench	C_4	S

Common name	Scientific name	Photorespiration type*	Photoperiodic reaction*
Sorghum (sorgo)	*Sorghum bicolor* (L.) Moench	C_4	S
Sorghum × sudangrass		C_4	S
Soybean	*Glycine max* Merrill	C_3.	S; N
Sudangrass	*Sorghum bicolor drummondi*	C_4	S
Sugar beet			
Sugarcane	*Saccharum officinarum* L.	C_4	S
Sunflower	*Helianthus annuus* L.	C_3	N
Teosinte	*Euchlaena mexicana* Schrad.	C_4	-
Timothy	*Phelum pratense* L	C_3	L
Tobacco	*Nicotiana tabacum* L.	C_3	N
Vetch, common	*Vicia sativa* L.	C_3	L
Vetch, hairy	*Vicia villosa* Roth	C_3	L
Wheat, common	*Triticum aestivum* L.	C_3	L; N
Wheat, durum	*Triticum turgidum* L.	C_3	L; N
Wheatgrass, crested	*Agropyron cristatum* (L.) Gaertn.	C_3	-
Wheatgrass, slender	*Agropyron trachycaulum* (Link) Malte	C_3	-
Wheatgrass, western	*Agropyron smithii* Rybd.	C_3	L

*Photorespiration type: C_3 = Low net assimilation rate (high compensation point); C_4 = High net assimilation rate (low compensation point)

*Photoperiodic reaction: L = Long day; S = Short day; N = Day neutral or indeterminate.

It can be observed that crop plants represent both C_3 and C_4 plants. The entire group of poaceous grasses is exclusively C_4. Several dicot families have C4 representatives but in leguminosae and compositae, no C_4 plant has been identified till date.

The important aspects of C_4 cycle are:

1. Mesophyll cells of C_4 plants contain few chloroplast.
2. The vascular bundles are surrounded by concentric cylinders called vascular bundle sheath, whose cells are packed with chloroplast and starch and the sheath is surrounded by mesophyll cells which are rich in chloroplast as well.
3. Mesophyll and vascular bundle sheath (VBS) cells take part in photosynthesis.
4. In C_4 plants the C_4 acids and their precursors have been identified in the mesophyll cells chloroplast whereas bundle sheath cells abound in Calvin cycle compounds.
5. Chloroplasts of VBS cells contain ribulose-1,5-diphosphate carboxylase (causes addition of CO_2 to C_5 sugars) while mesophyll chloroplast has PEP-carboxylase (CO_2 + PEP).

The significance of two photosynthetic carbon cycles operating in the same individual appears to assist the mechanism by which CO_2 in the environment is utilized most efficiently. The two types of cells (mesophyll cells and VBS cells) are associated with each other by a complicated transport process. Consequently malate is exported from the mesophyll and pyruvate is imported in turn. At present not much is known about the process precisely.

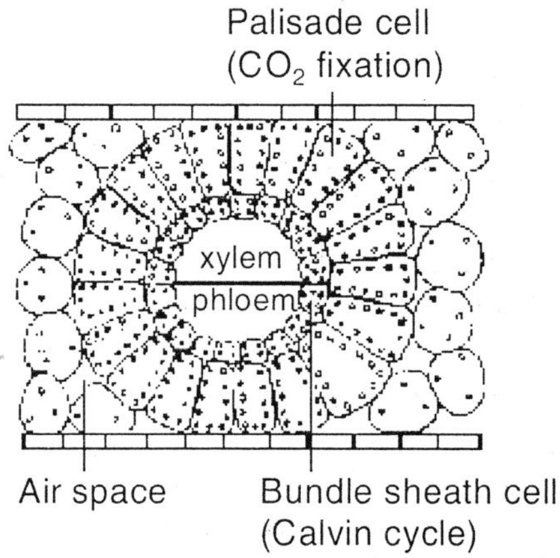

FIG. 7.8: *Diagrammatic transverse section of part of the leaf of a C_4 plant showing the close relationship between the palisade cells, which here form a circle round the vascular bundle, and the bundle sheath cells.*

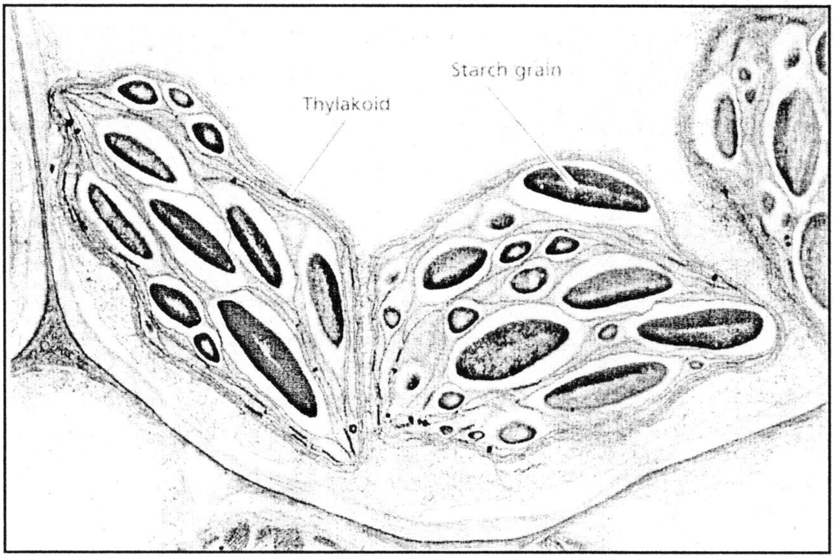

FIG. 7.9: *Electron micrograph of a bundle sheath cell from maize, showing the starch grains in the chloroplasts. (15,800x).*

PEP-carboxylase has strong affinity to combine with CO_2 than RuBP carboxylase. The CO_2 concentration for maximum photosynthesis in vivo is lower for PEP carboxylase. Therefore, the rate of photosynthesis is higher and CO_2 compensation point is lower in C_4 plants compared with C_3 plants. Thus C_4 plants can utilize CO_2 from weak concentration with the help of PEP- carboxylase. Obviously C_4 plants are more efficient for increased photosynthesis under low CO_2 concentration. The role of C_4 plants appears to concentrate CO_2 for RuBP- carboxylase of C_3.cycle.

Sucrose Synthesis in C_3 and C_4

Sucrose synthesis in cytosol is the main result of photosynthesis. The sucrose so produced has a number of possible fates. There are three major fates: translocation, storage in the vacuole or conversion into fructan. **Fructan** synthesis is particularly important in sucrose utilizing species especially at low temperatures. The rates of synthesis of sucrose, starch and fructan and the vacuolar storage and translocation of sucrose and their diurnal changes will thus determine the carbohydrate status of a leaf and its contribution to the plant.

Trioses phosphate exported from the chloroplast is converted to fructose-1, 6-bisphosphate (Fru 1,6-P_2) by aldolase and is subsequently hydrolyzed to fructose 6-phosphate (Fru-6P) by the cytosolic fructose –1,6- bisphosphatase (F-1, 6-Pase). The latter step is the first irreversible step in the committed pathway of sucrose synthesis and is believed to be an important control point in the pathway. Fru-6P is converted to glucose 6-phosphate (Glc6P), glucose 1-phosphate and finally to UDPG1c via UDPGlc pyrophosphorylase with the consumption of UTP and the release of the PPi. The proton-pumping pyrophosphatase in the tonoplast hydrolyzes most of the PPi produced during sucrose synthesis. Transport of PPi into chloroplasts, where it is hydrolyzed by the stromal inorganic pyrophosphatase, and a futile cycle of Fru 1,6 P_2 synthesis by pyrophosphate: fructose-6-phosphate 1-phospho-transferase and hydrolysis by fructose-1,6-bisphosphatsae (F1, 6-BPase might make a minor contribution to the turnover of PPi.

Sucrose is synthesized by the sequential action of sucrose-phosphate synthase (SPS) which catalyzes the synthesis of Suc6P from UDPGlc and Fru6P releasing UDP and sucrose phosphate phosphatase (SPP), which irreversibly hydrolyzes Suc6P to give sucrose and Pi. UDP is recycled to UTP by nucleoside diphosphate kinase with the consumption of ATP. There is some evidence for an association between SPS and SPP that might involve channeling of Suc 6P (Echeverria et al. 1997). An important aspect of the pathway is the release of Pi from the reactions catalyzed by the cytosolic F 1,6-BPase and SPP, and the turnover of PPi, which must be returned to the chloroplast via the phosphate translocator for the continued supply of triose phosphate to the cytosol.

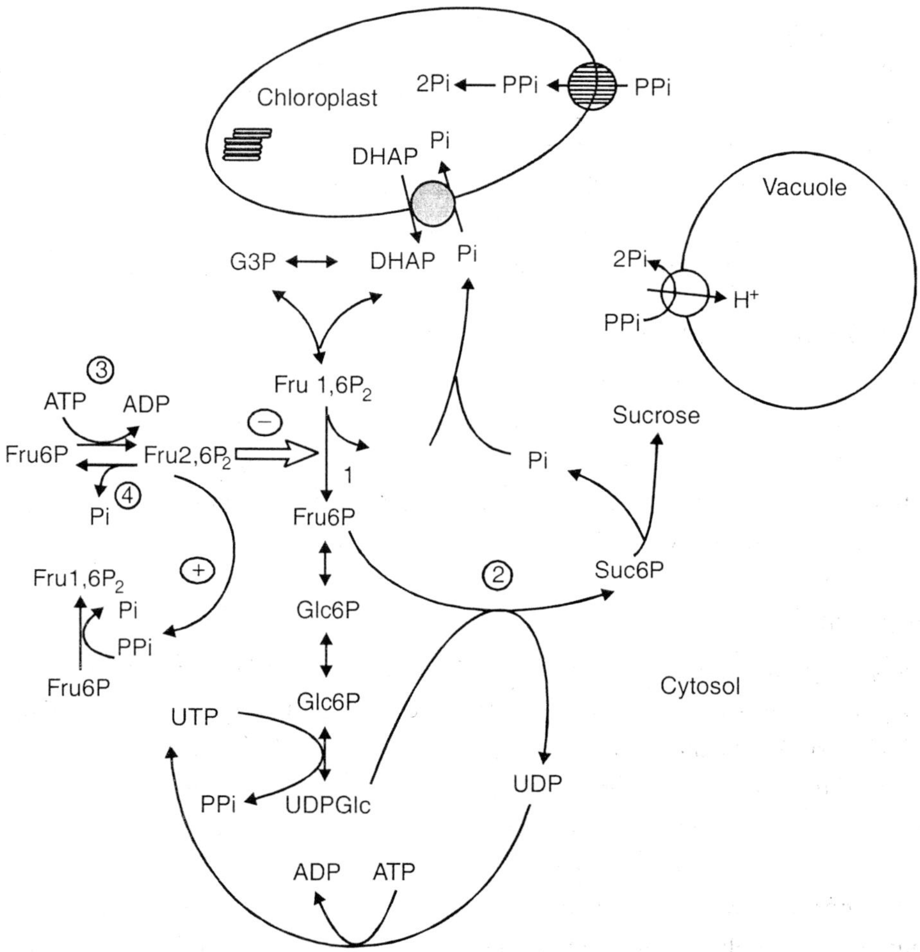

FIG. 7.10: *The pathway of sucrose synthesis in C_3 plants. 1. fructose-1-6-bisphosphate, 2. sucrose-phosphate synthase; 3. 6-phosphofructo-2-kinase; 4, fructose-2, 6-bisphosphatase, DHAP, dihydroxyacetone phosphate; G3P, glyceraldehydes 3-phosphate; GlclP, glucose 1-phosphate.*

In C_4 plants, there is the species–dependent localization of sucrose and starch biosynthesis in the mesophyll or bundle sheath cells (Lunn and Furbank, 1997). The key enzymes of starch biosynthesis are localized predominantly in the bundle sheath cell chloroplasts. It allows some recycling of Pi within the chloroplast and makes it easier to coordinate starch synthesis with the rate of CO_2 fixation by the PCR cycle. Compartmentation of the key enzymes of sucrose has been examined in a range of C_4 plants. Several studies have shown that SPS is present exclusively in the mesophyll cells and the incorporation of label from newly fixed CO_2 into sucrose occurs mostly in these cells. The result shows that sucrose is synthesized almost exclusively in the mesophyll cells. More recently, in a survey of ten C_4 species, the proportion of total leaf SPS activity in the mesophyll cells was found to range from 65% in *Sorghum bicolor* to 99% in *Atriplex spangiosa* (Lunn and Furbank, 1997).

Intracellular Transport in C_4 Plants During Photosynthesis

In leaves, the rate of photosynthesis is about 10-fold higher than the rate of dark respiration, which illustrates the potential for high rates of metabolite flux between cellular compartments during photosynthesis. For example, in C_3 plants there is a high rate of movement of CO_2 into the chloroplast, and of O_2 out of the chloroplast by free diffusion, equivalent to the rate of photosynthesis. The next most rapid exchange across the C_3 chloroplast envelope is P_i/triose-P on the phosphate translocator, which needs to function at rates near one-third of the rate of photosynthesis when sucrose is the major photosynthetic product. In C_4 photosynthesis, movement of metabolites across organelles occurs in association with the C4 cycle, the reductive phase of the C_3 pathway and photorespiration. As much as half of the PGA, the product of RuBP carboxylase, may be shuttled from bundle sheath to mesophyll chloroplasts for reduction, in which case this shuttle would operating at the same rate as CO_2 fixation.

Evidence that this transport must be occurring is based on studies of enzyme compartmentation and metabolism by isolated organelles. However, direct studies on metabolite transport, in order to identify carrier-mediated mechanisms and their kinetic properties, has been limited mainly to the mesophyll chloroplasts. The phosphate translocator is the only carrier protein identified in C4 photosynthesis.

Phosphate Translocator

A single phosphate translocator in the mesophyll chloroplast is considered to mediate PEP/P_i exchange associated with the C4 cycle and PGA/triose-P exchange associated with the C3 pathway. Thus the specificity of this translocator is different from that in C3 chloroplasts such as those of spinach, which can effectively exchange P_i/triose-P but not PEP/P_i. DIDS (4,4-diisothiocyanatostilbene- 2,2′-disulphonic acid), a disulphonic acid stilbene derivative, is a potent, irreversible inhibitor of the phosphate translocator. Incubation of chloroplasts with tritiated DIDS results in specific labelling of a polypeptide of 29 kDa in the envelopes of C3 and C4 mesophyll chloroplasts, which is considered to be the phosphate translocator protein.

Pyruvate Translocator

This translocator facilitates the uptake of pyruvate into the mesophyll chloroplast. It shows Michaelis-Menten type kinetics and is inhibited by analogs of pyruvate and by low temperature. There is light-induced uptake and accumulation of pyruvate by C4 mesophyll chloroplasts, which is driven by a proton or sodium gradient, depending on the species. In mesophyll chloroplats from monocots of NAD-ME and PEP-CK species there is pyruvate uptake in the light, addition of Na^+ substitutes for the light requirement. This light-dependent accumulation of pyruvate may be associated with an active Na^+/H^+ exchange, or active Na^+ efflux from the chloroplast which drives co-transport of Na^+ and pyruvate into the chloroplast. In NADP-ME type monocots there is enhanced uptake of pyruvate by mesophyll chloroplasts following a pH drop, suggesting a H^+/pyruvate import. Sodium does not stimulate uptake of pyruvate by chloroplasts of these species. In this case, the light-enhanced pyruvate uptake may be linked to light-dependent alkalization of the stromal compartment.

Oxaloacetate Translocator

There is a specific carrier in C4 mesophyll chloroplasts for uptake of oxaloacetate. It has a high affinity for oxaloacetate (K_m = 45 mm for maize chloroplasts), which is considered important since oxaloacetate is present at low levels in the cell.

Dicarboxylate Translocator

This translocator catalyses the exchange of dicarboxylic acids by the C4 mesophyll chloroplasts, as seen with *Digitaria sanguinalis* and maize. In vivo, its function may be linked in some way to the oxaloacetate translocator and involves the uptake of oxaloacetate and export of malate by the C4 mesophyll chloroplast. The dicarboxylate translocator(s) may also have other functions in association with movement of other dicarboxylic acids across the chloroplast envelope (for example, aspartate, 2-oxoglutarate and glutamate).

Glycerate Translocator

There are evidences for a glycerate/glycolate translocator in C4 mesophyll and bundle sheath chloroplasts which may function in the export of glycolate from the bundle sheath chloroplasts and import of glycerate into the mesophyll chloroplasts. Some stimulation of uptake of glycerate by light may be attributed to an alkalization of the stroma and a H^+/glycerate$^-$ import.

Intermediate Between C_3 and C_4 plants

More than 20 plant species exhibit photosynthetic traits that are intermediate between C_3 and C_4 plants (e.g., species in the genera *Alternanthera, Flaveria, Neurachne, Moricandia, Panicum* and *Parthenium*). These show reduced rates of photorespiration and CO_2-compensation points in the range of 0.8 to 3.5Pa, compared with to 4 to 5 in C_3 and 0 to

0.5 in C_4 plants. They have a weakly developed Kranz anatomy, compared with the true C_4 species, but Rubisco is located both in the mesophyll and the bundle sheath cells.

Two types of intermediates have been distinguished. In the first type (e.g., *Alternanthera ficoides, A. enella, Moricandia arvensis, Panicum milioides*) the activity of key enzymes of the C_4 pathway (i.e., pyruvate-Pi-dikinase, PEP carboxylase, NAD-malic enzyme, and PEP carboxykinase) is very low, and they do not have a functional C_4-acid cycle. Their low CO_2-compensation point is due to the light-dependent recapture by mesophyll cell released in photorespiration in the bundle sheath cells. The bundle sheath cells contain a large fraction of the organelles involved in photorespiration, compared with that in C_3 species. The CO_2-compensation point is negatively correlated with the percentage of photorespiratory organelles that occur in bundle sheath cells. These C_3-C_4 intermediates have evolved a system to utilize CO_2 that escapes from the bundle sheath cells but do not have the CO_2-concentrating mechanism of true C_4 species. In the leaves of these intermediate species, glycine decarboxylase, which is a key enzyme in photorespiration that releases the photorespiratory CO_2, occurs exclusively in the cells that surround the vascular bundle sheath. The mitochondria of the mesophyll cells, like those of true C_4 plants, have a low activity of glycine decarboxylase. Immunogold labeling shows a four times greater amount of one of the subunits of glycine decarboxylase cell. It is likely that products of the oxygenation reaction, including glycine, move to the bundle sheath cells. The products are presumably metabolized in the bundle sheath; therefore, serine can move back to the mesophyll. Due to the exclusive location of glycine decarboxylase in the bundle sheath cells, the release of CO_2 in photorespiration occurs close to the vascular tissue.

Evolution and Distribution of C_4 Species

C_4 species represent approximately 5 per cent of all plant species, C_3 species accounting for about 85 per cent and CAM species for 10 per cent. The C_4 pathway evolved nearly synchronously on several continents in the late Miocene 5 to 7 million years ago, probably in response to a reduction in atmosheric CO_2 concentration. This change decrease in atmospheric CO_2 concentration resulted from both the photosynthetic activity of plants and tectonic and subsequent geochemical events. To put it briefly, the collision of the Indian subcontinent caused the uplift of the Tibetan Plateau, causing new earthy crust that consumed CO_2 to become exposed over a vast area.

Low atmospheric CO_2 concentration would increase photorespiration and favour the CO_2-concentrating mechanisms and lack of photorespiration that characterize C_4 species. Considering the three subtypes of C_4 species and their occurrence in plants of at least 18 different families of diverse taxonomic groups, C_4 plants must have evolved from C_3 ancestors independently a number of times. The physiology of C_3-C_4

from C_3 ancestors independently a number of times. The physiology of C_3-C_4 intermediates suggests the mechanism to remechanism. Morphological and ecogeographical information combined with molecular synthesis has evolved in different lineages within the genus *Flaveria*. The phylogeny of *Flaveria* species deduced from an analysis of the nucleotide sequences that encode a subunit of glycine decarboxylase, suggests that C_4 species are derived from C_3-C_4 intermediates.

Apart from the decreasing atmospheric CO_2 concentration, high temperatures and water shortage may also have been the driving forces for the evolution of C_4 species. This makes it likely that C_4 species evolved primarily in tropical areas, which continue to be their centers of distribution e.g., tropical and temperate grasslands, with abundant warm-season precipitation, are dominated by C_4 species.

The high concentration of CO_2 at the site of Rubisco allows net CO_2 assimilation at relatively high temperatures, when photorespiration results in low net photosynthesis of C_3 species due to the increased oxygenating activity of Rubisco. This explains why C_4 species naturally occur in warm, open ecosystems, where C_3 species are less successful. At high temperature, C_4 plants have higher quantum yields than C_3 plants, despite the extra energetic cost of the C_4 pathway, mostly owing to the inefficiency imposed by photorespiration in C_3 plants. However, the carboxylase and oxygenase reactions of Rubisco are disproportionately affected by temperature, such that at low temperatures photorespiration is much decreased. This result in an increase in quantum yield in C_3 plants at lower temperature whereas the quantum yield of C_4 plants is relatively insensitive to temperature (Hatch, 1992). Thus, the selective advantage of C_4 plants at higher temperatures in terms of higher potential productively disappear as temperature decreases and the extra energy cost of C_4 pathway puts C_4 plants at a competitive disadvantage with C_3 plants at low temperature. This is probably a major factor in determining the geographical distribution of C_4 plants rather than any inherent cold-sensitivity of these plants. (Hatch, 1992). The high sensitivity to low temperature of pyruvate-Pi-dikinase, a key enzyme in the C_4 pathway, may be the main reason why C_4 species have rarely expanded to cooler places. Compatible solutes can decrease the low-temperature sensitivity of this enzyme; this could allow the expansion of C_4 species into more temperate regions in the future. On the other hand, rising atmospheric CO_2 concentration may offset the advantages of the CO_2-concentration mechanism of C_4 photosynthesis.

Breeding of C_3 and C_4 Plants

Because of the fundamental relationship between photosynthesis and yield, there has been considerable interest in enhancing photosynthetic capacity through breeding. Active research is presently going on in Plant breeding laboratories where incorporation of characters of C_4 plants to the C_3 plants is done and thereby augmenting the amount of metabolites they form. Probably limited number of genes are involved which need to be brought together. Incidentally, crosses between C_3 and C_4- *Atriplex* species have failed to yield progeny with integrated C_4 characters. In view of the non-availability of plants with C_4 characteristics closely related to cereals, attempt is being made to screen different seed plant for C_4 types. The possibility of breeding entirely new plants from the genetic pool of C_4 types can also be not ruled out.

C_2 PLANTS?

RuBP splits under O_2 consumption into the C_2 compound phosphoglycolate and a C_3 compound phosphoglycerate by the activity of an oxygenase. After dephosphorylation glycolate can be oxidized to glyoxylate either directly in the chloroplast or in the peroxisomes. This reaction also produces the aggressive H_2O_2 which is converted into water and O_2 by catalase, the key enzyme of the peroxisomes. The glyoxylate glycine transaminase produces the amino acid glycine. Two molecules of glycine can be condensed to generate CO_2 to form serine, either in the mitochondria or in the chloroplasts. Since these reactions are dominated by C_2 compounds, photorespiration is also called the C_2 cycle.

The significance of the C_2 cycle is not yet completely clear, since some plants have a photorespiration cycle while others do not. In addition to the production of amino acids it may have a role in photoinhibition. According to this concept the excessive oxygen produced at extremely high fluence rates would be deactivated by the C_2 cycle and the pigments and protein, especially in PSII, would be protected from oxidation. The disadvantage of photorespiration is the loss of fixed carbon which can amount to 25 per cent of the primarily fixed CO_2. Experiments to genetically eliminate or reduce photorespiration have not yet been successful.

REGULATION OF C_4 PEP CARBOXYLASE

The enzyme PEP (phosphoenol pyruvate carboxylase) is widely distributed in plant tissue, green algae and microorganisms but is absent is animals. It was first characterized in spinach leaves in 1953 and for a long time was considered a subsidiary plant carboxylase alongside Rubisco. However, the discovery of C_4 photosynthesis in the mid-1960s and the involvement of a specific isoform of PEP carboxylase in this pathway considerably boosted interest in this cytosolic enzyme. More recently, the extensive development and use of molecular and biochemical techniques has generated an impressive wealth of data and significantly advanced our understanding of the nature of the regulatory processes for this enzyme.

Experiments with recombinant C_4 PEP carboxylase from sorghum have confirmed that it is subject to opposing feedback inhibition by the end-product L-malate and allosteric activation by glucose-6-phosphate. Other sugar phosphates such as dihydroxyacetone-phosphate or amino acids are also able to alter enzyme activity. The L-malate appears to act as a competitive inhibitor. In contrast, glucose-6-phosphate increases the apparent affinity of PEP carboxylase for PEP such that, by lowering the K_m value for this substrate, this positive effector enhances the enzyme's ability to compete with L-malate. This is particularly striking at pH7. Thus, it appears that the activity of PEP carboxylase *in vivo* is fine tuned by the balance of negative/positive effector metabolites and this is further modulated by cytosolic pH.

The observation that plant PEP carboxylase was subject to phosphorylation/dephosphorylation-dependent regulation was made with the photosynthesis related isoform from maize. These important findings radically changed the existing dogma for PEP carboxylase regulation. It became evident that the phosphorylation of specific serine residue modulated the opposing effects of the metabolites L-malate and glucose-6-phosphate. These findings also implicated the involvement of a signal transduction controlling the phosphorylation state of PEP carboxylase in the cytosol of photosynthetic C_4 mesophyll cells.

The phosphorylation of PEP carboxylase is induced by light and final ratio of phosphorylated to dephosphorylated enzyme depends on the light fluence rate. Upon return to darkness, dephosphorylation occurs presumably by a type 2A protein phosphatase and PEP carboxylase-kinase catalyzes the phosphorylation of PEP carboxylase in light. Inhibition of PEP carboxylase-kinase activity correlates with the phosphorylation state of PEP carboxylase and ultimately CO_2 assimilation. These results suggest that PEP carboxylase phosphorylation has a critical regulatory impact on the overall functioning of C_4 photosynthesis.

In C_4 leaves, enzymes for carbon assimilation in photosynthesis are distributed in two concentric cell layers, the outer mesophyll, containing PEP carboxylse and several other C_4 enzymes and the bundle sheath in which Rubisco and the Calvin cycle are localized. The inhibitor-based experiments performed in vitro revealed that a functional Calvin Cycle in bundle sheath cells was required for the upregulation of C_4 PEP carboxylase in the mesophyll layer. This suggested that the signal transduction involves intercellular cross-talk between these neighbouring photosynthetic tissues. A working hypothesis was that light transduction was mediated through changes in the levels of a metabolites of photosynthesis and/or energy charge, as a consequence a transcellular message formed in the bundle sheath during C_4 photosynthesis would be delivered to the adjacent mesophyll cells where the physiological responses with respect to PEP carboxylase-kinase and its target protein were observed (Chollet et al., 1996). A cellular approach based on the isolation of mesophyll cell protoplasts from the C_4 grasses showed that in illuminated protoplasts, there is a marked decrease in L-malate sensitivity and an increase in catalytic activity of the endogenous C_4 PEP carboxylase. These changes were caused by light-induced stimulation of a Ca^{2+}-independent PEP carboxylase-kinase which as before, was found to be sensitive to cycloheximide in situ (Li and Chollet, 1993; Giglioli-Guivarch et al., 1996). Similar results were observed with intact mesophyll cells. Both mesophyll protoplasts and mesophyll cells isolated from a C_4 leaf retains the intrinsic ability to perform PEP carboxylase phosphorylation in a light-dependent manner.

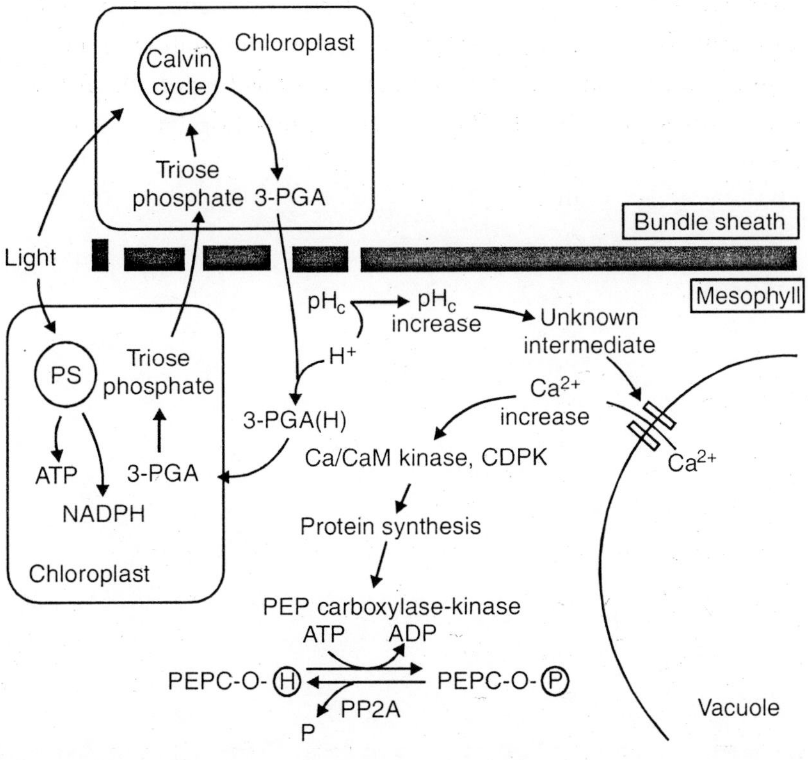

FIG. 7.11: *Schematic working model of the sequential organization of the various signaling elements of the C_4 PEP (phosphoenol pyruvate) carboxylase phosphorylation cascade in an illuminated leaf. Abbreviations: 3-PGA. 3-phosphoglyceric acid: PS. Photosystem: pH_c. cytosolic. PH: 3-PGA(H), partially protonated (2) form of 3-PGA: CaM. Calmodulin: CDPK. Calmodulin-like domain protein kinase: PP2A, type 2A protein phosphatase: P_1. inorganic phosphate.*

Intracellular alkalization in mesophyll cell protoplasts is implicated as an early signaling element in the C_4 – PEP carboxylase phosphorylation circuitry. A prime suspect is 3-PGA, a Calvin cycle intermediate, that diffuses into C_4 mesophyll cells for subsequent phosphorylation and reduction in the mesophyll chloroplast. When 3-PGA was substituted for the weak base in the mesophyll protoplast suspension medium, this metabolite elicited in situ changes in cytosolic pH and resulted in upregulation of PEP carboxylase kinase and the phosphorylation state of PEP carboxylase (Giglioli-Guivarch et al., 1996). Both light and alkalization of cytosolic pH are needed for the induction of C_4 PEP-carboxylase-kinase activity.

The working hypothesis is that the light induction pathway in an illuminated C_4 leaf involves sequentially: 3-PGA as an intercellular message indicating photosynthesis; increase in cytosolic pH in mesophyll cells, inositol triphosphate and corresponding tonoplast Ca^{2+} channels, cytosolic Ca^{2+} or a Ca^{2+} - dependent protein kinases and finally the upregulation of PEP carboxylase-kinase via stimulation of a cytosolic protein synthesis event. A model showing the light-signal transduction chain that controls C_4 PEP carboxylase-kinase and thus PEP carboxylase phosphorylation is illustrated in Fig. 7.11.

CHAPTER 8

CAM Plants

In addition to C_3 and C_4 species, there are many succulent plants with another photosynthetic pathway: Crassulacean Acid Metaboism (CAM). Primary CO_2 fixation is followed by a diurnal acid metabolism. This pathway is named after the Crassulaceae, which is a family in which any species show this type of metabolism. CAM, however also commonly occurs in other families, such as the Cactaceae, Euphorbiaceae, Orchidaceae, and Bromeliaceae (e.g., *Ananas comosus*, pineapple). There are 25 to 35 flowering plant families that have species showing CAM. Thus, CAM metabolism overall is an important feature of photosynthesis in flowering plants.

The unusual capacity of CAM plants to fix CO_2 into organic acids in the dark, which causes nocturnal acidification, with deacidification during the day, has been known for almost two centuries. A full appreciation of CAM as a photosynthetic process was greatly stimulated by analogies with C_4 species.

The productivity of most CAM plants is fairly low; however, this is not an inherent trait of CAM species because *Agave mapisaga* and *A. salmiana*, for example, may achieve an average productivity of 40 Mg ha^{-1} $year^{-1}$. Even higher productivity has been observed for both irrigated and fertilized and carefully pruned *Opuntia amyclea* and *O. ficus-indica* (46 Mg ha^{-1} $year^{-1}$). These are among the highest productivities reported for any species. In a comparison of two succulent species with similar growth forms, *Cotyledon orbiculata* (CAM) and *Othonna optima* (C_3), during the transition from the rainly season to subsequent drought, the daily net rate of CO_2 assimilation is similar for the two species. This shows that rates of photosynthesis of CAM plants may be as high as those of C_3, plants, if morphologically similar plants adapted to the same habitat are compared.

TABLE 8.1: Taxonomic survey of following plant families known to have species showing crassulacean acid metabolism

Agavaceae	Geraniaceae
Aizoaceae	Gesneriaceae
Asclepidiaceae	Labiatae
Asteranceae	Liliaceae
Bromeliaceae	Oxalidaceae
Cactaceae	Orchidaceae
Clusiaceae	Piperaceae
Crassulaceae	Polypodiaceae
Cucurbitaceae	Portulacaceae
Didieraceae	Rubiaceae
Euphorbiaceae	Vitacege

CAM plants, similarly to C4 plants, are divided into groups on the basis of decarboxylating enzymes, although sampling has not been broad. In malic enzyme (ME) – CAM plants and PEP carboxykinase (PEPCK) – CAM plants, there is sufficient RuBP carboxylase activity to account for active photosynthesis and chloroplasts actively show C3 – photosynthesis. Both types of plants have sufficient NADP and NAD malic enzyme activities. However, the ME – CAM plants are deficient in PEP carboxykinase leading to the conclusion that in PEPCK – CAM plants, the decarboxylating enzyme localized in cytosol or mitrchondria is PEPCK. In ME-CAM plants, either NADP malic enzyme localized in the cytosol or NAD malic enzyme localized in the mitochondria could be decarboxylating enzymes. The actual intracellular localization of the latter enzyme is difficult to state with certainty since isoenzymes of NADP malic enzyme are known to occur in several intracellular compartments.

DISTRIBUTION AND EVOLUTION OF CAM SPECIES

CAM is undoubtedly an adaptation to drought. They close their stomates during most of the day. There are two major ecological groupings of CAM plants:

- Succulents from arid and semiarid regions
- Epiphytes from tropical and subtropical regions

Although CAM plants are uncommon in cold environments, this may reflect their evolutionary origin in warm climates rather than a temperature sensitivity of the CAM pathway. Roots of some Orchidaceae, which lack stomata, also show CAM. CAM species are well adapted to dry environments and naturally occur in semiarid and tropical environments. Some submerged aquatics also show CAM photosynthesis.

Carbon Isotope Composition of CAM Species

The enzyme from CAM plants discriminates against $^{13}CO_2$ (similar to Rubisco from C_3 and C_4 plants). The fractionation is considerably less than that of C_3 plants and similar to that of C_4 species. This is expected, as the stomata are closed during malate decarboxylation and fixation of CO_2 by Rubisco. Hence, only a small amount of inorganic carbon diffuses back from the leaves to the atmosphere.

The carbon isotope discrimination consequently decreases, in accordance with the smaller extent of discrimination in CAM species.

ANATOMICAL, PHYSIOLOGICAL AND BIOCHEMICAL ADAPTATIONS

CAM plants are characterized by their succulence (*succulence is defined as the volume of water in the leaf at a relative water content of 100 per cent per unit leaf area*), but this is not pronounced in epiphytic CAM plants. The capacity to fix CO_2 at night via PEP-carboxylase, the accumulation of malic acid in the vacuole, and the subsequent deacidification during the day, when CO_2 is released from malic acid and fixed in the Calvin cycle using Rubisco.

CAM plants show a strong fluctuation in pH of the cell sap because of the synthesis and breakdown of malic acid. The concentration of this acid may increase to 100mM. By isolating vacuoles of the CAM plant *Kalanchoe daigremontiana*, it has been demonstrated that at least 90 per cent of all the acid in the cells is in the vacuole. In addition, the kinetics of malic acid efflux from the leaves of *K. daigremontiana* provide evidence for the predominant location of malic acid in the vacuole.

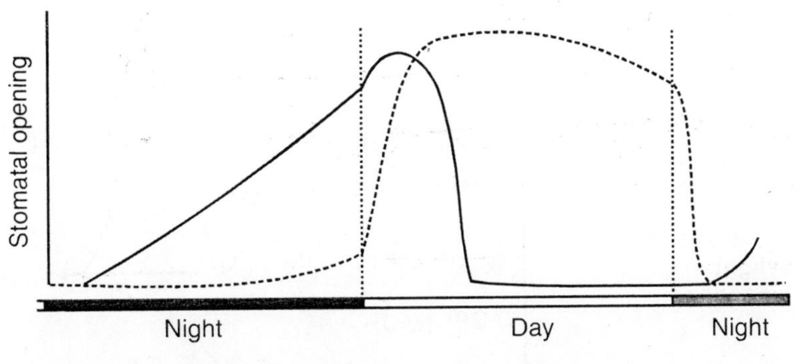

FIG. 8.1: *A diagram representing the daily changes in stomatal aperture in a succulent (continuous line), and mesophyte (broken line).*

At night, CO_2 is fixed in the cytosol which is catalyzed by PEP-carboxylase and produces oxaloacetate. PEP originates from the breakdown of glucose in glycolysis is formed from starch. Oxaloacetate is immediately reduced to malate which is catalyzed by malate dehydrogenase. Malate is transported to the large vacuoles in an energy-dependent manner, which involves a proton pump. An H^+-ATPase and a pyrophosphatase pump protons into the vacuole so that malate can follow passively.

The release of malic acid from the vacuole during the day is supposedly passive. Malate is then decarboxylated by malic enzyme (NAD- or NADP-dependent) or by PEP-carboxykinase. CAM species are subdivided like C_4 species, depending on the decarboxylating enzyme. The malic enzyme subtypes (ME-CAM) have a cytosolic NADP-malic enzyme, as well as a mitochondrial NAD-malic enzyme; they use a chloroplastic pyruvate-Pi-dikinase to convert the C_3 fragment that originates from the decarboxylation reaction into carbohydrate. Unlike C_4 species, ME-CAM species are not further subdivided into NADP- and NAD-malic enzyme subtypes. PEPCK-type CAM plants have very low malic enzyme activities (as opposed to PEPCK-C_4 plants) and no pyruvate-Pi-dikinase activity, but high activities of PEP-carboxykinase.

The C_3 fragment (pyruvate or PEP) that is formed during the decarboxylation is converted into starch, and the CO_2 that is released is fixed by Rubisco, much the same as it is in C_3 plants. The stomata are closed during the decarboxylation of malic acid and the fixation of CO_2 by Rubisco in the Calvin cycle. They are open during the nocturnal fixation of CO_2.

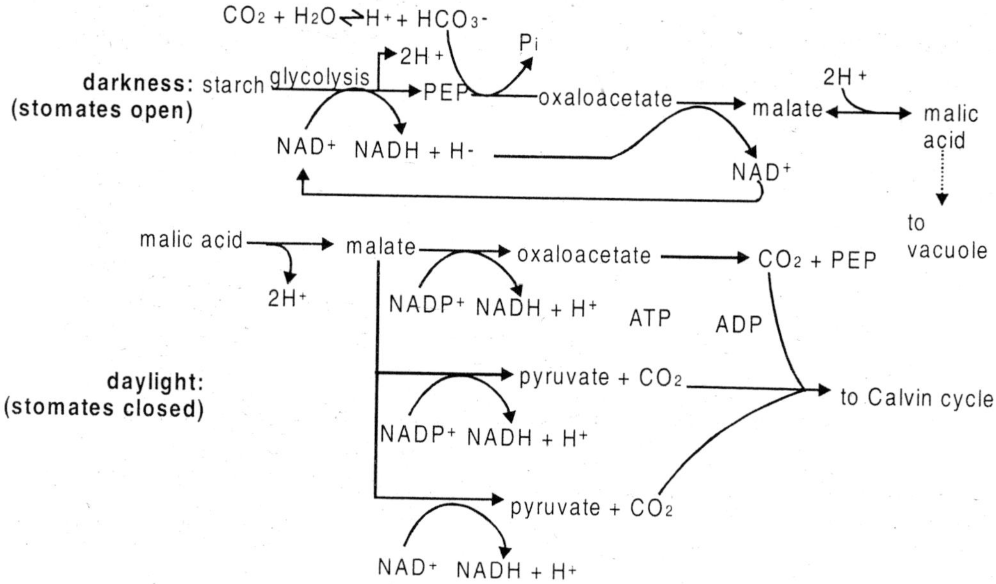

FIG. 8.2: *Crassulacean acid metabolism in light and dark.*

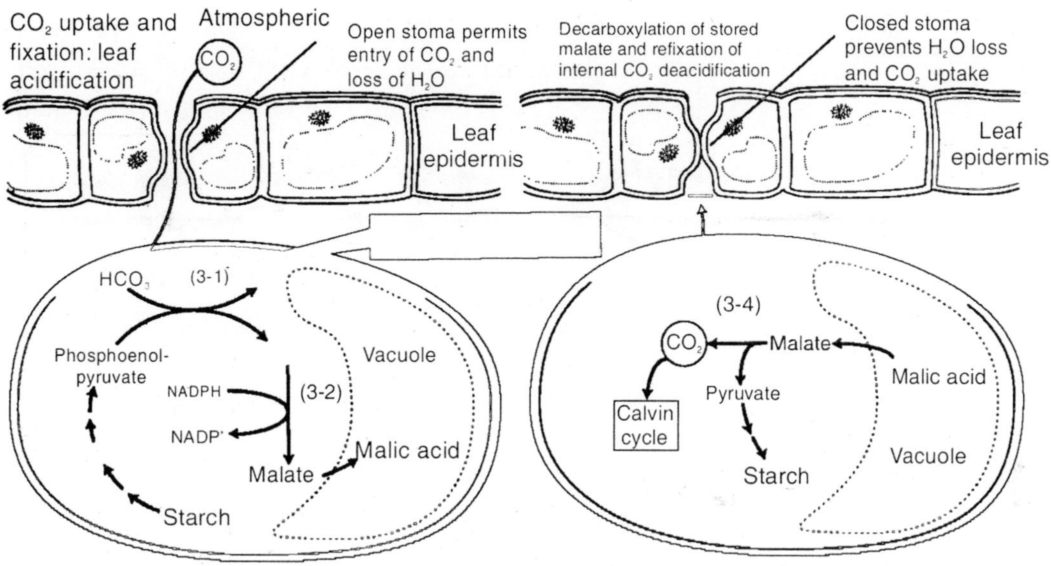

Fig 8.3: In crassulacean acid metabolism (CAM), the reactions of photosynthesis and CO_2 uptake are temporally separated : CO_2 uptake and fixation take place at night, and decarboxylation and refixation of the internally released CO_2 occur during the day. CAM is an adaptation primarily to minimize the quantity of water that is lost when stomata are opened to permit the entry of CO_2. In CAM plants, the stomata are opened in the cool of the night. CO_2 is fixed as malic acid, which is stored in the vacuole. As malic acid accumulates, the leaf vacuoles acidify in the dark. Upon illumination, the stomata close, and the leaf deacidifies. The malic acid is recovered from the vacuole and undergoes decarboxylation. The CO_2 that is released is prevented from escaping by stomatal closure and is assimilated via the Calvin cycle using photochemically generated ATP and NADPH.

The CAM traits can be summarized as follows:

- Fluctuation of organic acids, mainly of malic acid, during a diurnal cycle;
- Fluctuation of the concentration of sugars and starch, opposite to the fluctuation of malic acid;
- A high activity of PEP-carboxylase (at night) and of a decarboxylase (during the day);
- Large vacuoles in cells that contain chloroplasts;
- Some degree of succulence;
- The gas exchange of the leaves occurs predominantly at night.

Four phases in the diurnal patter of CAM are discerned (Fig. 8.5).

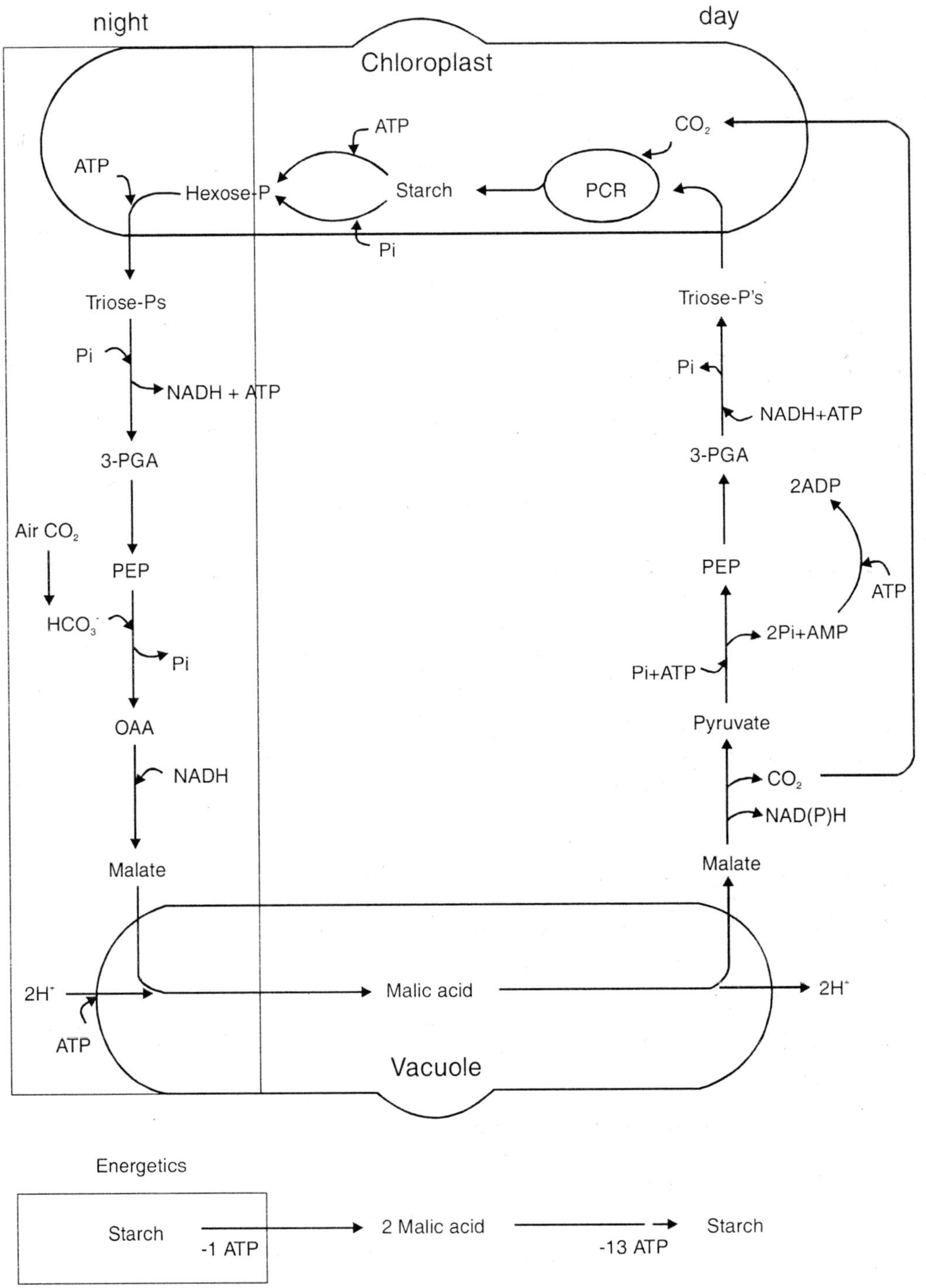

Fig 8.4: Metabolic pathway and cellular compartmentation of Crassulacean Acid Metabolism, showing the separation in night and day of carboxylation and decarboxylation.

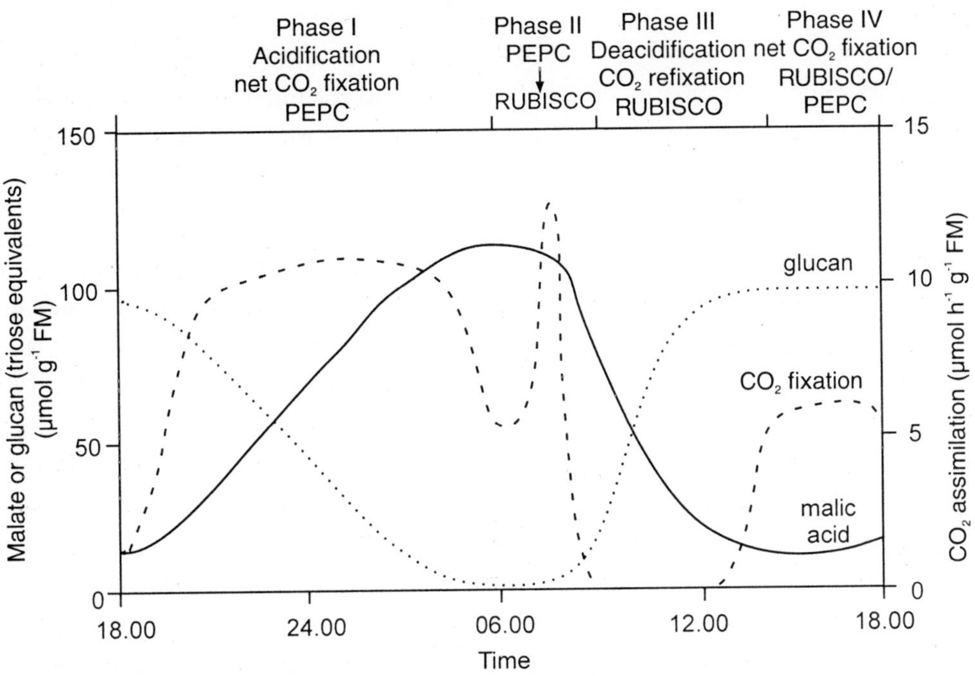

Fig 8.5: CO_2 fixation in CAM plants, showing diurnal patterns for stomatal conductance, net CO_2 assimilation, malic acid concentration, and carbohydrate concentrations; PEPC is PEP carboxylase.

Phase I

During the nocturnal phase I, phosphoenolpyruvate carboxylase (PEP carboxylase) assimilates atmospheric and respiratory CO_2 into oxaloacetate (OAA) by carboxylating phosphoenolpyruvate (PEP) derived from glycolysis. OAA is then reduced to malate by cytosolic and/or mitochondrial NAD (P) – malate dehydrogenase and stored in the vacuole.

Phase – II

In phase II, at dawn, both PEP carboxylase and Rubisco are active and malate is released from the vacuole.

Phase III

During most of the day, phase III, when stomata are closed, malate decarboxylation commences as the released CO_2 is refixed by Rubisco. The photosynthetic carbon reduction cycle incorporates the liberated 3-carbon compounds into storage carbohydrates via gluconeogenesis.

Phase IV

In phase IV, malate reserves are eventually depleted causing a decline in intracellular CO_2 availability. This results in stomatal opening and assimilation of atmospheric CO_2 until the onset of darkness. This phase resembles C_3 photosynthesis with Rubisco serving the primary carboxylase, although PEP carboxylase become active before the end of the light period.

CAM plants separate into two distinct metabolic groups based on the type of C4 – acid decarboxylases present. In malic enzyme (ME) – type CAM plants, malate is decarboxylated in either the cytosol or mitochondria utilizing NADP – ME or NAD – ME respectively; the reaction generates pyruvate which is subsequently converted to PEP in the chloroplast by pyruvate orthophosphate dikinase (PPDK) and to CO_2. In PEP carboxykinase (PEPCK) – type CAM plants, which lack PPDK activity, the OAA formed by NAD (P) – MDH is converted to PEP and CO_2 by PEPCK in the cytosol. In general, ME – type plants have higher ME activities than the PEPCK – type and completely lack PEPCK activity. These two decarboxylating systems are grouped within plant families but not at higher taxonomic levels. CAM plants also display a great variety of storage carbohydrate reserves which must be used to balance massive nocturnal acid accumulation. Some CAM plants accumulate predominantly starch or soluble sugars (e.g. sucrose, glucose, fructose) and polysaccharides (e.g. fructan, galactomannan) in extrachloroplastic compartments while other species store both starch and glucose. Diversity in carbohydrate accumulation patterns reflects evolutionary history rather than being governed by carbon flow constraints of the pathway (Christopher and Holtum, 1998).

How do CAM plants regulate the activity of the two caroboxylating enzymes and decarboxylating enzymes in coordinated way to avoid futile cycles? Rubisco is inactive at night for the same reason as in C_3 plant: This enzyme is light activated and therefore cannot function in the dark. In addition, the kinetic properties of PEP-carboxylase are modulated. In *Mesembryanthemum crystallinum* and in *Crassula argentea*, PEP-carboxylase occurs in two configurations: a *day-configuration* and a *night-configuration*. The night-configuration is relatively insensitive to malate (the Ki for malate is 0.06 to 0.9 mM, depending on pH) and has a high affinity for PEP (the Km for PEP is 0.1 to 0.3 mM). The day-configuration is strongly inhibited by malate (the Ki for malate is 0.004 to 0.07 mM, again depending on the pH) and has a low affinity for PEP (the Km for PEP has increased to 0.7 to 1.25 mM). When malate is rapidly exported to the vacuole at night in phase I, therefore, the carboxylation of PEP readily takes place, whereas it is suppressed during the day in phase III.

TABLE 8.2: Cumulative CO_2 and O_2 exchange and photosynthetic quotient (PQ) in the dark and in light [in mmol plant^{-1} (12 h)$^{-1}$] of one shoot of Ananas comosus (pineapple), measured over two consecutive days*

	Dark		Light		
	CO_2 assimilation	O_2 consumption	CO_2 assimilation	O_2 release	PQ
Day 1	10.6	6.4	10.4	27.1	0.99
Day 2	11.1	6.3	10.7	27.5	0.98

*PQ is the ratio of the total net amount of O_2 evolved in the light to total amount of CO_2 fixed in the light plus dark period.

The fluctuations in the kinetic properties of PEP-caroxylase do not require protein synthesis. In *Crassula argentea* the change in configuration involves a change from a dimer, with subunits of 100 kDa, during the day, to a tetramer, with subunits of the same molecular mass, at night.

TABLE 8.3: Effects of malate and glucose-6-phosphte (G-6-P) on the kinetic parameters of PEP-carboxylase

	Vmax mmolmg^{-1} (Chl) min^{-1}	Ratio	Km (mM)	Rate
Control	0.42	1.0	0.13	1.0
+ 1 mM G-6-P	0.45	1.07	0.08	0.61
+ 2 mM G-6-P	0.47	0.12	0.05	0.39
+ 5 mM malate	0.31	0.74	0.21	1.60
+ 5 mM malate and 2 mM G-6-P	0.34	0.81	0.05	0.39

Source : Kluge & Ting 1978. *The ratio gives the ratio of the values in the presence and absence of effectors.

The sensitivity for malate inhibition tends to be lost during storage PEP-carboxylase from the CAM plant *Kalanchoe daigremontiana*. This effect is stronger in the presence of inorganic phosphate, PEP, 3-phosphoglycerate (PGA), and glucose-6-phosphate. The presence of malate slows down the rate of loss of sensitivity to malate inhibition. In this case, there is no change from dimer to tetramer. There is some evidence that the modification of the kinetic properties involves the phosphorylation and dephosphorylation of PEP-carboxylase.

Though the modification of its kinetic properties, the inhibition of PEP-carboxylase is likely to prevent a futile cycle of carboxylation and concomitant decarboxylation reactions. Further evidence that such a futile cycle does not occur comes from studies on a labeling with ^{13}C of the first or fourth carbon atom in malate. This pattern of labeling is constant during de-acidification. If a futile cycle were to occur, then doubly labeled malate should appear because fumarase in the mitochondria would randomize the label in the malate molecule. Such randomization does occur during the acidification phase, which indicates rapid exchange of the malate pools of the cytosol and the mitochondria, before malate enters the vacuole.

Next to malate, glucose 6-phosphate is also an effector of PEP-carboxylase: It enhances the Vmax and decreases the Km. The physiological significance of this effect is that glucose 6-phosphate, which is produced from glucose, during its conversion into PEP stimulates the carboxylation of PEP.

Temperature has exactly the opposite effect on the kinetic properties of PEP carbxylase from a CAM plant and that from a C_4 plant. These temperature effect help to explain why a low temperature at night enhances acidification.

Circadian Rhythms

Detached leaves of several CAM plants kept in constant environmental conditions can exhibit persistent circadian rhythms of CO_2 exchange. These rhythms of CO_2 exchange are caused by periodic changes in the flux through PEP carboxylase. For example, in constant darkness and a stream of CO_2-free air, respired CO_2 is periodically fixed by PEP carboxylase so that little or no CO_2 escapes from the leaf, whereas when PEP carboxylase is inactive, respiratory CO_2 escapes. Thus troughs of CO_2 output correspond to periods of PEP carboxylase activity, whereas peaks correspond to periods during which PEP carboxylase is inactive. In constant light and normal air, the rhythm of CO_2 exchange also involves periodic activity of PEP carboxylase. The phosphorylation state of PEP carboxylase, the activity of PEP carboxylase kinase and the abundance of PEP carboxylase kinase transcripts, all exhibit a circadian rhythm in constant conditions. Light and temperature extremes can perturb the regular rhythms of CO_2 fixation through modifications of PEP carboxylase kinase activity. Fluctuations in malate concentrations in the vacuole and cytoplasm and processes that determine malate transport may play a role in generating and regulating circadian rhythmicity.

Metabolite Compartmentation and Transport

Transporters governing the intracellular compartmentation and the regulation of the diurnal organic acid/carbohydrate fluxes are central to the functioning of CAM. Vacuolar transport processes are energized by H^+-transporting, vacuolar type (V)–ATPase and/or H^+-PPiase activities. Increases in V-ATPase activity and subunit accumulation correlate with increased CAM capacity. Gene encoding vacuolar H^+-ATPase subunits, A, B, E and c from *M. crystallinum* as well as the c subunit from *K. daigremontiana* have been described. Diurnal changes in transcripts abundance for several of the subunit genes have been reported in leaves and may be related to CAM function.

Pyruvate import and PEP export via the chloroplast membrane is required to support diurnal cycling of malate and starch pools in the malic enzyme-type CAM plants. During the C_3 to CAM transition, light-dependent pyruvate uptake by the chloroplast is enhanced in *M. crystallinum*. Increased enzyme activities associated with glycolysis/gluconeogenesis and starch degradation and triose and hexose phosphate transport activities during CAM induction are accompanied by major changes in the profile of export products from chloroplast. Glucose-6-phosphate transport rates are reduced by illumination, suggesting that this transporter plays a role in the regulation of carbon flux.

SIMILARITIES BETWEEN CAM AND C_4 PLANTS

Recent studies have shown sufficient similarity between the CAM and C_4 plants in regard to carbon fixation and photosynthesis. In plants of categories, during photosynthesis CO_2 is fixed into C_4 dicarboxylic acid. However, they differ in respect to the place and time of carboxylation and decarboxylation. In CAM plants, carboxylation occurs during the night while decarboxylation takes place during the day time (temporal seggregation), however, in C_4 plants carboxylation reaction occurs in mesophyll cells while the carbohydrate synthesis takes place in the bundle sheath cells (spatial seggregation). Moreover, both these reactions occur during the day or night conditions. The organic acids formation in CAM plants is temporary during the night/dark.

CAM Genes

Characterization of CAM-specific genes, gene families and their expression pattern in facultative CAM plants has begun to shed light on the molecular mechanisms underlying the recruitment and evolution of genes important in CAM. In *M. crystallinum, K. blossfeldiana* and *Vanilla planifolia*, PEP carboxylase is encoded by a small multigene family of up to four members. CAM-specific isoforms exhibit enhanced expression in leaves of facultative or obligate CAM species. The remaining isoforms, which presumably fulfill anapleurotic housekeeping or tissue-specific functional roles, undergo little change in expression and generally are expressed at low levels in CAM-performing tissues. Multigene families with a least one CAM-specific isogene have also been reported for NADP-malic enzyme (ME), NAD(P) malate dehydrogenase (MDH), enolase and phosphoglyceromutase (PGM) in M. crystallinum and for the V-ATPase c subunit genes in *M. crystallinum* and *K. daigremontiana*. Cell-specific distribution patterns of PEP carboxylase kinase in the leaves of weak, CAM-inducible *Clusia apipoensis* suggest that one or more genes may encode this enzyme. Gene duplication, recombination or transposon-induced translocation events, followed by alteration of regulatory regions, are possible mechanisms to explain the recruitment of CAM-specific isogenes.

Alternatively, a single gene may fulfill both C_3 photosynthesis and CAM-specific functions, as appears to be the case for NAD-GAPDH (NAD-glyceroldehyde 3-phosphate) and PPDK (pyruvate orthophosphate dikinase). Other CAM enzymes are coordinately induced during CAM induction but are localized to distinct subcellular compartments including mitochondrial NAD-malic enzyme and cytosolie NADP-ME, cytosolic NAD-MDH and chloroplastic NADP-MDH. CAM induction in *M. crystallinum* also leads to large increases in the activities of starch-degarding enzymes and to increased rates of starch turn over. These increases also result in dramatic changes in the profile of starch degradation products exported from isolated chloroplasts. The CAM export mainly glucose-6-phosplate.

It has been demonstrated that mRNA accumulation occurs within hours of salinity or dehydration of plants or leaves of *M. crystallinum*. Transcription rates increase in salt-stressed M. crystallinum plants undergoing the C_3 to CAM switch. In the case of PEP carboxylase, increases in transcription are followed by subsequence increases in enzyme protein levels. The transcriptional activation of CAM-specific genes is mediated by the interaction of cis-acting DNA sequences and trans-acting factors. It has been suggested that Myb-related transcriptional factors participate during CAM induction. Multiple DNA-binding proteins interact with the Ppc1 (PEP carboxylase) 5′ flanking regions. One of them is PCAT-1, which binds differentially to two AT-rich sites. PCAT-1 shows increased DNA-binding affinity upon salt stress. Thus, PACT-1 and associated binding protein may play roles in the assembly of active transcription complexes during CAM induction. *In vitro*, PACT-2 and 3 show either reduced or enhanced abundance or binding affinity respectively upon salt stress. DNA-biding proteins that interact with the salt enhancer elements of the Ppc1 and Gap C1 (NAD-GAPDH) promoters, also show altered abundance or DNA-binding affinity during C_3 to CAM transition. Thus, the regulatory mechanisms that mediate transcriptional activation of CAM-specific genes remain unclear. The Ppc1 promoter in transgenic tobacco plants fail to exhibit salt inducibility. It suggests that tobacco lacks the requisite transcriptional machinery for recognizing the specific salt-responsive promoter elements. More detailed work characterizing *cis*-and *trans*-acting elements will be required in improve our understanding of transcriptions events leading to CAM induction and maintenance.

CAM idling

A few CAM plants may not open their stomata during the night when they are exposed to severe water stress. Under such conditions, they may continue to show a diurnal fluctuation in malic acid concentration and no gas exchange as first found in *Opuntia basilaris*. These plants recycle respiratory CO_2 through CAM pathway. The CO_2 they use to produce malic acid at night does not come from the air. It is entirely derived from respiration which is released during the day, which allows some Rubisco activity. This metabolism is termed CAM idling. It occurs typically in ordinary CAM plants that are exposed to severe water stress and have a very low stomatal conductance throughout the day and night.

Fluorescence measurements have indicated that the photosystems remain intact during severe drought. This has led to the suggestion that CAM idling prevents photodamage, in addition to recapturing respiratory CO_2. In the absence of an electron acceptor, damage of PSII might occur at the high irradiance to which these plants are exposed. CAM idling can be considered as a modification of normal CAM. The plants remain *ready to move* as soon as the environmental conditions improve, but they keep their stomates closed during severe drought.

CAM Cycling

For plants with weak CAM that appear "nearly-C_3" the term "CAM cycling" is used to describe C_3-type gas exchange patterns concomitant with diurnal fluctuations in C_4 acids but with little or no net noctural CO_2 assimilation. Plants that perform CAM-cycling typically grow in habitats with unpredictable daily water supply and may derive benefits by being poised to enter CAM rapidly when called for by prevailing environmental conditions. "Rapid-cycling CAM", a hypothetical variation of CAM in which both CO_2 acquisition and reduction phases of CAM are predicted to occur over time periods shorter than the normal diel cycle, has also been proposed.

INCOMPLETE AND FACULTATIVE CAM PLANTS

In a limited number of species, CAM only occurs upon exposure to drought stress. They are known as facultative CAM plants (e.g., *Agave desertil Clusia uvitana, Mesembryanthemum crystallinum, and Portulacaria afra*), Irrigation with saline water or drought can change from a virtually normal C_3 photosynthesis to the CAM mode. There is at least one genus that contains C_4 species that can shift from a normal C_4 mode under irrigated conditions to a CAM mode under water stress: *Portulaca grandiflora, P. mundula, and P. oleracea.* ABA can also induce CAM photosynthesis at a comparable rate. The transition from the C_3 or C_4 to the CAM mode coincides with an enhanced PEP-carboxylase activity and of the mRNA coding for this enzyme.

ECOLOGICAL SIGNIFICANCE

These plants are ecologically significant, because they reduce transpiration during daytime, and are well adapted to dry and hot habitats. The important features are as follows:

- The stomata remain closed during the day and open at night when water loss is little due to prevailing low temperature.
- CAM plants have parenchyma cells which are large and vacuolated. These vacuoles are used for storing malic and other acids in large amounts.
- CAM plants increase water-use efficiency of the plant and secondly through its enzyme PEP carboxylase, they are adapted to extreme hot climates.
- CAM plants can also obtain a CO_2 compensation point of zero at night and in this way accomplish a steeper gradient for CO_2 uptake compared with C_3 plants.
- They lack of a *real* photosynthesis during daytime and the growth rate is far lower than in all other plants (with the exception of pineapples).

CAM plants are not agriculturally important. The CAM mechanism is the characteristic fearure of many families of **Monocotyledons** (Agavaceae, Bromeliaceae, Liliaceae, Orchidaceae) and **Dicotyledons** (Aizoaceae, Asclepiadaceae, Cactaceae, Caryophyllaceae, Chenopodiaceae, Compositae, Convolvulaceae, Crassulaceae, Euphorbiaceae, Portulacaceae, Vitaceae). Pineapple (*Ananas comosus*) of Btomaliaceae family is an economically important CAM plant.

REGULATION OF CAM PEP CARBOXYLASE

CAM is a metabolic adaptation to arid environments: stomata are closed during much of the day and open at night. It is typified by a temporal seperation of two phases of CO_2 fixation. During the night the primary fixation of CO_2 is catalyzed by PEP carboxylase. This results in the formation of malic acid, which is stored in the vacuole. During the following day, the molate is released from the vacuole and decarboxylated – the resulting CO_2 is fixed via the Calvin cycle. It suggests that mechanisms must exist to permit flux through PEP carboxylase at night and reduce or eliminate this flux during the day to combat futile cycling between PEP and malate.

PEP carboxylase is an allosteric enzyme, inhibited by malate and activated by glucose 6-phosphate. These effects are modulated by phosphorylation, which activates PEP carboxylase in vivo by reducing its sensitivity to inhibition by malate. Many studies have shown that flux through PEP carboxylase in CAM plants is controlled by a circadian oscillator rather than by light-dark transition. Therefore, the control of PEP carboxylase has been studied both in the normal diurnal cycle and in constant conditions, in response to a circadian oscillator. These studies have provided new lights into the opration of protein phosphorylation in plants.

The first demonstration that higher plant PEP carboxylase is controlled by phosphorylation was obtained from study of the constitutive CAM plant *Kalanchoe fedtschenkoi* adapted to an 8 hours photoperiods. This work showed that PEP carboxylase was tenfold more sensitive to inhibition by malate during the light period than in the middle of the dark period. Purification of two forms shows that the night form is phosphorylated and the day form is dephosphorylated. Conversion of the night form of PEP carboxylase to the day form and vica-versa coincodes with the cessation and onset of malate accumulation respectively. It suggests that these interconversions are important in regulating the flux through PEP carboxylase. Both conversions occur during the dark period of the normal diurnal cycle. Conversions between the two forms of the enzyme are also observed in detected leaves kept in constant environmental conditions. It confirms that they are controlled by a circadian rhythm. The enzymes responsible for the dephosphorylation of PEP carboxylase in *K. fedtschenkoi* has been identified as protein phosphatase 2A (PP2A). The activity of this phosphatase does not change significantly during the normal diurnal cycle in *K. fedtschenkoi* leading to the suggestion that the phosphorylation state of PEP carboxylase could be regulated through changes in the activity of PEP carboxylase kinase.

The development of a method for measuring the amount of translatable mRNA for the kinase led to a significance advance. The method is based on *in vitro* translation of isolated RNA followed by direct assay of the translation products for kinase activity. Using this assay it has been shown that the level of PEP carboxylase kinase translatable mRNA in *K. fedtschenkoi* varies by some 20-fold, over the diurnal cycle, oscillates in constant conditions and determines PEP carboxylase kinase activity. Moreover, increases in kinase translatable mRNA are blocked by inhibitors of RNA synthesis. These results strongly suggest that PEP carboxylase kinase activity is regulated by the circadian control of kinase gene expression.

The cloning of the PEP carboxylase kinase gene has provided definitive evidence that flux through PEP carboxylase is controlled largely at the expression level of the PEP carboxylase kinase gene, which is controlled by the output from a circadian oscillator (Hartwell et al., 1999). The pathway connecting the circadian oscillator to PEP carboxylase kinase gene expression is not clear. There might be some direct connection, for example via circadian phosphorylation of a transcription factor required for the expression of the kinase gene, but there is increasing evidence that change in metabolic levels play a role (Nimmo, 2000).

Evolution of CAM

DNA and protein sequence data for PEP carboxylase has allowed modeling of molecular phylogenies. The most recent analyses using up to 75 PEP carboxylase amino acid sequence have resulted in phylogenetic reconstructions. Comparisons of prokaryotic and eukaryotic PEP carboxylase point to a common ancestral gene arising from the γ-proteobacterial lineage. Among the land plants, the precise relationships among various texa are not well resolved due to low sequence divergence and/or multiple speciation events. Nevertheless, phylogenetic reconstructions confirm the conclusion derived from comparative physiology and taxonomy that C_3 photosynthesis is ancestral to both CAM and C_4, with CAM appearing earlier than C_4 photosynthesis. Reconstructions with CAM-specific PEP-carboxylase isoforms support at least three independent origins of the pathways in the Orchidaceae, Crassulaceae and Caryophyllales.

The ecophysiological plasticity of CAM is demonstrated by nocturnal CO_2 uptake, CAM-cycling, CAM-idling and accumulation of organic acids by roots of epiphytic orchids, terrestrial fern species of the Isoetid family and large number of aquatic vascular species. These observations suggest that limited CO_2 may have provided a primary selective pressure for the evolution of CAM (Raven and Spicer, 1996).

SIGNAL TRANSDUCTION AND CAM REGULATION

Relatively little is known about the signaling mechanisms that induce or maintain CAM. It has been suggested that roots can perceive water stress and convey this information to leaves, triggering a switch from C3 and CAM. Although the nature of the signal in unknown, experiments with detached leaves suggest that a reduction in cytokinin content may derepress CAM expression. Treatment of detached leaves with the Ca^{2+} ionophore, ionomycin or with thapsigargin, a specific inhibitor of endomembrane Ca^{2+}-ATPases and stimulator of intracellular Ca^{2+} release, caused Ppc1 transcript accumulation in the absence of stress or ABA treatment. Similar results are observed for V-ATPase subunit. These results suggest that changes in intracellular Ca^{2+} concentration can stimulate Ppc1 and Atpc1.

Okadaic acid, an inhibitor of protein phosphatase 2A and 1 activities, eliminates the salt-responsive induction of Ppc1 and Imt1 transcripts but not the induction of Atpc1. Okadaic acid the protein phosphatase 2A inhibitor, cantharidic acid, are also effective inhibitors of Ppc1 transcript accumulation induced by dehydration and ABA. These results implicate protein phosphatase 2A and/or protein phosphatase 1 activity in CAM induction signaling. Treating detached *M. crystallinum* leaves with W_7, a specific inhibitor of Ca^{2+}/calmodulin-dependent protein kinases, blocked Ppc1 transcript accumulation in response to ionic, osmotic and dehydration stress and ABA treatments, suggesting that CA^{2+}/CAM protein kinases play a positive role in stress-induced increases in Ppc1 expression.

A Ser/thr protein kinase cDNA termed MK6 is preferentially expressed at night in plants performing CAM, suggesting that in may be involved din the diurnal or circadian expression of CAM enzymes. Additional protein kinases related to cyclic-nucleotide-dependent (PKA and PKG) or Ca^{2+} phospholipid-dependent kinases (PKC) have also been isolated from *M. crystallinum* and may function downstream of cAMP or cGMP second messengers in signal transduction cascades.

Physiological and molecular studies have documented the role of phytochrome in light-sensing and signaling in CAM plants. Phytochrome is likely to control short-day CAM induction since interrupting the dark period with red-light effectively inhibits CAM development. Phytochrome is also thought to enhance the appearance of a CAM-specific PEP carboxylase isoform in *Kalanchoe* and *M. crystallinum*. These effects, amplified in salt-stressed plants, suggest that phytochrome and NaCl or ABA may act along the same or parallel signal transduction pathways to promote CAM induction.

Recent work has demonstrated that CAM induction abolishes both white and blue-light-stimulated stromatal opening and light-dependent zeaxanthin formation, which suggests that light-dependent zeaxanthin formation in guard cell chloroplasts may an important component that mediates nocturnal stomatal opening.

CHAPTER 9

The Photorespiratory Carbon Oxidation Cycle

All land plants have an additional metabolic activity and lose a considerable amount of their photosynthate as CO_2 within a few seconds of its being fixed. It occurs when plants are illuminated. This process of CO_2 release is light-dependent and, because of its otherwise superficial resemblance to respiration, it has been given the name *photorespiration*. During the process O_2 is consumed and CO_2 is generated in light. It is independent of mitochondrial respiration but occurs in chloroplasts and peroxisomes and only a few reactions occur in mitochondria.. However, unlike true respiration (frequently called dark respiration) it performs no useful function in making energy available to the plant. Instead, its occurrence leads to a *loss* of energy and this is responsible for a considerable reduction in the yield of many of our crops.

Photorespiration results from the ability of Rubisco to catalyze the oxygenation of ribulose-1,5-bisphosphate as well as its carboxylation. This occurs because O_2 and CO_2 occupy the same catalytic site and compete in reacting with the sugar phosphate. The functioning of the RPP cycle is prevented by the production of one molecule of phosphoglycolic acid and only one molecule of phosphoglyceric acid. During photorespiration recycling of 75 per cent of the carbon in phosphoglycolate (25 per cent is lost as CO_2) and back to phosphoglyceric acid permit the continued operation of the RPP cycle to occur. The recycling occurs by the enzymes located in three separate organelles.

Photorespiration probably arose in evolution due to the gradual loss of CO_2 from the Earth's atmosphere to a level at which most land plants cannot take it up without losing some of what has already been fixed. However, there is evidence that some of the newer arrivals on the evolutionary scene, the so-called C_4 plants, have evolved a CO_2

concentrating mechanism which generates a high CO_2 concentration for their photosynthetic enzymes, and have in this way almost eliminated photorespiration.

The Rubisco in higher plants like rice and wheat is 100 times more likely to pick up CO_2 than O_2. But because the concentration of atmospheric O_2 is many times greater than that of CO_2, the greater affinity for CO_2 is largely canceled. As a result, 20% to 50% of the carbon fixed by photosynthesis is lost to photorespiration. The oxygenation reaction is - as far as we can tell, and a lot of researches over decades have gone into it – just a complete waste. It does not do any thing for the plants.

This striking inefficiency was no handicap when photosynthesis first evolved 3 billion years ago, because the atmosphere was almost devoid of oxygen. After photosynthesis filled the air with oxygen and Rubisco's weakness was revealed, it may have been too late for evolution to fix the problem, says Murray Badger, a Rubisco specialist at Australian National University *"It is somewhat general correlation that the more specific and discriminatory a reaction becomes, the slower it gets"*, he says. As a result, mutations that made Rubisco target carbon dioxide better might also have made it even slower.

TABLE 9.1: Differences between photorespiration and true respiration

Photorespiration	True Respiration
The Effect of Oxygen Concentration	
1. It increases steadily with oxygen concentration all the way from zero to one hundred per cent.	It is saturated at oxygen concentrations as low as two or three per cent and virtually no increase in rate is observed above these concentrations.
Requirement for the Photosynthetic Apparatus	
2. It is intimately linked to photosynthesis and occurs only in cells which are capable of photosynthesis.	It occurs in any living cell
The nature of the substrate	
3. The nature of the substrate is different.	The nature of the substrate is different.
Energy relationship	
4. Biochemistry of photorespiration differs fundamentally from that of the true respiration. Photorespiration, however, does not lead to a net production of ATP or reduced nucleotides. Glycolate arises as an unwanted byproduct of photosynthesis. The bulk of the carbon of glycolate, after oxidation, is converted to carbohydrate, but the remainder is converted to CO_2 which is released to CO_2 of photorespiration.	Biochemistry of true respiration differs fundamentally from that of the photorespiration. True respiration brings about the oxidation of its substrates to yield CO_2 and energy, mainly in the form of ATP and reduced pyridine nucleotides.

The Measurement of Photorespiration

During photorespiration green plants take up oxygen and release CO_2 in the light. Therefore, it can be measured by measuring either of these two activities. Unfortunately, most photorespiring tissues also show an oxygen uptake and CO_2 release in the light due to true respiration. This leads to an error in measurements of photorespiration, but this error is small, the rate of photorespiration for a tissue being normally several times greater than its rate of true respiration as measured in the dark. The rate of true respiration is reduced to an even lower level in the light. The main problem is that, under normal precise circumstances, photosynthesis and photorespiration occur simultaneously in the same tissue and carry out opposite overall reactions.

$$CO_2 + H_2O \xrightarrow[\text{Photorespiration}]{\text{Photosynthesis}} \text{Photosynthate} + O_2$$

Some of the simpler ways of measuring photorespiration are to inhibit photosynthesis and to measure O_2 uptake or CO_2 output. One means of doing this is to put a plant or leaf in the light into a CO_2-free air stream, so that photosynthesis is inhibited by lack of CO_2 and photorespiration can be measured as the rate of CO_2 output. This is not accurate measurement leaves some of the CO_2 released by photorespiration is refixed by photosynthesis so that the rate of photorespiration is underestimated. Although the degree of this underestimate can be minimized by using fast air flow rates to sweep as much as possible of the CO_2 away from the leaf before it is refixed, the error cannot be eliminated altogether by this means.

Photorespiration is also measured to turn off the light and to means the CO_2 released immediately afterwards. This is because photorespiration goes on for a minute or so after photosynthesis has stopped and takes the form of a sudden outburst of CO_2 at the onset of darkness which rapidly declines to the much lower level of CO_2 release due to true respiration. The peak rate of CO_2 release is taken as a measure of photorespiration. However, this too could lead to an underestimate if the cessation of photosynthesis is not as abrupt as assumed and some of the outburst is refixed.

A more sophisticated methods of measuring photorespiration is also known. It works without difficulties. Because of the criticism that the above two methods measure photorespiration when photosynthesis is absent, or at least greatly reduced, and that its rate may not normal under these circumstances, attempts have been made to measure photorespiration actually during active photosynthesis by the use of isotopic tracers. For example, photorespiration has been measured as the rate of $^{18}CO_2$ uptake from an atmosphere containing this isotope, but again this leads to an underestimate of photorespiration because some of the $^{16}O_2$ released by photosynthesis is taken up by photorespiration in addition to the supplied $^{18}CO_2$.

All the above methods for measuring photorespiration are somewhat inaccurate and most lead to an underestimate of the true rate. Despite this, it has been generally found that measured rates of photorespiration are very high, commonly three or four times the rate of true respiration measured for the same tissue in the dark.

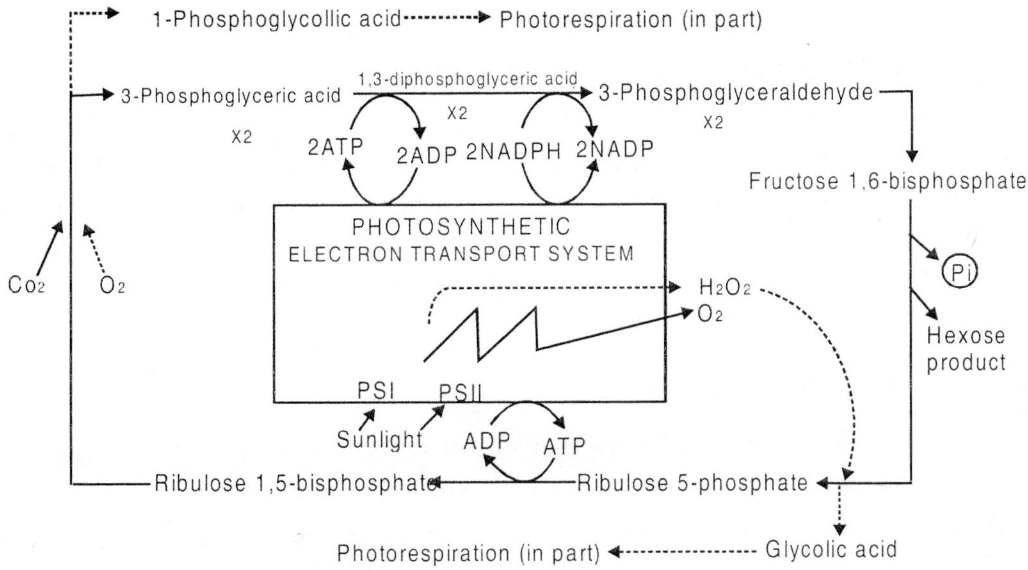

FIG. 9.1: *A simplified scheme of photorespiration.*

The Biological Function of Photorespiration is Unknown

Although the PCO cycle recovers 75 pert cent of the carbon originally lost from the Calvin cycle as 2-Phosphoglycolate, why does 2-Phosphoglycolate form at all? Formation of 2-Phosphoglycolate is a consequence of the chemistry of the carboxylation reaction. It requires an intermediate that can react with both CO_2 and O_2. Such a reaction would have had litter consequence in early evolutionary times if the ratio of CO_2 to O_2 in air were higher than it is today. However, the low CO_2-to-O_2 ratios prevalent in modern times are conducive to photorespiration, with no other function than the recovery of some of the carbon present in 2-phosphoglycolate.

It is also possible that photorespiration is important under conditions of high light intensity and low intercellular CO_2 concentration (for example, when stomata are closed because of water stress), to dissipate excess ATP and reducing power from the light reactions, thus preventing damage to the photosynthetic apparatus. *Arabidopsis* mutants that are unable to photorespire grow normally under 2 per cent CO_2, but they die rapidly if transferred to normal air. Furthermore, recent evidence with transgenic plants suggests that photorespiration protects C_3 plants from photooxidation and photoinhibition. Further work is needed to improve our understanding of the function of photorespiration.

THE METABOLISM OF PHOTORESPIRATION

Evidence for Glycolate as the Substrate for Photorespiration

Glycolate is the source of much of the CO_2. It is given off in photorespiration. The green leaves produced large quantities of glycolate in the light but not in the dark. The leaves could oxidize artificially supplied glycolate to yield, amongst other things, CO_2. The inhibition of an enzyme which oxidizes glycolate appears to inhibit photorespiration and causes a temporary increase in the apparent rate of photosynthesis in leaves which photorespire, but not in leaves which do not.

The Biosynthesis of Glycolate

Glycolate synthesis appears to be widespread occurrence in photosynthetic organisms and has been demonstrated in many of the algae and in higher plants.

It is an unwanted metabolic excrete of many algae into the surrounding water, however, higher plants cannot excrete it in this way, but they oxidize it instead by the process of photorespiration. In algae and higher plants the origin of the glycolate seems to be similar under similar environmental factors. Glycolate production in both the groups occurs only in the light and is stimulated by high oxygen and low CO_2 concentrations and the pattern of labeling in glycolate, when $^{14}CO_2$ is being photosynthesized is also similar.

Site of Glycolate Production

The isolated chloroplasts are capable of synthesizing glycolate. It is a early product of photosynthesis and becomes labeled within seconds when $^{14}CO_2$ is being photosynthesized. The mechanism has been proposed by a group of workers in Illinois (Ogren and Bowes, 1971; Bowes *et al.*, 1971). They obtained that a derivative of glycolate, namely, 2-phosphoglycolate, was produced as a result of the apparent malfunctioning of the enzyme RuBP carboxylase. Normally, it does this by adding a molecule of CO_2 to the 5-carbon compound ribulose-1, 5- bisphosphate (RuBP) to form the two 3-carbn molecules of phosphoglyceric acid (PGA) which are further metabolized into sugars etc. by the Calvin cycle. However, these workers showed that oxygen was able to function as an alternative substrate for this enzyme in place of CO_2. Using purified preparations of RuBP carboxylase they showed that when oxygen underwent the reaction in place of CO_2, *phosphoglycolate* was produced. The exact mechanism of the reaction was not investigated but it is probably an oxidative splitting of the 5-carbon RuBP into a 2-carbon and a 3-carbon unit:

$$\text{RuBP (5-carbon)} + CO_2 \xrightarrow{\text{Normal}} \text{PGA (3-carbon)} + \text{PGA (3-carbon)}$$

$$\text{RuBP (5-carbon)} + O_2 \xrightarrow{\text{Malfunction}} \text{Phosphoglycolate (2-carbon)} + \text{PGA (3-carbon)}$$

Phosphoglycolate, produced by this means is converted within the chloroplasts to free glycolate by phosphoglycolate phosphatase. This enzyme is shown to be located in the chloroplasts and is irreversible, highly active and specific for phosphoglycolate. The enzyme produces glycolate from phosphoglycolate. This indicates glycolate arises in nature from phosphoglycolate. The above proposed mechanism of glycolate biosynthesis explains the following:

- High oxygen and low CO_2 concentrations stimulate glycolate production. If glycolate arises due to oxygen competing with CO_2 for an enzyme, then anything which increases the O_2/CO_2 ratio would be expected to increase glycolate synthesis.
- Glycolate is produced only in the light. The enzyme RuBP carboxylase, when in its natural state in the chloroplasts, is only active in the light (Jensen and Bassham, 1968).
- Glycolate becomes labeled when $^{14}CO_2$ is being photosynthesized after approximately the same time interval as RuBP.
- The capacity to produce glycolate has been retained during evolution, even though it appears to be detrimental to the plant. Naturally, the loss of the enzyme which leads to the production of glycolate, namely RuBP carboxylase, would also lead to the loss of the plant's ability to photosynthesize.
- The mechanism of glycolate production is backed by experimental evidence.

It remains possible that there may also be other mechanisms, but these have yet to be demonstrated in quite such unequivocal terms.

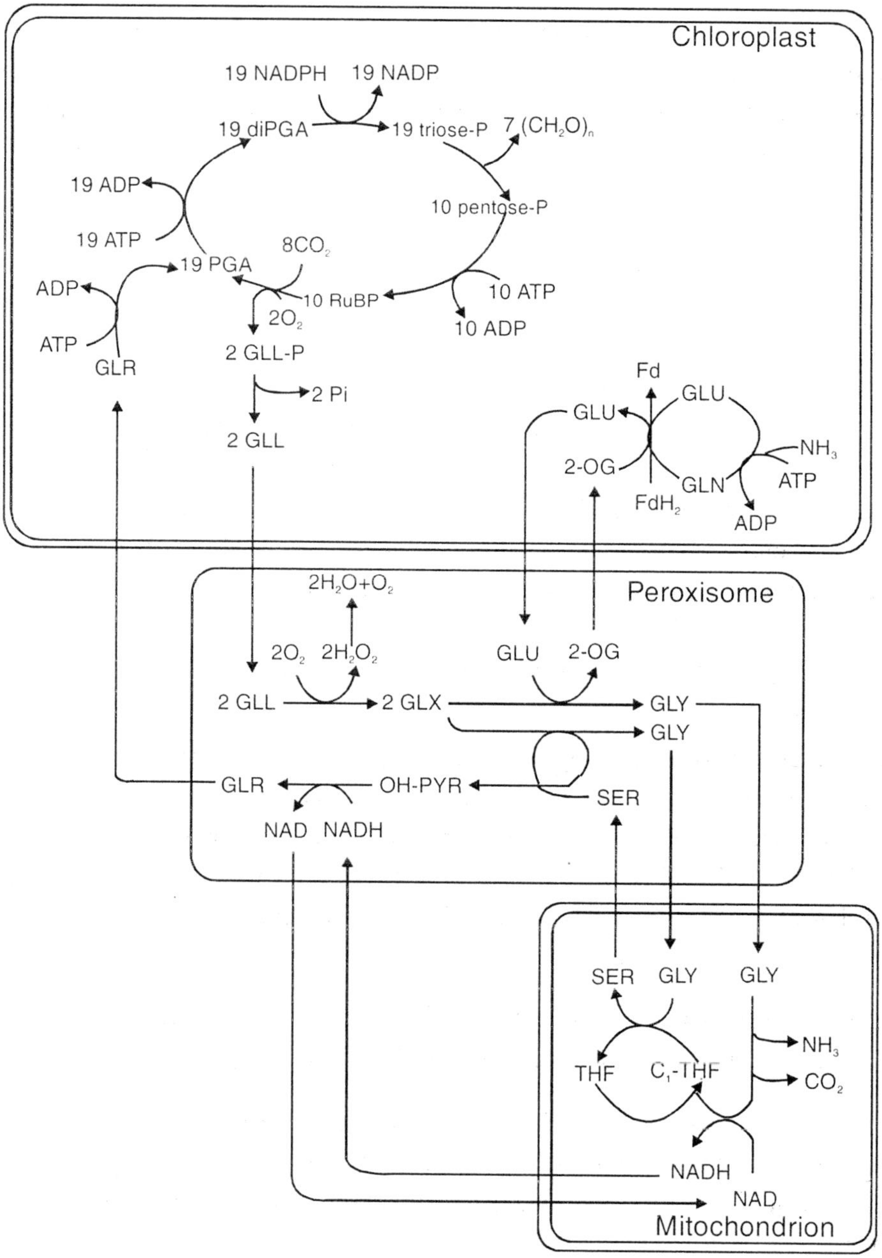

FIG. 9.2: *Reactions and organelles involved in photorespiration. In C_3 plants, at 20° C, 20,000 Pa O_2 and 25 Pa CO_2 in the chloroplast, 2 out of the 10 RuBP molecules are oxygenated, rather than carboxylated. The oxygenation reaction produces Phosphoglycolate (GLL-P), which is dephosphorylated to glycolate (GLL). Glycolate is subsequently metabolized in peroxisomes and mitochondria, in which glyoxylate (GLX) and the amino acids glycine and serine (SER) play a role. Serine is exported from mitochondria and converted to hydroxypyruvate (PYR) and then glycerate (GLR) in the peroxisomes which it returns to the chloroplast.*

Metabolism of Glycolate by the Peroxisomes

Glycolate leaves the chloroplasts and enters small cytoplasmic organelles called peroxisomes. The conversion of glycolate to the amino acid glycine takes place in the peroxisome. The perioxisomes are microbodies bounded by a single membrane and are found closely associated with chloroplasts and mitochondria in the leaf. The first step in glycolate metabolism is its oxidation to glyoxylate and hydrogen peroxide. In higher plants, this oxidation is catalyzed by the enzyme glycolate oxidase, a flavoprotein. The H_2O_2 is decomposed to water and O_2 by the enzyme catalase found in abundance in the peroxisomes.

$$H_2O_2 \longrightarrow H_2O + \tfrac{1}{2}O_2$$

Thus the net O_2 uptake in the peroxisomes during glycolate oxidation is one atom of oxygen for every molecule of glyoxylate formed. It has been suggested that a very small fraction of glyoxylate formed may undergo a non-enzymatic reaction with H_2O_2 forming formate and releasing CO_2. This reaction could theoretically be a source of photorespiratory CO_2. However, there is an abundance of catalase in the peroxisomes to decompose all the H_2O_2 thus minimizing such oxidation of glyoxylate to produce formic acid, a deleterious product.

Glyoxylate is then transminated to glycine by two amino transferases found in the peroxisomes.

$$\text{Glyoxylate} + \text{L-Glutamate} \longrightarrow \text{Glycine} + \text{2-Oxoglutarate}$$

(catalyzed by glutamate-glyoxylate aminotransferase)

$$\text{Glyoxylate} + \text{Serine} \longrightarrow \text{Glycine} + \text{Hydroxypyruvate}$$

(catalyzed by serine-glyoxylate aminotransferase)

Physiologically these reactions are irreversible.

Metabolism of Glycine by the Mitochondria

The conversion of glycine to serine and CO_2 takes place in the leaf mitochondria catalyzed by the enzymes glycine decarboxylase and hydroxymethyltransferase found as a multienzyme complex in the mitochondrial membrane.

$$\text{Glycine} + \text{Tetrahydrofolate} + NAD^+ \longrightarrow \text{5, 10 Methylene tetrahydrofolate} + NH_3 + CO_2 + NADH_2$$

$$\text{5,10 Methylene tetrahydrofolate} + \text{Glycine} \longrightarrow \text{Serine}$$

Reaction (1) is liked to mitochonrdiral electron transport, the $NADH_2$ being subsequently oxidized with ATP/O ratio of about 3. This CO_2 which is formed during the oxidative decarboxylation of glycine in the mitochondria is the main source of photorespiratory CO_2. The NH_3 released is mostly refixed to glutamate by mitochondrial glutamate dehydrogenase or to glutamine by the cytoplasmic glutamine synthetase. Using ^{15}N glycine it has been demonstrated that the rate of ^{15}N released to the atmosphere by wheat leaves in the presence of methionine sulfoximine (a specific inhibitor of glutamine synthetase) is four-fold higher than the rate in the absence of the inhibitor. Thus, photorespiration involves recycling of nitrogen pool.

Serine formed from glycine provides substrate for hydroxypyruvate in the peroxisomes. The hydroxypyruvate is reduced to glycerate by the enzyme glycerate dehydrogenase. The glycerate is subsequently phosphorylated to 3-phosphoglycerate by a chloroplast phosphokinase, which completes the photorespiratory cycle.

Conversion of Serine to Carbohydrate

In tobacco leaves, some of the label from photosynthesized $^{14}CO_2$ first entered glycine and serine and then became incorporated into carbohydrate. Similarly, artificially added serine-3-^{14}C also became incorporated into carbohydrate, but only in the light.

The conversion of serine to carbohydrate, is an energy requiring process which can use the energy of light. However, since the carbohydrate products of serine metabolism accumulated both inside and outside the chloroplasts, it could not be certain whether the formation of carbohydrate from serine was confined to the chloroplasts. There is evidence that at least some of the stages in this reaction sequence can take place in peroxisomes since these particles have been shown to contain some of the necessary enzymes.

Peroxisomes derived from spinach leaves contain an aminotransferase which converts serine to hydroxypyruvate and a NAD-linked dehydrogenase which converts hydroxypyruvate to D-glycerate. The exact location of the remainder of the pathway is not clear, but a D-glycerate kinase enzyme has been demonstrated in higher plant leaves which uses ATP to convert D-glycerate into 3-phosphoglycerate. This compound (PGA) is, of course, an intermediate of the Calvin cycle and would, if it entered the chloroplasts, become converted to carbohydrate, using ATP and NADPH, by the normal processes of photosynthesis.

PHORORESPIRATION AND CROP PLANTS

Photorespiration is common in those crops (all C3 plants) that are an important part of the world's food supply such as wheat and rice, many legumes, and sugar beets. In some of these, photorespiration reduces as much 50 per cent of the net assimilation of CO_2 by photosynthesis. Therefore, it is essential to find out the means to inhibit photorespiration to increase the crop yield. Other equally important food crops- corn, sorghum, sugarcane exhibit much less photorespiration. This difference can be explained, in part, by the fact that these species are C_4 plants carrying out C_4 photosynthesis. In this process, the high concentration of CO_2 in the bundle-sheath chloroplasts will selectively favour **carboxylation** rather than **oxygenation** and little phosphoglycolate will be produced.

Photorespiratory Oxygenation

The oxygenation of ribulose-1,5-bisphosphate results in the incorporation of one atom of molecular O_2 into one of the products of the reactions, 2-Phosphoglycolate. The ability to catalyze the oxygenation of ribulose-1,5-bisphosphate is a property of all Rubiscos regardless of taxonomic origin. Even the Rubisco from anaerobic, autotrophic bacteria catalyzes the oxygenase reaction when exposed to oxygen. As alternative substrates for Rubisco, CO_2 and O_2 compete for reaction with ribulose-1,5-bisphosphate, since carboxylation and oxygenation occur within the same active site of the enzyme. Offered equal concentrations of CO_2 and O_2 in a test tube, angiosperm Rubiscos fixes CO_2 about 80 times faster than they oxygenate.

Evidence in support of the PCO cycle stems from several types of experiments. Using mutant *Arabidopsis* plants lacking Phosphoglycolate phosphatase, glycolate synthesis *in vivo* is dependent on O_2 and is inhibited by CO_2 in a competitive manner, as expected from the primary role of rubisco in the cycle. Several studies have shown that, as with the Calvin cycle, the enzymes of the PCO cycle catalyze all the proposed reactions *in vitro* at rates higher than the *in vivo* rate of photorespiration. In addition, the participating enzymes are localized in the choroplast (rubisco, Phospho glycolate phosphatase, glycerate kinase), the peroxisome (glycolate oxidase, catalase, hydroxypyruvate reductase, and the two aminotransferases), and the mitochondrion (glycine decarboxylase and serine hydroxymethyl-transferase).

Competition between Carboxylation and Oxygenation

Since photorespiration is concurrent with photosynthesis, it is difficult to measure the rate of photorespiration in intact cells. Two molecules of 2-Phosphoglycolate (4 carbon atoms) are needed to make one molecule of 3-phosphoglycerate, with the release of one molecule of CO_2; so theoretically one-fourth of carbon entering the PCO cycle is released as CO_2. The rate of *true* photosynthesis is approximately 120 to 125 per cent of the rate of net photosynthesis. The ratio of carboxylation to oxygenation in air at 25°C is computed to be between 2.5 and 3.

Interlocking of Carboxylation and Oxygenation

Photosynthetic carbon metabolism in the intact leaf reflects the integrated balance between two mutually opposing and interlocking cycles. The Calvin cycle is capable of independent operation, but the PCO cycle depends on the Calvin cycle for a supply or ribulose-1,5-bisphosphate. The balance between the two cycles is determined by three factors: the kinetic properties of rubisco, the concentrations of the substrates (CO_2 and O_2) and temperature.

As the temperature increases, the concentration of CO_2 in a solution is in equilibrium with air decreases more than the concentration of O_2 does. Consequently, the concentration ratio of CO_2 to O_2 decreases as the temperature rises. As a result of this property, photorespiration (oxygenation) increases relative to photosynthesis (carboxylation) as the temperature rises. This effect is enhanced by the kinetic properties of rubisco, which also result in a relative increase in oxygenation at higher temperatures. Overall, then, increasing temperatures progressively tilt the balance away from the Calvin cycle and toward the photorespiratory carbon oxidation cycle.

THE ENERGETICS OF PHOTORESPIRATION

The total amount of energy lost by the plant will depend on the amount of glycolate produced. Zelitch (1959) measured the rate of glycolate synthesis by following the rate of its accumulation when its further metabolism was blocked by an enzyme inhibitor. He found that in tobacco about half of the carbon fixed by photosynthesis was passing into glycolate. Other measurements, based on the rate of CO_2 release by photorespiration suggest that this proportion might be even higher. It can be seen, therefore, that photorespiration must represent a considerable loss of the plant's available energy.

The conversion of two molecules of glycolate into carbohydrate generates two molecules of ATP during conversion of glycine to serine and CO_2, but the subsequent conversion of serine to carbohydrate uses two molecules of ATP and two of reduced pyridine nucleotide (NADH and NADPH). If all the carbon of the original two glycolate molecules is to be recovered, then the CO_2 released during serine formation must be refixed by normal photosynthesis and this requires two further molecules of NADPH and three of ATP. Thus, to restore to the level of carbohydrate the four carbon atoms lost in the form of a pair of glycolate molecules, the plant must use a net total of four molecules of reduced pyridine nucleotides and three ATP.

IS PHOTORESPIRATION ADVANTAGEOUS TO THE PLANT?

It is estimated that the energy consumed in the release of one molecule of CO_2 by the glycolate oxidation pathway is more than three times the energy required to fix a molecule of CO_2 into sugar by the Calvin Cycle. Although part of the photorespiratory energy loss is regenerated as ATP by the glycine oxidation-linked electron transport in mitochondria, overall, photorespiration is a wasteful process. Various assays show that at 25°C the extent of CO_2 loss by photorespiration in many C_3 species is up to 50% of the net CO_2 assimilation; this percentage will be higher at higher temperature. The question has been often asked as to whether photorespiration is an essential requisite for photosynthesis or whether it is an unavoidable consequence of the biochemistry of Rubisco enzyme.

It has been suggested that a function of photorespiration is to protect the chloroplast against photo-oxidative destruction of its components in a natural environmental of high light and low CO_2 such as is often found during the day when stomatal closure occurs in periods of water stress. Another function suggested for photorespiration is the provision of glycine and serine pools for the plant through the glycolate pathway. But a small fraction of the carbon flux into the glycolate cycle is enough to provide all the required glycine and serine. Also there are other metabolic pathway available to the plant to synthesize these amino acids. Another proposed role for photorespiration is the transfer of reductant from the chloroplast to the mitochondria via the glycine oxidation reaction. Again, there are less complicated mechanisms, such as the malate-oxaloacetate cycle, available to the leaf for the import of reductant from chloroplast to mitochondria. Thus, at present one has to come to the conclusion that photorespiration is the result of the competitive affinity for CO_2 and O_2 to the same site of Rubisco and is the inevitable result of the metabolic function of this enzyme.

REGULATION OF PHOTORESPIRATION

One mechanism for regulating photorespiration would be to change the characteristics of the enzyme so that the affinity of Rubisco for CO_2 is greater and there is relatively less oxygenation. The findings of Jordan and Ogren (1983) indicate that it might be possible to increase the CO_2/O_2 specificity by appropriate selection for genetic changes or by use of site directed mutagenesis to alter the properties of the enzyme in transgenic plants.

A mutant of the alga *Chlamydomonas* has been described in which the Rubsico specificity ratio was decreased compared with the wild-type. Although the specificity was not altered in the direction needed to reduce photorespiration. These results demonstrate that CO_2/O_2 specificity is not necessarily fixed within a species.

Raising the CO_2 concentration increases net photosynthesis and decreases photorespiration and the efficacy of CO_2 enrichment in increasing yields in closed environments, such as green house, is well documented. An analogous example of the regulation of photorespiration exists in leaves of all C_4 plants in which specialized bundle-sheath cells surrounding the vascular elements decarboxylate a C_4 compound to pyruvic acid and CO_2. This system raises the level of CO_2 in the same cellular compartment in which the Rubisco is located and the CO_2 is then fixed as in C_3 plants but in the absence of photorespiration.

Another example of compartmentation in plants that partially regulates photorespiration occurs in some species known as the C_3-C_4 intermediates. These plants have as a common feature a rudimentary bundle sheath system in their leaves and show less inhibition of photosynthesis by oxygen than do C_3 plants and more than C_4 plants. Their CO_2 compensation points are also intermediate between C_3 and C_4 values. Many of the photosynthetic features of the C_3-C_4 intermediates, including a decreased rate of photorespiration, can be explained by the finding that the glycine decarboxylase reaction of the photorespiratory pathway that produces CO_2 is confined to mitochondria of the bundle sheath cells. There the Rubisco carboxylase reaction can more efficiently recapture photorespiratory CO_2.

It appears that the photorespiration is regulated in C_3-C_4 species but it seems puzzling that under moderate temperatures and atmospheric levels of CO_2, photosynthesis rates are often lower in intermediate species. It appears that the recapture mechanism is generally not efficient enough to bring about greater net photosynthesis when photosynthesis is rapid because the apparent reduction in photorespiration requires released CO_2 to be refixed without C_4 photosynthesis.

There are few reports of greater net photosynthesis in some intermediate species of *Flaveria*. Schuster and Munson (1990) described enhanced photosynthesis in the intermediate *F. floridana*, which has a C_3-like photosynthesis, temperature response compared to the C_3 *F. cronquistii*. Their observations are qualitatively similar to the increased photosynthesis observed in a tobacco mutant with decreased photorespiration, although the mutant has a different biochemical mechanism that decreases the production of photorespiratory CO_2 rather than conserving it by recycling.

The plants with enhanced catalase activity produce less photorespiratory CO_2 and thus show some oxygen resistant because the peroxidative decarboxylation of ketoacids is slowed. Such plants might show increased net photosynthetic CO_2 assimilation. Tobacco plants with oxygen-resistant photosynthesis has 40-50% more catalase activity than wild type, whereas other peroxisomal enzymes such as glycolate oxidase and NADH-hydroxypyruvate reductase are the same as wild-type. Mutants and wild-type plants also have similar superoxide dismutase activities. The five isozymes that comprise CAT-1 (major isoforms of catalase with similar antigenic and enzymatic properties) are all present in greater amounts in mutant leaves than in wild-type. The results indicate that the mutant has greater net photosynthesis under conditions where there is a high ratio of photorespiration/net photosynthesis such as higher temperatures, high oxygen and low CO_2. Cloning of catalase gene and the production of transgenic plants with different levels of catalase activity may help to establish a biochemical linkage.

PHOTORESPIRATION AND EVOLUTION OF PHOTOSYNTHESIS

Chemical and immunological studies of Rubisco in a wide range of organisms show that the larger subunit of the enzyme molecule, which contains the catalytic and activation sites for carboxylation and oxygenation of RuBP, has changed very little during the course of evolution. The universal occurrence of this enzyme enables us to trace the evolutionary route followed by photosynthetic organisms. The photosynthetic bacteria would have evolved in the earth at least three billion (3×10^9) years ago, presumably from the anaerobic fermenters, to compete with the fermenters for limited amounts of reduced carbon substrates. The C_3 carbon reduction cycle and the enzyme Rubisco should have coevolved with the photosynthetic bacteria. These photosynthetic anaerobes functioned well in the primitive earth atmosphere rich in CO_2 and with very little O_2-the oxygenase activity of Rubisco being completely suppressed. With the advent of blue-green algae, which could use water as a substrate, the atmosphere began gradually to accumulate O_2. With the increasing O_2 in the air the oxygenase activity of the blue-green algal Rubisco began to express itself with the production of glycolate, which algae then excrete. The glycolate oxidation pathway would have evolved later, probably before the origin of land plants as a means of recycling some of the fixed carbon in glycolate from the CO_2 formed during its breakdown. The final stage in this sequence would have been the developmental of the CAM and C_4 plants with their metabolic ability to concentrate internal CO_2 which then almost eliminates photorespiratory CO_2 loss.

CHAPTER 10

Carbon Partitioning and Crop Productivity

Plants are the reliable source of food; therefore, there has been intense interest to improve crop productivity. Attempts to improve crop yields have been increased largely through efforts to increase plant adaptability to environmental conditions, resistance to diseases and pests, improve agronomic and horticultural practices and improvement of genetic yield potential. Efforts are also made to increase agricultural productivity with the help of basic researches on biological determinants of yield:

1. Photosynthetic efficiency
2. Harvest index

Of all the biological yield determinants, photosynthesis continues to be considered a high-priority area of research.

Carbon partitioning is influenced by both the supply and demand for photosynthate in plant. It is moderated by vascular connections and the storage capacity of the leaves and pathway tissues. The transport of photosynthate may be either for short distance (within the source or the sink) or for long-distance transport (in the phloem).

Carbon partitioning in plants is influenced by a range of factors. If a particular organ or tissue is adequately supplied with photosynthates, then the organ can fully express its growth potential (a source or sink limitation) and depend on the direct supply and demand for carbon throughout development. The hormonal or nutritional factors indirectly control the growth of an organ. The vascular constraints may favour the growth of one organ over another. The buffering effect of carbon storage at different sites in the plant and in some instances photosynthetic rate (source activity) may also regulate the rate of carbon utilization (sink activity). The supply of photosynthate in various plant parts can also be modified either experimentally, or through the action of environmental stress and disease. The above aspects show that a series of factors, and not a single limiting event, control carbon partitioning in higher plants.

In thermodynamic terms the efficiency of crop production is defined as the ratio of energy output (carbohydrate) to energy input (solar radiation). Temperature and water supply are the main climatic constraints on efficiency. In India, the radiation and thermal climates are not uniform and rainfall is the main discriminate of yield between regions.

Total dry matter production by barley, potatoes, sugar beet, and apple is strongly correlated with intercepted radiation and these crops form carbohydrate at about 1.4 g per MJ solar energy, equivalent to 2.4% efficiency. Crop growth may, therefore, be analyzed in terms of (*a*) the amount of light intercepted during the growing season and (*b*) the efficiency with which intercepted light is used. These variables are discussed in terms of the rate and duration of development phases.

PHOTOSYNTHETIC EFFICIENCY

About 90 per cent of a plant's dry matter is the result of photosynthesis. In other words there must be a correlation between net photosynthesis and dry matter content of the whole plant, which vary among cultivars. This relationship is very complex, and for many crops a positive correlation is not obtained. The reasons for a poor correlation between leaf photosynthesis and yield are varied. Before a crop canopy achieves full interception of light, leaf area is more important than the photosynthetic rate per unit leaf area in determining crop growth rate. After canopy closure, the *carbon exchange rate* (CER) per unit leaf area becomes the most important factor. High CER has always been considered a desirable characteristic. Still the evidences are needed which can reveal that CER per unit leaf area has increased during the domestication of important crops like wheat, maize, sugarcane, and cowpea. Actually, the highest measured CER for wheat, cotton, and pearl millet are found in the wild relatives, not in modern cultivars (Gifford and Evans 1981). However, a correlation seems to exist between yield of forage crops and their respective CER.

Several attempts have been made to select for a specific photosynthetic component that would result in a higher CER. For example, increasing chlorophyll content does not seem to have much impact on CER. No good correlation has been shown between Hill reaction activity or phosphorylation per unit leaf area and CER. Nevertheless, a correlation between ribulose bisphosphate (RuBP) carboxylase activity and CER does seem to exist for some crops. Efforts to breed out photorespiration have failed so far, i.e., there is no clear evidence that a C_3 plant can have low oxygenase activity with normal carboxylase activity (Gifford and Evans 1981). Considering the low success rate for all of these approaches, an alternative might be to breed directly for overall CER, rather than for any one specific photosynthetic component.

Still we can reach to the final conclusion which can express the correlation between yield and photosynthesis. We need more standardized system for measuring photosynthetic variables. More important, at the whole plant level, a lack of correlation between photosynthetic rates and economic yield may result from the chemical partitioning of carbon between starch and sucrose, as evidenced in the work of Marini and Barden (1981). According to them there is a linear correlation between net photosynthetic rates and specific leaf weight (a possible indicator of leaf starch content) of apple leaves. Since the relationship between CER and final yield is weak, this area of research does not appear to be over promising.

Harvest Index

There are two major determinants of high productivity in a crop:

1. The ability to produce high levels of photosynthates over a wide range of environmental conditions,

2. The efficiently transport and partition of high proportion of those assimilates into economically important organs.

Yield increases have resulted mainly from empirical selection for high harvest index. Therefore, increasing the harvest indices of crops seems to be the most effective means for future yield improvements.

If a crop plant are not able to allocate a major portion of the fixed carbon into yields, high photosynthetic rates would not translate into increased food production. Therefore, bioregulation for more efficient allocation of fixed carbon to harvestable parts may hold the key to assuring adequate world food supplies in the future.

Since, agricultural yield (the economic product) is not necessarily synonymous with biological yield (total biomass), the concept of an "efficiency" index was first proposed by E.S. Beaven in 1914 . In 1962, Donald proposed the more explicit term harvest index (HI), which is defined as the ratio of economic yield to total biomass production (HI = Economic Yield/Biological Yield).

Harvest index, by definition, has a value less than unity, and it ranges between 0 and 55% for most crops. Although HI indicates how the products of photosynthesis are partitioned among different plant parts, a positive correlation between economic yield and HI is not, by itself, unequivocal evidence of the value of the harvest index as a selection criterion (Donald and Hamblin, 1976; Charles-Edwards, 1982). Charles-Edwards (1982) argued that HI integrates phonological, physiological, and environmental factors and, unless each factor is well defined, HI does not provide much aid in understanding crop performance. Therefore, direct analysis of the main parameters of economic yield is a more realistic means of evaluating crop yield than is analysis of HI. This argument is consistent with previous findings that HI values of many crops are sensitive to factors such as crop density, water status, and nitrogen availability (Donald and Hamblin 1976). They showed, in cereal grains, that up to a point as the population density was increased, a greater grain yield per unit area was achieved, but HI declined before the maximum yield was attained. Likewise, under water stress conditions, the harvest indices of several grain species dropped (Poostchi *et al.* 1972). Nitrogen application usually increases biomass production but reduces HI for some crops (Donald and Hamblin 1976). On the other hand, McNeal *et al.* (1971) reported that the increase they noted in biomass production greatly exceeded the decline in HI, the net result being increased grain yield.

Of the two alternatives to further improve crop yield-increasing photosynthetic rates and/or increasing harvest index-the potential for eventual success seems to be much greater if HI is manipulated by modifying carbon partitioning and allocation patterns. Although some description of how assimilates are partitioned in plants is emerging, no single aspect of it is fully understood. Crop plants are highly integrated systems consisting of multiple sources and sinks. Presumably, the whole system is so integrated that it maintains a balance of growth. Although this balance determines the harvestable yield, its mechanism is poorly understood.

Partitioning of the Products of Photosynthesis and Regulation of Feedback

Most of the products of photosynthesis are exported out of chloroplast to the cytosol as triose phosphate in exchange of inorganic phosphate (P_i). Triose phosphate is the substrate for the synthesis of sucrose in the cytocol and for the formation of cellular components in the source leaf. Sucrose is exported to other parts (sinks) of the plant via the phloem. Some species export oligosaccharides or sugar-alcohols rather than sucrose. Partitioning of the products of the Calvin cycle within the cell is largely controlled by the concentration of P_i in the cytosol. If this concentration is high, then the rapid **exchange** for triose-phosphate allows export of most of the products of the Calvin cycle. If the concentration of P_i drop, then the exchange rate will decline, and the concentration of triose-phosphate in the chloroplast increases. Inside the chloroplast, the triose-phosphates are used for the synthesis of **starch** within the chloroplast. The partitioning, therefore, of the products of photosynthesis between export of the cytosol verses **storage** in the chlorplasts is largely determined by the availability in the cytosol. In intact plants the rate of photosynthesis is also reduced when the plant demand for carbohydrate (**reduced sink strength**) is decreased, for example by the removal of part of the fruits or girdling of the petiole. Partitioning within the whole plant involves production of carbohydrates in photosynthetic organs (**sources**), phloem loading, and subsequent translocation and unloading at regions of growth or storage (**sinks**).

When the rate of incorporation and /or export of the products of photosynthesis is low, the rate of photosynthesis may become restricted by **feedback inhibition**. Under such conditions, phosphorylated intermediates of the pathway leading to sucrose accumulation, inexorably decreasing the cytosolic Pi in the chloroplast, the formation of ATP is reduced and the activity of the Calvin cycle is decline. That is less intermediates are available and less RuBP is regenerated, so that the carboxylating activity of Ribisco and hence the rate of photosynthesis is dropped.

How important is feedback inhibition in plants whose sink has not been manipulated? To answer this question, the oxygen sensitivity of photosynthesis must be determined. The rate of CO_2 assimilation normally increases when the O_2 concentration is lowered from normal 21 per cent to 1 or 2 per cent because of the suppression of oxygenated reaction.

The Gap

Although, individual biochemical and physiological processes relating to photosynthesis and growth are well known but the factors primarily responsible for the control of carbon partitioning in plants still need more understanding. One has to understand the factors which control organ initiation and development (source and sink formation and potential size), Whether growth is sink or source (including possible sink-controlled photosynthesis) a detailed assessment of the role of storage in buffering developmental and environmental changes in sink and source activity. The knowledge is still needed on the role of hormonal and nutritional factors in regulating source and sink activity (organ interactions not directly associated with carbon transfer).

In unicellular green algae, each cell is capable of carbon fixation by photosynthesis in the light, the uptake of nutrients, growth in size and reproduction. In land plants a division of labour accompanies these processes between different cell types, tissues and organs, with an additional requirement for a suitable structural framework and an interconnecting transfer and transport system. The chlorophyll of the leaf (the site of CO_2

fixation) is spatially separated from the nutrient absorbing root system and the reproductive organs, yet each tissue and functional cell is part of a closely integrated whole plant system. Considerable variation has developed in these systems, that allows plants to grow and survive under a wide range of environmental conditions. This diversity of function and the control of partitioning of dry matter to different plant organs is the platform on which crop yield is based and in many agricultural plants improved yields have been associated, not with an increase in total production, but a greater partitioning of the available carbon to the organs being harvested (Gifford and Evans 1981).

SOURCE-SINK RELATIONSHIPS IN PLANTS

Sink Demand and Photosynthetic Capacity

Substantial evidence shows that alterations in the source: sink ratio influence the rate of photosynthesis by a leaf. Partial or complete removal of sinks generally leads to a reduction in net photosynthesis. Associated with this phenomenon is the hypothesis that the concentration of assimilates in leaves alters the net photosynthetic rates of those leaves; however, evidence for this is equivocal.

A photosynthetic rate response to altered sink activity is not necessarily a universal phenomenon. Certain conditions must be present for such feedback to be effective. Herold (1980) described some requirements for demonstrating when sink demand is communicated to the source. She proposed that, in order to demonstrate a positive correlation between photosynthetic rate and sink demand, three conditions should be fulfilled:

1. The activity of the sink components (other than manipulated sinks) should remain constant.
2. The direct, vascular link between the supplying organ and the receiving sink should be maintained.
3. A potential must exist and be maintained for enhanced photosynthetic rate (dependent on environmental conditions).

She further suggested that two kinds of sinks may be defined in studies of carbon partitioning: **primary** sinks, which are the cytoplasm of photosynthetic cells, and **secondary** sinks, which refer to distant, nonphotosynthetic cells such as roots or storage tissues. In this view, the chloroplast is considered to be the source; therefore, the ultimate barrier between source and sink is the chloroplast envelope, and changes in photosynthetic rate (given that environmental conditions are constant) must be initiated by some message that crosses the chloroplast membrane.

It appears that an increase in photosynthetic rate as sink demand is increased is not a typical response in short-term studies. However, since there is need for adjustment within the plant, this response is generally observed in longer-term experiments. Whether the adjustment is a direct result of sink demand or an incidental effect of associated changes in hormone synthesis or availability of crucial nutrients for the photosynthetic apparatus is difficult to establish (Herold 1980).

Local Transport

Studies of source-sink relationships have been largely empirical in nature and is mainly dependent on the consequences of change to either the supply or demand for photosynthate in plants. Phloem transport can occur over large distances, but a sink is generally supplied with photosynthate from a nearby source, with the association between specific sources and sinks continuously changing; as growth proceeds; new sources develop and the photosynthetic rate of older leaves declines (Wardlaw, 1968; Rawson and Hofstra, 1969). In jute, a typical distribution pattern has been studied by Palit (1985). His studies reveal that the [^{14}C]photosynthates (carbon source) is supplied form the top leaves to the apical bud and to the young growing leaves; the centrally placed leaves supply carbon to the stem and the basal leaves supply carbon to the lower stem and roots. Similar studies have been done by various workers (Khan and Sagar, 1967; Wardlaw, 1968; Kriedemann, 1970; Singh and Pandey, 1980) in a wide range of fruits species where carbon is received from nearby leaves.

Hierarchy of Sinks

In higher plants there is a hierarchy of sinks. The fruit and seed growth dominate the growth of vegetative tissues, although flowers in contrast to fruits appear to be poor competitors under source-limiting conditions. Shoot growth generally dominates that of the roots, but many underground storage organs have the same ability as fruits. This concept is supported by experiments in which the supply of photosynthate, or the size and number of growing organs were manipulated (Wardlaw, 1968; Starck, 1969, 1971).

Sucrose-Cleaving Enzymes and Carbon Partitioning

One of the crucial function of source leaves is the synthesis of energy-rich molecules for the transport of carbon, whereas heterotrophic sink organs such as developing fruits, seeds, roots and tubers are dependent on the import and utilization of these compounds. In most plant species, assimilated carbon is transported as sucrose, cleavage of sucrose initiates utilization and in plants this reaction is catalyzed by two enzymes with entirely different properties: invertase and sucrose synthase. Invertase is a hydrolase and cleaves sucrose into two monosaccharides. By contrast, sucrose synthase is a glycosyl transfrase, which in the presence of UDP, converts sucrose into UDP-glucose and fructose. Cleavage of sucrose by invertase in generally correlated with growth and cell expansion. Sucrose synthase is located in the cytoplasm. In general, cleavage of sucrose by sucrose synthase is associated with anabolic process in which UDP-glucose is the precursor of numerous compounds. It is also located in the sieve tube- companion cell complex of source leaves where it is involved in the catabolism of sucrose to generate ATP for phloem loading.

It had been suggested that sucrose transport from source leaves into sink organs are controlled by sink strength. Experiments to unravel the molecular nature of this force have identified the cleavage of sucrose by cell wall invertase at the site of phloem unloading as one of the key steps. By contrast, sucrose synthase is responsible for feeding assimilated carbon into sink metabolism. Studies on maize and fava bean (*Vicia faba*) seeds and carrot tap roots support this model.

It is evident that a turgor pressure gradient based on sucrose concentration is the driving force for assimilate transport from source leaves into sink organs. The sucrose concentration gradient depends on the rate of sucrose utilization in sink tissues. Thus, cleavage of sucrose by all sink-located sucrose cleaving enzymes might contribute to sink strength. In fava bean seeds, sink strength is generated by sucrose synthase activity rather than by cell wall invertase. The conversion of an osmotically active sugar into an insoluble polymer such as starch creates a strong sucrose sink, making cell wall invertase activity unnecessary. This hypothesis is further supported by developing potato tubers and tomato fruits which have little acid invertase activity but high sucrose synthase levels. A reduction in sucrose synthase activity in potato tubers by anti-sense technology leads to marked reduction in their starch content and to a marked reduction in tuber yield (Zrenner et al., 1995). During tomato fruit development, sucrose synthase but not acid invertase is positively correlated with fruit relative growth rate and with starch content in the pericarp tissue (Wang et al., 1993).

In seeds of maize and fava bean, expression of cell wall invertase is restricted to only a few cells at the transition zone, at the point where sucrose exits the phloem and enters the storage tissue. In fava bean and cells of *Chenopodium rubrum*, expression of cell wall invertase is coupled to the expression of a plasmamembrane-located hexose transporter, facilitating the entry of hexoses into the storage tissue (Ehness and Roitsch, 1997). Thus, in these examples, both cell wall invertase and a sugar transporter appear to contribute to sink strength and also to control assimilate partitioning by sink-located transfer and transport processes (Weber et al., 1997). After hexose import into the storage tissue, sucrose is partially resynthesized.

During maturation of tomato fruits, starch deposited during early fruit development is reconverted into sucrose and depending on the activity of an acid invertase in the vacuole, is stored mainly as sucrose or as a mixture of glucose and fructose. Antisense repression of the fruit vacuolar invertase converts hexose storing fruit into sucrose-storing fruit without affecting the allocation of assimilated carbon to the sink organs (Klann et al., 1996).

Post-harvested cold treatment of potato tubers leads to enhanced starch degradation and resynthesis of sucrose. Subsequently, a cold-inducible vacuolar invertase hydrolyzes the disaccharide to glucose and fructose (Cold Sweetening) Antisense repression of this cold –inducible invertase in transgenic tubers does not change the total quantity of soluble sugars that are released in response to low temperatures. However, it does alter the ratio of hexose to sucrose in favour of the disaccharide, confirming the role of vacuolar invertase in the control of sugar composition.

Improved Canopy Structure and Light Interception

Shoot density and leaf geometry improve photosynthetic input in a plant without a changing the photosynthetic potential of individual leaves. Trenbath and Angus (1975) showed the importance of leaf angle in improving plant productivity in cereal crops. They concluded that upright leaves have an advantage for yield improvement in rice. However Pendleton *et al.* (1968) obtained increased yields in near isogenic lines of maize with upright leaves and also demonstrated a rise in yield when leaf angle was increased by mechanical means. Although Shibles and Weber (1966) found that the dry matter production of soybean was a function of interception of solar radiation regardless of the planting pattern.

Atmospheric CO_2 Concentrations

Many ecological processes affect photosynthesis through their impact on plant demand for carbohydrate. The processes that increase carbohydrate demand cause an increase in photosynthetic rate, whereas factors that reduce carbohydrate demand reduce photosynthesis. Although defoliated plant in reduction of photosynthetic tissue, causes a compensatory increase in photosynthetic rate of remaining leaves.

Increasing the CO_2 concentration of the air is one way to increase net CO_2 exchange (NCE). It provides a possible experimental procedure to determine how plants may respond to increased leaf photosynthetic rates. Considerable work is being done in this area with the current concern over rising atmospheric CO_2 concentrations and the consequences of the *greenhouse effect*, However, these studies cannot be interpreted in terms of carbon partitioning.

Accumulation of carbohydrates has been observed in many studies on plant growth under CO_2-enrichment. This is because the photosynthetic rate exceeds the sink capacity to utilize the photosynthate for growth. Therefore, much attention has been paid to whether there is a causal relationship (s) between carbohydrate accumulation and the suppression of photosynthesis. One of the possible mechanisms for the suppression of photosynthesis is a feedback inhibition of carbohydrate synthesis by a high carbohydrate level. Stitt (1991) proposed the existence of another feed-back mechanism(s) by which the accumulation of carbohydrates directly or indirectly leads to a decrease in the amount of key components of the photosynthetic apparatus. In fact, many studies of long-term CO_2-enrichment have shown a decrease in Rubisco content (Osborne *et al.*, 1998; Sims *et al.*, 1999). The transcripts levels of several photosynthetic genes including rbcS decrease in the plants grown at elevated CO_2 levels (Moore *et al.*, 1998; Cheng *et al.*, 1998). However, increased carbohydrates do not directly inhibit the expression of such genes and it seems to be associated with the metabolism of hexoses derived from sucrose hydrolysis by acid invertase (Moore *et al.*, 1998). Although the phosphorylation of hexoses by hexokinase may involve the repression of photosynthetic gene expression, there are still many unknown aspects between the metabolism of hexoses and the repression of gene expression at elevated CO_2 (Moore *et al.*, 1999).

In many species the increase of starch content by CO_2-enrichment is greater than that of soluble sugars. There have been some reports describing that extreme enlargement of starch grains may lead to the physical damage of the chloroplasts (Pritchard *et al.*, 1997) but it is possible that starch accumulation hinders CO_2 diffusion in the chloroplast (Nakano *et al.*, 1999). The extent to which starch and soluble sugars accumulation responds to CO_2 enrichment greatly depends on species. For example, bean, cotton, soybean and clover preferentially accumulate starch whereas in wheat, sunflower and rice the absolute amount of soluble sugars accumulated is greater than that of starch. For some of these plants, the decrease in photosynthesis by CO_2-enrichment can be found to be more strongly correlated with starch accumulation. Difference in the down-regulation of photosynthesis among species might be closely related to a difference in the absolute amount of starch accumulating in leaves (Nakano *et al.*, 1998).

The down-regulation of photosynthesis by CO_2-enrichment is observed when the photosynthate exceeds the carbohydrate utilization for growth. However, there are some species showing little or no down-regulation of photosynthesis even when they are grown during long term CO_2-enrichment and have excess carbohydrates. For examples, potato and radish do not show any down-regulation of photosynthesis whereas they genetically belong to starch-accumulating species. These species have tubers or roots which can act as a large sink for carbohydrates. In radish, the biomass of the storage roots is markedly enhanced under CO_2-enrichment and consequently no over-accumulation of carbohydrates is observed in leaves (Usuda and Shimogawara, 1998). Similarly in rice, the leaf sheaths can act as a temporary sink for carbohydrates and the absolute amounts of carbohydrates in the leaf blades are considerably small. Therefore, the down-regulation of photosynthesis in rice may be small compared with that in bean, cotton or soybean. These species accumulate a great deal of carbohydrates in the leaves especially as the starch in chloroplasts.

There is a decrease in leaf N content by CO_2-enrichment and is commonly found in many species. Makino *et al.*, (1999) reported that the decrease in leaf N content is the result of a change in N allocation at the morphogenic level of the whole plant. During long-term growth under conditions of elevated CO_2, plants reallocate N away leaf blades to leaf sheaths and roots.

Differences in the long-term response to CO_2-enrichment is also related to differences in the sink-source status of the whole pant depending on the developmental stages. CO_2 enrichment leads to a greater stimulation of biomass production in young seedlings than in matured plants such as soybean, cotton, alfalfa, tobacco and rice. Similarly, CO_2-enrichment promotes the development of tillers especially during early stage of growth. In tobacco carbohydrates levels are lower in young seedlings than in older plants (Geiger *et al.*, 1998). The growth during the seedling stage is source-limited and the carbohydrates increased by CO_2-enrichment can be efficiently utilized for additional sink such as the development of new tillers or secondary shoots.

Elevated CO_2 levels accelerate development of the whole plant such as radish and tobacco. On the other hand, the development of rate of wheat and rice is net affected by CO_2 enrichment (Makino *et al.*, 1997; 1999). Miller *et al.*, (1997) reported that the down-regulation of photosynthesis by CO_2-enrichment in tobacco appears to be the result of a shift in the timing of the natural ontogenetic decline of photosynthesis to an earlier onset associated with leaf senescence. An acceleration of leaf senescence has been observed in other species such as wheat, sunflower and rice. In rice, although leaf senescence is accelerated by CO_2-enrichment, there is no shift in the timing of the maximum photosynthetic stage of leaf ontogeny (Makino *et al.*, 1999). Similar results can also be found in wheat. Thus, different responses of photosynthesis to CO_2 observed at the level of a single leaf may result from different growth strategies of the whole plant under conditions of CO_2-enrichment (Makino *et al.*, 1999).

The Timing of Development

Cereals

In the cereals (Evans and Wardlaw, 1976; Wardlaw, 1987), as in other species (Brouwer, 1983), there is a short period following germination when the root is the major sink for seed reserves, but this dominance is lost as the shoot is exposed to light and the leaves become the major source of carbon through photosynthesis. However with time this main shoot dominance is lost and the laterals (tillers) commence to grow rapidly, firstly as parasites for carbon and mineral nutrients from the main shoot and subsequently as independent units. The rate of tillering is known to be under genetic and environmental controls and involves a complex balance between mineral nutrition, the supply of carbon and the concentration of specific growth regulators (Harrison and Kaufman, 1980; Michael, 1980; Richards, Mueller and Mott, 1988).

A major change in development is the switch from leaf production at the shoot apex in cereals to the formation of floral organs. This switch is also followed by a marked increase in stem growth and there is an overlap in the growth of the remaining young leaves, the stem and the ear. This is a critical stage in development of cereals as it covers spikelet and floret development and therefore the potential of seed production.

Stem growth may be dependent on the production of gibberellins by the developing ear and in this way stem and ear growth may be both competitive for a limited supply of carbon, but still show integrated growth (Stoy and Hagberg, 1967; Michael, 1980). In the dwarf wheat containing the *Rht* gene for GA insensitivity the improved harvest index (grain weight: above-ground dry matter) could possibly result from a modified association between ear and stem growth (Brooking and Kirby, 1981). Stems and leaves of cereals have a distinct pattern of growth, with the stem below the point of insertion of the leaf only growing after the leaf is fully expanded (Percival, 1921). Following anthesis grain growth is dominant and there is generally little production of new roots or tillers.

Woody Species

In woody species the seasonal nature of development also allows periods of little as well as intense overlap between growing organs and tissues. The pattern of growth and development is derived mainly from an analysis of conifers in the higher latitudes of the northern hemisphere (Kienholz, 1934; Krueger and Trappe, 1967; Dickmann and Kozlowski, 1968), but the general sequence is applicable to many species, including deciduous forest trees (Derr and Evert, 1967; Donnelly, 1974) and fruit trees such as apple (Mochizuki, 1962; Priestley, 1970; Hansen, 1977).

Maximum root growth occurs in the spring and autumn when the growth of other organs is less pronounced. Priestley (1964) has stressed the importance of autumn foliage for root growth and carbohydrate storage in apple. The late stages of flower development and pollination may also occur in the less competitive early parts of the growing season, but subsequent fruit development overlaps that of the elongating shoot and the formation of woody tissue. Heavy fruiting, which reduces shoot growth, will also reduce the formation of new flowers and result in poor fruit production in the following year (biennial bearing).

THE RESPONSE TO STRESS

Carbon partitioning is sensitive to environmental changes that may differentially influence the relation between growing organs and between growth and photosynthesis.

For example low nitrogen nutrition results in a much greater partitioning of dry matter to the roots of soybean (Rufty, Raper and Huber, 1984; Tolley-Henry and Raper, 1986), thus changing the recognized pattern away from shoot dominance. Similarly, in wheat grown under low phosphorus conditions the grain is less dependent on the supply of photosynthate from the flag leaf (Batten and Wardlaw, 1987). Low nutrition often results in the accumulation of reserve carbohydrates in the stems and leaves, which indicates that growth is retarded more than photosynthesis. However a further complication may be that carbohydrate accumulation under low nutrient conditions may in turn result in reduced photosynthetic rates (Selga *et al.*, 1983).

Low temperature

Low temperature can also result in a change in the partitioning of dry matter between growing organs (Wardlaw, 1968) and also an accumulation of starch and soluble carbohydrates (Wardlaw, 1985; Farrar, 1988). In peanut (Bagnall, King and Farquhar, 1988), reduction in leaf photosynthesis with a fall in temperature (30 to 20°C) is related to a rise in storage of photosynthate resulting from slow growth and a low demand for photosynthate at the lower temperature, rather than a direct effect of temperature on leaf function.

The accumulation of sucrose is a common response to low temperature in many species. For example, cold-induced sweetening in potato tubers is a well-known phenomenon of major economic importance. Higher sucrose concentrations lower the freezing point of the cells content, helping to prevent ice crystal formation at moderate sub-zero temperature. In addition to a cryoprotective function, sucrose reverses allow the cell's metabolism to recover quickly when more favorable conditions return.

Some C_3 plants but very few C_4 plants, develop freezing tolerance after previous exposure to low but not freezing, temperatures. This accumulation distinguishes winter from spring varieties of crop species. The activities of the cytosolic F-1, 6-BPase and SPS increased in winter varieties but not spring varieties of wheat, rape and rye, when plants are exposed to chilling temperatures. These findings suggest that they are part of the acclimation response, allowing continued photosynthesis and growth in the cold. SPS activity also increases in potato tubers in the cold and is brought about by an increase in the activation state of SPS and by the appearance of new form of the enzyme with a slightly higher molecular mass (Deitging et al., 1998). Monroy et al. (1998) found that protein phosphatase 2A activity was almost completely inhibited when alfalfa cells were exposed to low temperature. This suggests that the cold-induced activation of SPS is unlikely to occur by a type 2A protein phosphatase and therefore differs from the activation of SPS by light.

Drought Conditions

Under drought conditions the level of water-soluble carbohydrates in the leaves of species such as sunflower (Bunce, 1982) and cotton (Timpa *et al.*, 1986) is increased. This accumulation in the leaves of plants under **water stress** also probably results from a reduction in growth and a slowing in carbohydrate utilization prior to a reduction in photosynthesis (Munns, 1988). However the response to drought is dependent on the stage of plant development and in cereals the developing grains continue to grow under severe water stress, with the result that there is a relatively greater use of both the limited supply of current photosynthate and stored carbohydrates.

The imposition of water or salt stress led to sucrose accumulation in the leaves of several C_3 plants. A stress-induced change in photosynthetic partitioning in the favour of sucrose, was observed in several plants and was correlated with an increase in SPS activity and the rate of sucrose synthesis. In contrast, water stress decreased the activity of SPS and the rate of sucrose synthesis in *Phaseolus vulgaris* but decreased the rate of starch synthesis even more, so a change in partitioing in favour of sucrose was also seen in this species. Other factors, such as decreased export or turnover of sucrose, might also contribute to sucrose accumulation. In resurrection plant, *Craterostigma plantagineum*, which can withstand extreme dissication, dehydration induces the transcription of one of the two SPS genes, leading to increased levels of SPS protein and activity (Ingram et al., 1997).

In contrast with most C_3 plants, the sucrose contents of maize and sugarcane leaves rose very little or even decreased when plants were subjected to water stress (Foyer et al., 1998) At the same time there was an increase in hexoses associated with higher activity of soluble acid invertase whereas the activities of the cytosolic F-1, 6-BPase in sugarcane and of SPS in both species decreased. The increases in free hexose levels are too low to make a significance contribution to osmotic adjustment within the cell but could be important in signaling as part of the adaptive response to the plants.

Control at the Source

Types of sources

Leaf photosynthesis is generally the main source of carbon for growth and storage in other parts of the plant. However since other chlorophyll containing organs such as stems, inflorescence parts and fruits also contribute significantly to the overall supply of photosynthate and thus influence the pattern of carbon partitioning. The formation of new leaves and their longevity is a major factor in the supply of photosynthate for growth that can influence the source-sink relationships.

Regulation in the Leaf

Availability of carbon for translocation

Carbon incorporated into the leaf during photosynthesis faces a range of alternative metabolic pathways, transfer through several organelle and cell membranes, temporary storage as starch, oligosaccharides and sugars, and possible recycling back to the atmosphere following respiration, all of which are factors either directly or indirectly linked to the availability of carbon for translocation out of the leaf via the phloem.

Spatial and biochemical partitioning

C_3 species

In C_3 species the first step in carbon fixation in the light is the carboxylation of ribulose 1,5-bisphosphate (RuBP) to from two molecules of P-glycerate. However, ribulose bisphosphate carboxylase/oxygenase (Rubsico) the enzyme responsible for carbon fixation also converts RuBP to P-glycolate in the presence of oxygen, resulting in the photorespiratory pathway and loss of CO_2. Net CO_2 exchange by the leaf is therefore largely a balance between the carboxylase and oxygenase (photorespiration) activity. P-glycerate in turn forms triose-P which may continue as part of the photosynthetic carbon reduction cycle (PCR cycle) to produce further substrate (RuBP) for CO_2 fixation, or be diverted into starch in the chloroplast, or transferred from the chloroplast into the cytosol where under normal conditions a large part of the carbon would be incorporated into sucrose.

In many plants, carbon is translocated through the phloem of the vascular system from the leaf to other parts of the plant, however, sucrose and other sugars are stored in vacuoles of mesophyll cells. Based on this pattern it should be evident that there are many processes that may be involved in regulating the movement of carbon from the chloroplast to the vascular tissue. The enzyme, sucrose phosphate synthase (SPS) is also involved in the synthesis of sucrose from uridine 5′-diphosphate glucose (UDPG) and fructose 6-phosphate (F6P). It also regulate F6P formation from fructose 1,6-bisphosphate (F-1, 6-P_2) by fructose 2,6-bisphosphate (F-2, 6-P_2). The role of the latter may be important in the regulation of photosynthesis when the demand for carbon in growth is low. Slowing the transfer of triose P out through the chloroplast membrane to the cytosol and hence lowering the reverse exchange of P_1 into the chloroplast. This event is expected to favour starch synthesis in the chloroplast.

C_4 species

In C_4 species, where the first products of carbon fixation in the mesophyll are the C_4 acids, the system of carbon transfer is perhaps even more complex than in C_3 species (Hatch, 1987). In **maize**, for example malate is the main C_4 acid transferred from the mesophyll to the bundle sheath cells where it is decarboxylated and reassimilated in the PCR cycle. The carbon from P-glycerate may continue in the PCR cycle to regenerate RuBP, or be siphoned off into starch. However, part of the P-glycerate returns to the mesophyll cell where it is converted to triose phosphate which in turn is either converted to sucrose or returned to the bundle sheath cells to re-enter the PCR cycle in the chloroplasts (Fekete and Vieweg, 1984). Several processes could operate to regulate photosynthesis and carbon transfer in this system. The photorespiratory pathway is not active in C_4 species, but other controls could be similar for both C_3 and C_4 plants. However, in addition it is known that the C_4 acids control the activity of PEP carboxylase and malate accumulation could be expected to reduce photosynthesis in maize (Hatch, 1978).

Vein Loading and Vascular Limitations

The path of transfer

In sugar beet, *Amaranthus, Vicia* and pea the movement of sugars from the mesophyll to vascular tissue in the leaves is thought to involve an apoplast step prior to loading into the companion cell-sieve element complex, because of selective transfer of leaf metabolites to the sieve elements, the low number of plasmodesmatal connections between the companion cells or sieve tubes and adjacent parenchyma cells, the high osmotic potential of the sieve tube-companion cell complex in comparison with either the mesophyll or adjacent phloem parenchyma cells (i.e. maintenance of a concentration gradient) is energy dependent. This apoplast step could provide another important control in the export of carbon from the leaf, governed by the rate of uptake of sugars across the companion cell-sieve tube membrane and the area of membrane available for uptake. In cucumber and maize the argument for an apoplast step in the transfer of photosynthate from the mesophyll to the veins is less convincing. Concentration gradients between the photosynthetic tissues and the vascular system are smaller, discrimination between metabolites transferred is not so apparent, there are numerous plasmodesmatal connections between the conducting elements and adjacent cells and free space exudates collected from $^{14}CO_2$ fed leaves are low in ^{14}C-labelled photosynthate.

In C_4 species, like maize the path of movement of carbon is further complicated by rather complex pattern of transfer of carbon between the mesophyll and the bundle sheath cells. The enzymes necessary for sucrose synthesis occur mainly in the mesophyll and not in the bundle sheath cells as a result the carbon fixed by the mesophyll must be transferred first to the bundle sheath cells for incorporation into 3-PGA and then returned to the mesophyll for reduction to triose phosphate and conversion to sucrose before being transferred back through the bundle sheath cells to the vascular tissue. This final transfer is not fully understood, but presumably must involve movement through the plasmodesmatal connections between the mesophyll and bundle sheat, because of the suberized wall between the two cell types. These additional transfer steps would appear to negate, at least in part. It claims that C_4 species have an advantage over C_3 species in relation to photosynthate export because of the shorter distance of transfer and fewer cells between the mesophyll and the vascular system.

Uptake by the transport tissue

1. The minor veins of sugar beet are 13 times as extensive as the major veins and a 10 μm length of minor vein services approximately 8 or 9 mesophyll cells; the minor veins are considered to play a major part in the initial uptake of sugars from the mesophyll (Geiger and Cataldo, 1969). The sieve elements of these fine veins are very narrow and in both sugar beet and sunflower. There is evidence that the initial site of uptake and short distance transfer to a better developed vascular pathway is by the relatively large intermediary cells (a special form of companion, or transfer cell?).

2. In Gramineae, the smallest veins in the grasses are the transverse veins croos-linking the larger parallel veins running the length of the leaf. These cross veins could act as a direct loading point for photosynthate from the mesophyll, a transfer that in rice and wheat may involve an apoplast step. Therefore, there also appears to be a division of labour with the smaller veins acting as a collecting system for photosynthate, which is then transferred to the larger veins for export.

Vein patterns

It has been suggested that both venation patterns and the dimensions of the vascular system (sieve-tube cross-sectional area-leaf area) may be important in providing a transport system capable of matching the potential photosynthetic rates of a leaf. There is considerable variation in the interveinal distance of leaves between grass species ranging from around 90 μm in the C_4 species maize and sorghum to more than 300 μm in some C_3 species. C_4 species have closely spaced veins, a high rate of assimilate transfer and a high rate of net CO_2 exchange.

Response to carbon input

Detailed analyses of carbon assimilation and export by leaves have resulted in models that include spatially and metabolically separated carbon pools both in series (direct export path) and in parallel (storage) which vary in size and exchange capacity. The complexity of this system is such that it is often impossible to establish precise relations between individual factors and the rate of transfer of carbon into the vascular system, although this has been attempted in relation to the rates of net CO_2 exchange and amounts of sugar and starch in the leaf.

Photosynthetic rate: The export of carbon from leaves has been positively correlated with the rate of day time photosynthesis, but seldom with the amount of sucrose in the leaf, even though the latter is the main form of carbon exported. There is a strong positive relation between photosynthesis and carbon export in tomato, above a threshold rate of 2 mg C dm^{-2} h^{-1}. Similar correlations between photosynthesis and the export of carbon have been reported for a range of species including cotton, sugar beet, *Sorghum* and *Lolium*. However the correlations within a species do not necessarily hold between species, where there can be big differences in the proportion of carbon that is stored in the leaf or translocated.

In leaves, high light immediately follows darkness and there may be an initial period during which translocation lags behind the rapid increase in photosynthesis. During the day there can be changes in storage and translocation within the leaf without changes in net CO_2 exchange. The failure to correlate sucrose concentrations in photosynthesizing leaves with the rate of carbon export is not difficult to understand if the pool of sucrose associated directly with transport is small, mobile and continuously balanced by storage in the vacuole. Thus any correlation between carbon export and sucrose in a leaf during photosynthesis might be expected only at the very early stage of carbon fixation following a dark period, or under low light conditions. There are however some indirect evidences supporting the importance of sucrose synthesis in transport with a correlation between sucrose export in the light and the sucrose phosphate synthase activity in the leaves of a number of species, including soybean and cotton.

Translocation of carbon from a leaf is frequently determined from the loss of [^{14}C] photosynthate following pulse labeling with $^{14}CO_2$. Although this method has been used with apparent success (Hofstra and Nelson, 1969; Lush and Evans, 1974) under certain conditions caution is needed in the interpretation of these data as the rate of loss of ^{14}C from the leaf depends on pool size and respiratory losses as well as the rate of carbon export through the phloem in *Lolium* (C_3) and *Sorghum* (C_4) species.

The mobilization of leaf storage carbon: In pepper and spinach, with a fall in light intensity towards the end of a day and a lowering of photosynthesis there is a point of time when the supply of current photosynthate is inadequate to maintain translocation, and carbon reserves built up during the day may be mobilized and exported. Grange (1985), Servaites *et al.* (1989), and Ho (1978) postulated a base rate of export of carbon in the tomato leaf, which if not maintained by current photosynthesis will require the mobilization of leaf reserves. If starch reserves are still available in the morning, as light gradually increases, starch can continue to be mobilized until photosynthesis is adequate to maintain translocation.

With a sinusoidal change in light during the day, net carbon export can therefore remain steady despite changes in photosynthesis (Servaites *et al.*, 1989). There are differing opinions on the amount of turnover of starch in illuminated leaves; according to Porter, Martin and Bird (1959) there is simultaneous synthesis and degradation of starch in the leaves of tobacco under low light conditions. Similar results have been obtained by in isolated chloroplasts from spinach (Stitt and Heldt, 1981). However, in pea and pepper variance with this view has been observed.

When plants are transferred abruptly from light to dark it has been shown for a range of species that there is an initial export of sucrose prior to the mobilization of starch. In spinach leaves, with a gradual decrease in light towards the end of the day there is an export of accumulated sucrose prior to the mobilization of starch has been observed. The rate of export of carbon during the night is much lower than during the day and may or may not be related to sucrose concentration in the leaf. The maximum export of carbon during the night can also be related to the rate of starch mobilization and sucrose synthesis in a leaf in soybean (Mullen and Koller, 1988).

The response to daylength: Some continuity in the supply of carbon from the leaf to growing organs over a diurnal cycle is maintained by the build up and mobilization of reserves. However this may be more than just a simple response to a changing photosynthetic input, as it has been shown that starch accumulation in the leaf of a range of species can be increased up to five fold when they are shifted from long to short days. The mobilization of these reserves during darkness may be slower under short days. Soybean is one of the species in which starch storage in the leaf is inversely related to daylength but this response is less evident during seed filling when the demand for photosynthate is high.

The duration of the high light (photosynthetic) period, rather than the photoperiod *per se*, regulates the rate of starch synthesis in the leaf of soybean. Under long days, in comparison to short days, the lower rate of starch accumulation in the leaf of soybean was found to be associated with a greater sucrose phosphate synthase activity, higher leaf sucrose and a higher rate of photosynthate export. In the leaf of cucumber, sucrose phosphate synthase has also been implicated in the relation between daylength and starch storage, the total radiation does not appear to be the only factor influencing the partitioning of photosynthate in this leaf.

In *Digitaria decumbens*, using low intensity white light and far-red irradiation for daylength extension and night breaks, there is at least a partial photoperiod control of assimilate partitioning into starch. These endogenous rhythms found to differ between sugars (approx. 26 h) and starch (approx. 23 h) may also be a factor in the photoperiodic regulation of starch and sugar metabolism in leaves.

The response to demand: Individual leaves do not necessarily respond to a change in the demand for photosynthetic assimilates resulting from changes in sink capacity or changes in other photosynthetic inputs, as storage outside the leaf and the development of alternative sinks may moderate the response. However there are many reports where a change in assimilate demand has resulted in either a change in the partitioning of carbon between storage and translocation in the leaf, or a change in the rate of photosynthesis, or both. Although artificially induced changes in the demand for photosynthate can clearly influence the rate of leaf photosynthesis. There has been some doubt concerning the operation of this control under natural conditions.

The defoliation of maize increased the net CO_2 exchange (NCE) of the residual source leaf in genotypes with a large source-sink ratio (i.e. when the initial demand was relatively low), while ear removal decreased NCE only in genotypes with a small source-sink ratio. In rice, a high NCE in cultivars with a low source-sink ratio and a low NCE in cultivars with a high source-sink ratio and concluded that it should be possible to improve the yield of many rice cultivars by improving the sink. These two examples refer to reproductive growth and probably the strongest case for sink limited photosynthesis and provides the relation between high photosynthetic rates and fruiting (for example in apples, pea, peach, pepper and soybean).

In potatoes and tuberous roots, a similar response is also associated with other rapidly growing storage organs. In soybean, rather anomalous results have been obtained when pod set and the growth rate were increased by an early thinning treatment and this was accompanied by an increase in the photosynthetic rate of the young leaves and no change in the mature leaves. However decreasing the demand by pod removal did not result in a decrease in photosynthesis.

Based on a simulation model for the vegetative growth of tobacco the photosynthetic rate of tobacco leaves was commonly suppressed due to a shortage of demand. The interpretation of these source-sink interactions is often clouded by the possibility that changes in demand may influence photosynthesis not only directly, but indirectly through factors other than the accumulation of photosynthate, possibly hormonal. Mechanisms are available to explain the end product inhibition of leaf photosynthesis based on the accumulation of sucrose in the mesophyll cytosol.

Genetic variation in net CO_2 exchange

Although yield usually is limited by photosynthetic rates under field conditions, genetic improvements in photosynthetic potential have not been achieved to date.

The rates of photosynthesis relate directly to the photosynthetic potential of a leaf, or this is governed by other factors in the plant like genetic variation in photosynthetic rates, dark respiration in lines or related species. These factors have been directly concerned with the consequences of the differences in relation to the partitioning of carbon.

In wheat, it appears that with increased ploidy and the change from primitive to modern lines there has been a decrease in the rate of photosynthesis per unit leaf area, although there has been a compensation with larger and thinner leaves in the modern lines. Although, plant form has changed, total biomass has not greatly altered and improved yields are associated with a greater harvest index. In a study with 6 cultivars and 120 progeny of wheat, no relation between the peak photosynthetic rate of leaves at different positions on the culm (third and flag leaf blades) and no correlation between photosynthetic rate and yield. In this study it was concluded that the major determinants of photosynthetic inputs were leaf area and maintenance of photosynthesis (rate of senescence).

In some of the legumes a better relationship has been observed between leaf net CO_2 exchange and yield. The high yielding varieties of soybean tended to have high photosynthetic rates and the seed yield and canopy photosynthesis are significantly related. However, in 36 cultivars of soybean, where photosynthesis ranged from 12 to 24 mg CO_2 dm^{-2} h^{-1}, there is no correlation between photosynthesis and yield. In 20 cultivars of black gram, a positive correlation has been observed between net CO_2 exchange after anthesis and total dry matter, pod yield and harvest index. Thus there is correlative evidence for an improved biomass and an improved harvest index in response to increase in photosynthesis.

The Utilization of Photosynthate: Sink Characteristics and Limitations

An important factor in the control of carbon partitioning, and one that ensures balanced development, is the *timing of organ initiation and growth.* Once a potential sink is established, for example following floral initiation, or the formation of an axillary bud, the success or competitive ability of that organ is dependent on the development of an adequate vascular link for the supply of carbon and possibly more importantly the characteristics of the growing organ itself. There are also less direct controls such as the amount of mineral nutrient that is delivered to the growing tissue through the phloem and xylem and the often quoted, but not understood, controls imposed by growth regulators.

There appear to be a number of features within a growing organ that give a competitive advantage in terms of controlling carbon partitioning:

1. The size of the surface (membrane) area across which metabolites are transferred from the vascular system to the zone of utilization (unloading area).
2. The efficiency of transfer of carbon (rate per unit area) from the vascular system to the sink.
3. The spatial or biochemical isolation of assimilates in growing (or storage) organs once they leave the vascular system.

The first attribute would make size a major factor in sink dominance. However, it has been suggested that organs with a high relative growth rate (RGR), rather than size *per se*, will be competitive for a limited supply of photosynthate. This suggests that the efficiency of transfer of photosynthate and the spatial and/or biochemical isolation of photosynthate in sinks must also be important in sink dominance. The underlying basis of sink strength (assuming a pressure flow mechanism of translocation) is an ability to effectively lower the concentration of photosynthate in the sieve elements servicing the sinks and thus establish a favourable concentration gradient between the sink and the source.

Vegetative Tissues

Transport and transfer limitations

The growth of the vegetative shoot apex is dependent on the movement of photosynthate over considerable distances, presumably through the still undifferentiated procambial strands linked to the mature vascular system below. In young leaf primordial of *Coleus* the sieve elements are not differentiated until the leaves are longer than 0.4 mm and it is not until the leaf reaches a length of 1.4 mm that the sieve tubes are closer to the leaf tip than 142 μm.

The young leaf primordial may be limited by the supply of photosynthate. The first differentiated sieve tubes in the protophloem of a developing leaf may be formed in the absence of recognizable companion cells and appear to have plasmodesmatal connections to the surrounding cells, which is a cellular organization that may favour vein unloading and the supply of photosynthate to the apex via the symplast.

In dicotyledonous leaves, which are also capable of their own photosynthesis as they expand, the maximum import of photosynthate occurs at about 15-20% of full size. However this will vary with conditions and in the presence of a pathogen such as *Uromyces fabae*, which acts as an additional sink for photosynthate, import continues for a much longer period.

As growth proceeds the terminal part of the leaf matures first and the apparent anomaly can arise where photosynthate is being imported into the growing base of the leaf and simultaneously exported from the apex in the leaf of **eastern cottonwood**. Such apparent bi-directional movement of photosynthate through the petiole of a developing leaf is likely to result from movement in different bundles in the petiole and not within a single sieve element. In eastern cottonwood differentiation of the photosynthetic tissue and stomata proceeds basipetally in the leaf lamina, while simultaneous differentiation of the major vein network proceeds acropoetally. Maturation of the minor veins (generally associated with vein loading) proceeds basipetally and laterally in line with their export role.

The mature leaves are not considered to be parasitic on the rest of the plant and will not import carbon even under very low light conditions at the base of the canopy, they can be induced to become minor sinks under extreme conditions This transition occurs without any apparent anatomical changes. In mature leaves of *Gomphrena globosa* induced to import photosynthate that there is an activation of soluble invertase in the leaf. In soybean the symplast is the major supply route of photosynthate entering rapidly expanding stem tissue.

In the leaves and stems of grasses growth occurs at an intercalary meristem; the growing tissue is enclosed by leaf sheaths and unlike many dicotyledonous plants growth and the source of photosynthate are spatially separated. Actively growing tissue therefore occurs between regions of mature tissue and the question is not only one of photosynthate supply to the region of growth, but also transport through the growing region. Although the rate of differentiation of sieve elements may be sufficient to maintain phloem continuity across the intercalary growth region of the stem or lamina, the tenuous nature of this connection would be expected to favour retention of assimilates in this tissue.

Roots are of special interest as they often appear to be the poor relations when it comes to the allocation of a limited supply of photosynthate. In millet, the number of both phloem pole and pericycle cells is proportional to the diameter of the stele at the base of the primary root, an indication that the vascular capacity may match root size. Therefore, the growth of roots is not limited by long-distance phloem transport and the rate of translocation and growth of part of the root system can be increased experimentally if the internal competition between roots is reduced. These responses are probably the basis of the adaptation to drought where part of the root system is inhibited by lack of water, but the remainder will grow to a greater depth in the moist soil. The root associate such as mycorrhiza, nematodes and nodulated legume roots capable of nitrogen fixation all stimulate the transport of carbon through the root system, apparently by creating a greater demand for photosynthate. If transport processes are involved in limiting the growth of roots this may involve transfer from the young developing vascular tissue to the actively expanding and meristematic tissue at the root apex rather than long distance transport. The transfer of photosynthate in the root from the phloem to the pith or cortex behind the apex may follow a symplastic path.

In roots sucrose does not undergo hydrolysis during unloading, although for example in corn there is an effective cell wall invertase in the apoplast. The earlier autoradiographic studies indicate that there could be an apoplast step in the movement of carbon from the end of the vascular channels in the root to the root apex a symplastic transfer of photosynthate from the vascular system to the expanding apical tissues of both root and shoot appears to be the favoured mechanism.

Metabolic limitations

The rate of growth is positively correlated with acid invertase activity. There is a good relation between the activity of acid invertase and the growth of sugar cane, in growing. and stem tissue. Acid invertase is an essential enzyme in the metabolism of sucrose for use in both growth and storage, however, it is difficult to confirm that the activity of this enzyme is more than just an expression of growth, and the rate of cell elongation can also be correlated with other enzyme systems such as β-glucosidase. The activity of sucrose synthase may be more important than invertase as a measure of sink strength in the roots of several species and the activity of this enzyme has also been shown to parallel the rate of leaf expansion in maize.

Turgor and cell wall relaxation

Turgor and cell wall relaxation play important role in controlling growth. Cosgrove (1986) stated that *'cell growth starts with a reduction in wall stress because of irreversible yielding of the wall'* and that cell turgor is the link that coordinates water uptake with cell wall yielding. Thus the concept has emerged of a turgor threshold that is essential for growth, with a major control by cell wall extensibility. An effect on cell wall extensibility may explain in light of the stimulation of stem elongation by gibberellins and auxins. We still need the proper identification for this kind of control and the difference in growth between or within species.

Physical constraints

The possible role of physical constraints in regulating growth is also important. This has been suggested as a factor in the growth of the rice kernel and in the expansion of tillers in wheat. The possible mechanical control of leaf expansion by the epidermis. It is not difficult to implicate a physical limitation on growth as an important control in the pattern of plant development, but in a finely tuned system this concept needs further experimental support.

Fruit growth

The dominance of fruit and seed growth over vegetative organs has been recognized for a long time. Although many of the basic processes involved in the growth of fruits and seeds have been identified, the basis of this dominance still needs clarification. There is a definite contrast between kernel of cereals and the seed of legumes. In kernel of cereals the growth consists largely of storage in an endosperm tissue where cell development is complete by the time. The kernels are about a third of their maximum dry weight and only the outer layers of cells of the endosperm (the aleurone layers) are still viable at maturity. In legumes seeds, storage largely occurs in the cotyledons of the embryo with cells that remain viable throughout development.

Seed composition also varies between cereals and legumes and the cost of carbon in producing 1 g of seed is inherently greater in legumes, which have a high protein content, than in the cereals where starch is the main form of storage.

From Flower to Fruit

In many species the potential for seed production is very high, but this potential is not realized because of premature flower drop, the failure of fertilization or early abortion of the young developing seed. Experiments in which light, leaf area or the number of developing fruits are varied indicate that the supply of photosynthate may be critical at these stages. The yield of dry bean is correlated with the rate of photosynthesis at the time of pod set for eight out of nine cultivars tested. However the actual requirement for photosynthate by flowers and small fruits at this critical stage is relatively low.

This apparent anomaly, where a threshold level of carbon is required to stimulate both fertilization and the initial stages of pod development. Once a pod is rapidly growing it will, according to the model, lower the availability of carbon to a critical level in parts of the shoot and prevent at least some of the remaining flowers from developing fruits. This interaction can also involve vegetative tissue. In French beans the removal of the apical and axillary vegetative shoots at flowering can increase seed number. It may be due to the competition for photosynthate. Binnie and Clifford (1980) showed that mild defoliation, which reduced the supply of photosynthate, did not result in a parallel reduction in seed number.

There is the possibility of less direct effects of photosynthate supply on flower and fruit development. For example through a change in the movement of nutrients or hormones to the shoot from the roots, the roots being the organs most likely to respond to source limitation or a form of apical dominance where a more advanced fruit could regulate the growth and development of later or less robust fruits. This is a critical stage of development in many species and a knowledge of the controls that operate over this period could provide the key to harvest index and yield.

Vascular Constraints and Temporary Storage

Vascular links

The role of the vascular system in regulating the partitioning of carbon in plants is still unclear, although there are evidences to suggest that the carrying capacity of the phloem is not a major factor in determining yield. The vascular links do appear to govern specific source-sink relationships. Also temporary storage of photosynthate along the path of movement acts to moderate the effect of a change in the activity of either a source or a sink.

Retention of photosynthate and lateral transfer from the phloem

During long-distance phloem transport between a source and sink the loss of assimilates from the sieve element-companion cell complex is small compared with the amount transported, a characteristic that is central to the role of the vascular system in carbon partitioning.

In *Vicia faba* and *Phaseolus vulgaris* the phloem unloading in the stem operates by passive leakage with the net transfer depending on the balance between membrane permeability and active uptake. In *Ricinus communis* the uptake of sugars by the phloem involve a sugar-proton co-transport system. Based on structure and the frequency of plasmodesmatal connections, the transfer of photosynthate from the sieve tube-companion cell complex to other tissues in the stem of *Phaseolus vulgaris* could be either by the symplast or the apoplast, while in sugar cane the apoplast is clearly involved. In dicotyledonous plants (e.g., willow) it has been shown that tangential movement of photosynthate can occur around the stem if the sink demand in a longitudinal direction is low. In monocotyledonous plants such as wheat, the vascular bundles of the stem running between the nodes are clearly isolated and there are no cross-connecting veins, there appears to be very little tangential movement. However in both dicotyledons and monocotyledons the exchange of carbon between stem traces and leaf traces entering the stem is largely governed by linkages in the nodal regions.

Proximity of source-sink and vascular connections

The studies on between source and sink have led to the conclusion that the distance over which the transfer of assimilates has to occur is not a major limiting factor. In grapes, where the leaves and fruits are generally in close proximity, an appropriate defoliation treatment showed that photosynthate could move readily from a leaf to a fruit cluster 3.8 m distance The increasing distance between the source leaves and fruits of apple had no significant effect on fruit growth. A large sinks dominate the overall supply of current photosynthate with the smaller sinks having to rely on local supplies and storage. Although the proximity of source and sink is an important factor in regulating the pattern of carbon partitioning in plants, vascular links and phyllotaxis also have a part to play in this transfer.

In sunflower the movement of photosynthate has been shown to follow the phyllotaxic link between leaves and growing organs. The phyllotaxic links between the leaves and the tubers of potatoes also became less noticeable in relation to assimilate distribution if the source is restricted to a single leaf.

Carrying capacity

Evidences are available on translocation where measurements and estimates were made of the *specific mass transfer* (SMT) of dry matter through the phloem there appeared to be an upper limit for transport of about 3-5 g cm^{-2} phloem h^{-1}, or 15-25 g cm^{-2} sieve tube area h^{-1}. This suggests that the carrying capacity of the phloem may have been a factor limiting growth. However, more recent data have shown that specific mass transfer can be much greater than the earlier estimates suggested. Confirmation for an excess carrying capacity in the phloem of cereals can be seen in the failure to observe a reduction in grain growth of either sorghum.

The xylem as an alternative transport system

If phloem continuity in the peduncle of wheat is destroyed by a heat treatment there is an inhibition of the movement of photosynthate from the flag leaf to the ear, although phloem mobile mineral nutrients may bypass this block by transferring to the xylem. This clear distinction in the role of phloem and xylem in transporting sugars and mineral nutrients is however less evident in deciduous trees where there is considerable storage of sugar in the xylem. Although the subsequent movement of these stored sugars is often uncertain.

Path Storage

The storage of photosynthate in the leaf and in tissues along the path of transport between the source and sink provides an important control in relation to the partitioning of carbon in plants. Storage is often considered a rather passive response to an excess supply of photosynthate, with remobilization occurring when current photosynthesis is low, or the demand is increased. However, the control of storage is more complex than this simple model suggests and its regulation and cost in terms of respiration and carbon turnover, as well as its influence on carbon partitioning needs more careful consideration. Pathway storage, like that in the leaf, can have a buffering action on the diurnal fluctuations in the supply of photosynthate available for growth. Little is known about the control of pathway storage, the capacity for storage, or the extent to which storage tissues are competitive sinks for carbohydrates. Probably the most detailed analysis of pathway storage in stems is that for sugarcane, where the stored sugar can reach as much as 50 per cent of the total dry matter of the stalk. Following are the steps that are associated with sugars storage in sugarcane:

1. Sucrose moving through the sieve elements in the vascular strands of the stem is released into the adjacent apoplast (free space), where it is hydrolysed by invertase to fructose and glucose.

2. The hexoses are then in turn taken up by the parenchyma cells where they are phosphorylated and condensed to form sucrose P, which is transferred through the tonoplast into the vacuole and dephosphorylated for storage as sucrose.

3. In the storage parenchyma sucrose may reach 80 per cent of the total dry matter and uptake occurs against a concentration gradient and is sensitive to low oxygen and low temperature.

4. Morphological as well as biochemical factors may be important in sugar storage.

5. Environmental conditions also influence the degree of carbohydrate storage and also the form of that storage.

The form of storage in trees can also depend on temperature, as for example in *Populus* where an increasing degradation of starch and the accumulation of sugars in the late autumn to mid winter is reversed when temperatures fall to zero the resynthesis of starch begins in the spring as the temperature rises above 5°C. However, in contrast in the ray cells of other genera e.g., *Quercus, Fraxinus* and *Larix* there is little starch degradation at low temperature.

The utilization of stored carbohydrate

Carbohydrates stored in tissues adjacent to the translocation pathway are used for growth when remobilized. However the maintenance of storage tissue, the transfer of metabolites across membranes and the elaboration of new materials require the use of energy and hence must involve a 'cost' to the system in terms of carbon lost in respiration. The *economic ratio* (i.e. the efficiency of conversion of reserve substances into growing organs) for the formation of a new shoot in a potato tuber was 0.4-0.5 and compares with a similar ratio for germinating rice. Also the economic ratio for the utilization of reserves in mulberry for 10 d following pruning. Thus it appears that less than half of any reserve carbohydrate may be effective in new growth.

To date there is not firm indication that either short distance transfer out of leaves, or into growing organs, or long distance transport between the source and sink are major factors in regulating the overall growth of the plant, although actual patterns are influenced by vascular linkages and these may determine the growth of individual organs under source-limited conditions.

There is often considerable debate as to whether dry matter production is source or sink limited and this will certainly vary with genotype, stage of development and environmental conditions. However there would still appear to be considerable scope for improving yields, both through an increased efficiency of canopy or single leaf photosynthesis and also through improved partitioning of dry matter to the organs required for harvest (an increased harvest index).

Understanding the control of organ initiation and development and a recognition of the appropriate balance between source and sink tissues is basic to the manipulation of carbon partitioning in plants. There is also a need to determine the components most limiting sink development and the factors that influence the expression of sink potential. It is tempting to suggest that an apoplast step in the transfer of carbon from the vascular tissue to the region of growth is associated with sink dominance, but there are notable exceptions to this rule, such as the potato tuber (Oparka, 1986). The role of temporary storage in modifying sink-source relationships has also been largely neglected and the importance of this needs to be assessed under both near optimal and stress conditions. In the final analysis there appear to be more questions than answers in relation to the control of assimilate partitioning and yield in plants; careful experimentation should allow many of these questions to be resolved to give a firm base for future genetic manipulation.

It is therefore, concluded that carbon partitioning in plants is controlled by a number of factors which include the supply of photosynthate, the number and size of competing sinks and their location on the plant, vascular connections and the potential for temporary storage in the leaves and along the path of transport. However many functions are closely interrelated, with the growth of one organ dependent on another either directly through the supply of essential metabolites or nutrients, or less directly through the action of plant growth regulators. Our understanding of the latter is far from adequate and needs further careful analysis and experimental evaluation.

Although there is considerable information on the operation of individual processes in plants, such as photosynthesis, starch metabolism, translocation and cell expansion (at the sub-cellular, cellular and tissue levels of organization) it appears that the controls which actually regulate the partitioning of photosynthate in the integrated whole plant systems are still only poorly understood.

CHAPTER 11

Transgenic Plants and Photosynthetic Carbon Metabolism

Transgenic plants refer to those genetically modified plants in which functional foreign gene(s) have been inserted in their genome. Recent advances in plant molecular biology, tissue culture and recombinant DNA technology have opened this new avenue in transformation of higher plants. Thus, transgenic plants have a piece of alien DNA that are introduced by procedure other than sexual crossing and thus are genetically modified. Crops with transgenic plant are now common feature of the agricultural landscape (Dunwell, 1999).

For improvement in crop yield genetically modified crops have been developed because carbon metabolism in photosynthetically active cells of higher plants is of central importance for the allocation of carbohydrates formed to tissues that are dependent on the supply of carbon and energy. As discussed earlier, many control mechanisms influence carbohydrate formation and partitioning into different pools.

Photosynthetic process that occur exclusively within the chloroplast as stated earlier can be divided into light-dependent and light-independent reactions. The light-dependent reactions convert electromagnetic energy into biologically convertible energy forms such as ATP and NADPH. These are used to reduce CO_2 in the reductive pentose phosphate cycle (RPP cycle; Calvin cycle) reactions, which are chemically light independent, but nevertheless can be light regulated.

MANIPULATION OF CHLOROPLAST

The manipulation of organelle DNA has become a realistic possibility now. The first report of chloroplast transformation was a result of cocultivation of *Nicotiana tabacum* and *Agrobacterium*. Some of the transformants obtained with pGV 3850 : pNosCAT7 showed a maternal inheritance of the introduced chlorophenicol acetyl transferase (CAT) activity. Although this activity could be localized in the chloroplast, southern blots reveled that the integrated DNA fragment was not stably maintained. This instability was considered one of the major problems and many laboratories started to develop more sophisticated vectors. The first step was to fuse chloroplast promoters to the selection marker to ensure a high level of expression in the chloroplast. As a next step the vectors were equipped with plastid DNA sequences that allow homologous recombination without disturbing functional regions. With a similar construct, PGV2260 containing the NPTII gene fused to the psbA promoter and flanked by a 3 rbcL gene fragment, *Agrobacterium* co-cultivation experiments with tobacco protoplasts were performed. However, due to unexpected nuclear activity of this construct, it was not possible to detect any putative chloroplast transformation event.

Current methods for DNA delivery into chloroplast include particle bombardment and polyethylene glycol. The particle bombardment is very expensive and polyethylene glycol method requires preparation of protoplasts and is less efficient. Knoblauch et al. (1999) described a novel galinstan expansion femtosyringe method that allows microinjection of foreign DNA and others substances into prokaryotic cells and eukaryotic organelles. The essence of the method is the heat induced expansion of a liquid metal called galinstan (an alloy of gallium, indium and tin) within a glass syringe to expel sample through a capillany tube with a diameter of about 0.1 μm. Khan and Maliga (1999) described the use of a new fluorescent antibiotic resistance enzyme conferring resistance to spectinomycin/streptomycin (FLARE-S) to detect chloroplast transformation. FLARE-S was created by fusing the antibiotic resistance enzyme to the GFP from the jellyfish *Aequorea victoria*.

Recently, commercial useful chloroplast transgenic plants expressing high levels of herbicide resistance (Daniell et al., 1998) or insect resistance (Kota et al., 1999) have been engineered. Guda et al., (2000) reported exploitation of the chloroplast transformation approach for molecular farming by the genetic engineering of a value added trait that is not advantageous or required by transgenic plants. Bioelastic protein-based polymers (PBP) have several medical (prevention of post-surgical adhesions) and non-medical (biodegradable plastic) applications. They compared the expression levels of PBP genes (synthetic) integrated into the nuclear genome or the large single-copy (LSC) or inverted repeats (IR) region of the chloroplast genome in transgenic tobacco plants. Higher levels of polymer transcripts were observed only in the IR transgenic plants.

A Better Rubisco

By interacting with atmospheric carbon dioxide, Rubisco initiates the chain of biochemical reactions that creates the carbohydrates, proteins and fats that sustain plants and animals. But the enzyme also has another distinction. According to T. John Andrews, Australian National University, Canberra, "*Rubisco is nearly the world's worst, most incompetent enzyme- it is almost certainly the most inefficient enzyme in primary metabolism that there is*". First, it is the slowest known enzyme and its turnover in higher plants is as little as two or three reactions per second. Second, Rubisco triggers an additional reaction that interferes with the first. Rubisco catalyzes a second, opposing reaction. In what is called photorespiration, the enzyme combines with oxygen rather than carbon dioxide to create a compound that is subsequently converted partly into carbon dioxide. In other words, Rubisco catalyzes one reaction that incorporates carbon into plants and another that ultimately strips them of carbon.

Rubisco's ineffectiveness has been a spur to scientists since it became fully apparent in the 1970s. According to George Lorimer, the quest for a better Rubisco is *a Holy Grail in Plant Biology*. Everyone always goes in with the hope of changing the face of agriculture. Despite more than 20 years of effort, the hopes have not yet paid off. But recent advances in molecular biology and the unexpected discovery of more efficient Rubisco in red algae have given now impetus to the long struggle to modify the enzyme. Laboratories across the world are trying to improve the Rubisco in food crops by either replacing the existing enzyme with red algae form or bolting on what could be thought of as molecular superchargers.

Hopes reawakened in 1992, when F. Robert Tabita and B.R. Read of Ohio State University in Columbus discovered that some diatoms and red algae have more-specific Rubisco that in higher plants. In 1997, a team led by Akiho Yokota at the Research Institute for Innovative Technology for the Earth in Japan found red algae with Rubisco is about three times more efficient. The scientists are attempting to insert Rubisco genes from red algal chloroplasts into the chloroplasts of higher plants using techniques for manipulating chloroplastic DNA developed by Rutger university biochemist Pal Maliga. The scientists hope to see result in about "10 years". To alter Rubisco, scientists must work with a 16-part molecule that is encoded by many genes in both the cell nucleus and the chloroplast where photosynthesis takes place. The enzyme also depends on supporting cast of other enzymes, some of which will probably need to be revamped if Rubisco is changed.

Supercharging Photosynthesis

While most researchers trying to modify the Rubisco, a few are trying another, even more ambitious strategy. Just as small engines can go faster if they are equipped with a supercharger which force-feeds them with fuel, some plants have their own photosynthetic supercharger, known as the C_4 cycle . In C_4 plants the bundle sheath cells where photosynthesis takes place are surrounded by specialized mesophyll cells, which temporarily fix carbon dioxide and jam it into the bundle cells at such high concentration that the oxygen reaction is effectively blocked. The C_4 cycle requires so much energy that C_4 plants can not grow in dim light but in well illuminated conditions, C_4 crops like sugarcane photosynthesize more efficiently. About 5% of all terrestrial higher plant species use the C_4 cycle.

A joint team at Japan's National Institute of Agrobiological Resources and Nagoya university led by M. Matsuoka is now attempting to reproduce the C_4 cycle in rice. His team is trying to identify non-functional equivalents of C_4-type gene in rice and selectively replace them with their active counterparts from maize. It includes phosphoenol pyruvate carboxylase (PEPC) which catalyze the beginning of the C_4 cycle. Rice has mesophyll-like cells that are not photosynthetically active and these may have to be activated. The transgenic rice with only the C_4 enzyme NADP-malic tends to die quickly. A preliminary evidence suggests that active PEPC in rice cuts the distructive reaction by about a third.

Rubisco activity in Transgenic Plants

Tobacco plants with reduced Rubisco content were analysed with respect to their photosynthetic activity, their growth, and their carbon allocation under different environmental conditions. Under an ambient light regime (310 μmol photons m^{-2} s^{-1}) it was shown that Rubisco content can be decreased down to 50-60 per cent of wild-type levels before any effects on the photosynthetic activity of tobacco leaves are observed, clearly indicating that Rubisco is present in excess. It is now established that Rubisco is present in excess. Experimental evidences indicate that photosynthetic activity was not affected over a wide range of Rubisco reduction, therefore, neither the production of biomass was also not affected, nor was starch synthesis over the same range of inhibition. When Rubisco was reduced in a way that photosynthetic rates declined, lower biomass production and lowered starch synthesis in leaves were observed.

Rubisco activity is positively regulated through carbamylation by Rubisco activase. Furthermore, it is possible that the activase promotes the release of the inhibitor carboxy-arabinitol 1-phosphate from the enzyme in tobacco plants. Interestingly, lower Rubisco activation was compensated by higher Rubisco levels in the transgenic plants.

Triose Phosphate Translocator Activity in Transgenic Plants

It has been a matter of debate for a long time whether photosynthates are exported as triose phosphate or as hexoses from the chloroplast into the cytosol. Furthermore, it is an open question whether this export is rate limiting for photosynthesis. Biochemical studies with spinach chloroplasts clearly indicate that triose phosphate export is the major route for carbon channeled into sucrose for further transport into sink organs, since other transport activities could not account for the sufficient translocation of metabolites across the chloroplast membrane during photosynthesis.

Photoassimilates enter the cytosol as triose phosphates and two major possibilities exist for the subsequent utilization. Triose phosphates can be used either in respiratory reactions or for sucrose biosynthesis. Several attempts have been made to analyze the influence and function of key enzymes involved in these pathways by creating transgenic plants. The interest behind these experiments is to understand the mechanism controlling partitioning and allocation of carbohydrates in higher plants. To increase crop productivity it would be useful to shift the direction of triose phosphate flow from glycolysis to sucrose synthesis, as the amount of sucrose present in the source leaf is a rate-limiting factor with respect to the capacity of a source leaf.

Manipulation of Glycolytic Pathway

After transport into the cytosol triose phosphate may enter the glycolytic pathway. This pathway ends in the cytosol at the level of pyruvate which subsequently enters the mitochondria for further metabolism, such as respiration, amino acid or fatty acid synthesis. Pyruvate kinase (PK) catalyses the formation of pyruvate and ATP from phosphoenolpyruvate (PEP) and ADP.

It is known that there are multiple isozymes of PK in plastids (PK_p) and the cytosol (PK_p) and that there are complex regulatory mechanisms controlling their expression. In a recent study the potato PK_c was fused to a Rubisco transit peptide to target the protein to plastids in transgenic tobacco plants. Surprisingly, expression of both the resident and transgenic PKc was eliminated in some plants as a result of the phenomenon of co-suppression. Plant lacking any PKc grew and reproduced normally. Lack of PK would be fatal in onther organisms and it is assumed that plants have some means of bypassing the reaction catalysed by PKc, perhaps through the action of PEP phosphatase or by the combined action of PEP carboxylase, malate dehydrogenase, and malic enzyme. A better understanding of the glycolytic pathway can be achieved through the modification of the levels of the various PK forms in transgenic plants and subsequent biochemical and physiological analysis.

Manipulation of sucrose biosynthesis

The major controlling steps of sucrose biosynthesis in higher plants are the interconversion of fructose 1,6-bisphosphate and fructose 6-phosphate, the formation of sucrose 6-phosphate from UDPG and fructose 6-phosphate, and the dephosphorylation of sucrose 6-phosphate yielding the final product, sucrose. Three enzymes are involved in catalyzing the first reaction: the forwards reaction is catalysed by fructose 1,6-bisphosphatase (FBPase), the backward reaction towards glycolysis is controlled by phosphofructokinase (PFK), and the third enzyme, pyrophosphate : fructose 6-phosphate-1-phosphotransferase (PEP), is able to catalyse the reaction in both directions. Each of these enzymes is allosterically regulated by intermediates of carbohydrate metabolism, furthermore FBPase and PEP are highly regulated by the regulatory molecule fructose 2,6-bisphosphate (F2, $6P_2$).

To date, two approaches using transgenic plants have been employed to manipulate the interconversion of F1,6P and F6P using transgenic plants. As the precise functional significance of PEP is still unclear and controversial. The transformed potato plants with antisense constructs to the genes encoding the ∝- and β-subunits of PEP. In several transgenic plants PEP expression was decreased by up to 99 per cent. These plants did not show any phenotypical changes, reduced growth, or a decrease in tuber yield. The reduced PEP activity leads to a small reduction in starch and to a small increase in sucrose in potato tubers. These results were unexpected, because the essentially complete reduction of an enzyme that is highly regulated and expressed in most tissues was expected to influence plant carbohydrate metabolism significantly.

UDP-glucose (UDPG) is derived from glucose 1-phosphate by the action of the enzyme UDP-glucose pyrophosphorylase (UGPase) and represents a branchpoint in sugar metabolism: UDPG is needed for sucrose biosynthesis and as a precursor of cellulose. Transgenic plants with only 4 per cent of the wild-type UGPase activity showed no phenotypic changes, and sugar metabolism was not substantially affected (Zrenner et al., 1993). The fact that neither the suppression of PEP nor suppression of UGPase leads to major phenotypic effects could have two meanings. Either plants express certain enzymes in such an excess that even the low remaining activity is sufficient for normal development and growth, or these enzymes even when they are expressed in a more or less constitutive way are only needed under certain, as yet unknown, environmental or stress situations.

The reversible reaction catalysed by PEP may be shifted towards sucrose synthesis by removal of the phosphate group donor inorganic pyrophosphate (PP). In a recent study, *E. coli* inorganic pyrophosphatase (PPase) was introduced into the cytosol of transgentic tobacco and potato plants (Sonnewald, 1992). The expression of PPase in transgenic plants leads to a dramatic increase in the level of soluble sugars in source leaves. Although an accumulation of starch is observed in older leaves, the ratio between soluble sugars and starch is increased up to three times. This result could mean that expression of PPase leads to the expected alteration of carbon partitioning. The irreversible of F1,6P and F6P is shifted towards F6P in the direction of sucrose synthesis. In addition, the formation of UDPG catalysed by the enzyme UGPase becomes irreversible due to removal of PPi and based on an increase in free Pi in the cytosol, the export of triose phosphate from the chloroplast is triggered. Growth and development of these plants are changed drastically. The plants are characterized by stunted growth and loss of chlorophyll. Potato plants showed an increase in tuber number, arguing for a reduction of sink capacity. As the PPase was expressed in all tissues several explanations are possible for the observed changes. Expression in sink tissues may prevent the utilization of UGPG which is produced by the cleavage of sucrose by the enzyme sucrose synthase. Additionally, PPase activity in companion cells may block phloem loading (Sonnewald, 1992).

A growing body of work indicates that sucrose phosphate synthase (SPS) regulates the synthesis of sucrose in the cytosol and, therefore, may play a major role with respect of the amount of photoassimilates available for export from source leaves (Stitt and Quick, 1989). The maximal activity of SPS is regulated in a light-dependent manner through phosphorylation or dephosphorlation, and furthermore through G6P and Pi (Huber and Huber, 1992). Different species show a variation with respect to light activation of SPS (Huber et al., 1989). Maize SPS activity, for example, is strongly down-regulated during the dark, but the enzyme does not respond to G6P or Pi concentrations in the light (Huber et al., 1989). The expression of the maize SPS in transgenic tomato plants caused a reduction of starch and increased the amount of sucrose in mature leaves (Worrell et al., 1991). This result clearly demonstrates that SPS is involved in the regulation of carbon partitioning in leaves.

Manipulation of Sucrose loading into the phloem

In photosynthetically active tissues a surplus of fixed carbon is produced, which is transported to sink tissues, e.g. roots, seeds, tubers, and sink leaves. Solutes are transported within the phloem from source to sink tissues. In this section we will concentrate on the transport of sucrose from mesophyll cells into the **sieve element/companion cell complex** (SECCC). Whether phloem loading is exclusively symplastic or if an apoplastic transport step is involved has been the subject of many discussions (Lucas and Madore, 1988; Bush, 1933). The distribution of plasmodesmata has provided evidence for symplastic transport, at least between mesophyl! cells. In certain plant species, e.g. Laminaceae, a high symplastic connectivity between mesophyll cells and the SECCC (type 1 plants) may allow phloem loading via plasmodesmata, whereas in species like potato, sugar beet, and tobacco (type 2b plants), with comparatively low symplastic connectivity (van Bel et al., 1993), an apoplastic step is proposed.

The apoplastic loading model predicts an energy-dependent transport step of sucrose across the SECCC plasma membrane, generating a sucrose gradient with a tenfold difference in concentration. Transport is driven by a proton gradient generated by plasmalemma H^+-ATPase. In a variety of experiments a sucrose / proton symport activity was undoubtedly detected in plasma membrane vesicles and protoplasts of several plant species (Bush, 1993). There are two explanations for such a transport activity. First, the transport protein is part of a general sucrose retrieval system. Secondly, the protein is responsible for the active transport of sucrose into the SECCC. Recently, transgenic plants were generated which gave a deeper insight into the *in vivo* processes related to the questions concerning phloem loading.

Yeast invertase has been expressed in various organs and different subcellular compartments of transgenic plants (Sonnewald et al., 1991) such as in the apoplast of potato leaves (Heineke et al., 1992). Because phloem loading in these leaves involves an apoplastic step, sucrose is cleaved before long-distance transport. As a consequence, export is blocked and high levels of soluble sugars and starch accumulate, leading to the repression of photosynthetic genes. At the same time, the sink organs are starved of carbon and the plants produce fewer tubers.

Unloading and transport of sucrose in potato tubers is thought to occur symplastically. Surprisingly, expression of yeast invertase in the apoplast of transgenic tubers lead to a reduction in tuber number per plant and an increase in tuber size (Sonnewald et al., 1997). Total tuber yield per plant on a fresh weight basis increases whereas production of tuber biomass (dry weight) remains constant. Expression of the yeast enzyme in the cytosol of transgenic potato tubers is detrimental to both tuber size and yield. Together, these studies reveal that the activity of a sucrose-cleaving enzyme at the wrong time and place has profound effects, drastically altering carbon metabolism and partitioning as well as gene expression and plant development.

Reduction of the expression of sucrose transporter in transgenic plants

The transgenic plants show reduced enzyme activity due to the *antisense* technology allowing new insights into the mechanisms present *in vivo*. A prerequisite for this approach is the isolation of the gene to be suppressed. Recently sucrose transporter cDNA clones from spinach and potato were isolated by complementation of a yeast mutant (Riesmeier et al., 1993). The transporter proteins from both spinach (SoSUT) and potato (StSUT) showed characteristics consistent with those reported for the sucrose / proton symport in isolated plasma membrane vesicles. RNA in situ hybridization data show that the transporter gene is especially expressed in the minor veins of mature leaves (Riesmeier et al., 1993).

To further investigate the physiological function of the isolated cDNAs, potato plants were transformed with an *antisense construct* (Riesmeier et al., 1992). The antisense inhibition leads to drastic phenotypic effects. The growth of the plants is strongly retarded; mature leaves are curled, bleached, and accumulate anthocyanines at the rims, whereas sink leaves showed no phenotypic changes. Measurement of the export of soluble sugars through the petiole provides direct evidence that the flow of sucrose is reduced in the antisence plants. The reduced efflux strongly affects the supply of carbohydrates to sink organs, clearly visible as a reduced root system. No tubers are formed in the case of plants showing the highest reduction of StSut mRNA. The plants were analysed more intensively with respect of physiological changes. The soluble carbohydrate content in mature leaves increased up to twentyfold and the starch amount up to tenfold. The maximal rate of photosynthesis decreased to 50 per cent in strongly affected plants as compared to wild-type.

Starch Synthesis in Transgenic Plants

It is regulated in a complex manner. Probably only part of the controlling mechanisms have been identified. Starch is produced in the chloroplasts, where part of the carbon fixed by the Calvin cycle is channeled along the pathway starting at fructose 6-phosphate (F6P). F6P is converted to ADP-glucose by the action of the enzyme phosphoglucoisomerase (PGI), phosphoglucomutase (PGM), and ADP-glucose pyrophosphorylase (AGPase). ADP-glucose serves as a direct precursor for starch production, which involves a complex array of enzymes including different starch synthases (soluble and granule-bound isoforms), branching enzyme, and, more than likely, other unidentified propteins. During the day, i.e. when temporary storage of starch occurs, only a part of the recently fixed carbon is immediately available for regeneration of RuBP in the Calvin cycle or for conversion to sucrose in the cytosol. At night, degradation products are converted to sucrose, allowing transport of carbon metabolites to heterotrophic tissues. A precise control of leaf carbohydrate metabolism is needed to organize these processes efficiently and to ensure efficient growth of the plant.

ADP-glucose pyrophosphorylase

ADP-glucose pyrophosphorylase, a major enzyme controlling starch synthesis, has been demonstrated in many different plant species, including lower plants, monocots, and dicots . The enzyme is also present in bacteria, where ADP-glucose is the substrate for glycogen synthesis. Two long-known features of plant AGPase are the substrate for glycogen synthesis. Two long-known features of plant AGPase are the allosteric activation by 3-phosphoglycerate (3-PGA) and the inhibition by inorganic phosphate. In contrast, the enzyme from enteric bacteria is mainly regulated by fructose 1,6-bisphosphate (F1,6P; activator) and AMP (inhibitor) (Ghosh and Preiss, 1966; Plaxton and Preiss, 1987).

Following data underscores the importance of AGPase in starch synthesis:

- AGPase is localized in chloroplasts and amyloplasts, i.e. in starch-storing organelles.
- The in vitro, activity of AGPase play an important role in *in vivo* rate of starch accumulation.
- Kinetic models led to the conclusion that 3-PGA and Pi, which are known to influence AGPase activity in vitro, play an important role in controlling starch synthesis *in vivo*.
- A *Chlamydomonas reinhardtii* mutant with an unregulated AGPase (no activation by 3-PGA and no inhibition by Pi) produces only 5 per cent of starch seem in the wild type.
- Mutants deficient for AGPase also have reduced levels of starch.
- Transgenic plants in which AGPase activity was modulated showed altered levels of starch.

Plant AGPase has been described as a heterotetrameric protein composed of two different but similar polypeptides. The size of the two smaller subunits ranges from 50 to 55 kDa, the size of the larger subunit from 51 to 60 kDa, depending on the plant. At least ten different AGPase polypeptides from angiosperms, e.g. wheat, maize, rice, and potato, have been characterized at the DNA level (Anderson et al., 1989). Based on biochemical and DNA sequence data, the large subunit polypeptides appear to be more variable than the small subunit polypeptides exist in the potato tuber (Okita et al., 1990).

Transgenic plants with altered levels of AGPase activity

In order to down-regulate starch levels in transgenic plants, AGPase activity was decreased by the antisense RNA technique. The cDNA encoding the small subunit of the potato enzyme (Müuller-Röber at el., 1990) was inserted in its reverse orientation behind the cauliflower mosaic virus (CaMV) 35S RNA promoter and transformed into potato. The leaf AGPase levels of some of the transgenic plants were decreased to 5-30 per cent of the wild-type activity. A good correlation was observed between the amount of residual mRNA and protein expression and the degree of inhibition in AGPase.

GENETIC ENGINEERING OF C_4 PLANTS

There is considerable interest in the genetic engineering of C_4 plants for improved agronomic performance, which has led to the development of transformation procedures for major crops such as maize and sugarcane, and the model C_4 dicot *Flaveria bidentis*. Modern high-yielding varieties of sugarcane have been produced by conventional breeding but there is a limit to further improvements in yields by this approach. The accumulation of sucrose in sugarcane stems is determined by the relative activities of the vacuolar acid invertase and sucrose phosphate synthase (Zhu et al., 1997). Therefore, a decrease in vacuolar acid invertase activity by anti-sense or the over expression of SPS in the stem would seem to be a promising approach to increasing sucrose accumulation. Over expression of SPS activity in the leaves might also lead to higher rates of sucrose synthesis in the leaves and export to the stems (Grof et al., 1998).

Rocher et al (1989) reported a strong positive correlation between leaf SPS activity and growth rate in maize plants and Sarquis et al (1998) reported an increase in SPS activity as maize plants were selected for higher yield. However, attempts to over-express SPS in species other than tomato have met with limited success due to inactivation of the over-expressed enzyme probably by protein phosphorylation. In the transgenic tomato plants the maize SPS was not recognized by the endogenous tomato SPS protein kinase, so the overexpressed enzyme remained active. One approach to overcoming the problem of enzyme inactivation would be to express a prokaryotic SPS gene in plants, such as from the cyanobacterium *Synechocystis* sp. PCC 6803, which encodes an SPS protein that lacks the phosphorylation – site motifs found in the higher plant enzyme (Lunn et al., 1999).

Another possible area for engineering sucrose metabolism in C_4 plants is stress-tolerance. Transgenic tobacco plants, engineered to accumulate compatible solutes (such as trehalose, glycine betaine and sugar alcohol) showed some increase in water stress tolerance over wild-type plants (Hare et al., 1998). These findings suggest that the diversion of photosynthate from sucrose synthesis towards the synthesis of compatible solutes might also improve water-stress tolerance in other species. However, improved water-stress tolerance might be more difficult to achieve in C_4 plants, given their generally higher water use efficiency than C_3 plants. Cold tolerance is a multifactorial trait that is unlikely to be affected by changes to sucrose metabolism (Sarhan and Danyluk, 1998). Therefore, the diversion of photosynthate from sucrose to more valuable products such as fructan or polyhydroxybutyrate could be worthwhile goal.

The bundle sheath cells of C4 plants possess thickened and suberized cell walls that act as a diffusion barrier to help decrease the loss of CO_2 from the bundle sheath cells. A consequence of this anatomical adaptation is that the bundle sheath cell walls of C4 pasture grasses are resistant to degradation in the digestive tract of grazing animals. In addition, because most of the starch in the leaves of C4 plants is within the bundle sheath cells it is also inaccessible to digestion. An improvement in the digestibility of C4 pasture grasses might be achieved either by bringing about a higher accumulation of soluble carbohydrates such as sucrose or by transferring starch synthesis from the bundle sheath cells to the mesophyll cells.

Selected Bibliography

1. Akie Fujii, J. A. 1984. Photosynthesis and photorespiration in apples. Ph.D. Thesis, Washington State Univ., Pullman.
2. Albertsson, P-A., E. Andreasson, and P. Svensson 1990. The domain organization of the plant thylakoid membrane. FEBS Lett. 273: 36-40.
3. Avery, D. J., Priestley, C. A. and Treharen, K. J. 1979. Integration of assimilation and carbohydrate utilization in apple. In: Photosynthesis and plant development. R. Marcelle, C. Chjsters, and M. Van Pouke (eds.). W. Junke, Hague, The Netherlands.
4. Bangerth, F. 1980. Some effects of endogenous and exogenous hormones and growth regulators on growth and development of tomato fruits. In: Aspects and prospects of plant growth regulators. B. Jeffcoal, (ed.). Monograph 6. British Plant Growth Regulator Group. Wessex Press, Wantage, Oxfordshire.
5. Beery, J. A., Downtown, W. J. S. and Tregunna, E. B. 1970. The photosynthetic carbon metabolism of *Zea mays* and *Gomphrena globosa*: the location of the CO_2 fixation and carboxyl transfer reactions. Can. J. Bot. 48: 777-86.
6. Bendall, D. S. and Manasse, R. S. 1995. Cyclic phosphorylation and electron transport. Biochim Biophys. Acta. 1229, 23-38.
7. Bendich, A. J. 1987. Why do chloroplast and mitochondria contain so many copies of their genome ? Bio Essays 6, 279-82.
8. Bennett, A. B. and Spanswick, R. M. 1983. Solubilization and reconstitution of an anion-sensitive H^+ ATPase from corn roots. J. Membrane Biol. 75:21-31.
9. Bennett, A. B., Sweger, B. L. and Spanswick, R.M. 1984. Sink to source translocation in soybean. Plant Physiol. 74:434-436.
10. Benson, A. A. and Calvin, M. 1950. The path of carbon in photosynthesis VII. Respiration and photosynthesis. J. expt. Bot. 1: 65-8.
11. Bhagsari, A. S. 1981. Relation of photosynthetic rates to yield in sweet potato genotypes. HortScience 16:779-780.

12. Bird, I. F., Cornelius, M. J., Keys, A. J. and Whittingham, C. P. 1972. Oxidation and phosphorylation associated with the conversion of glycine to serine. Phytochemistry 11: 1587-94.

13. Bishop, P. M. and Whittingham, C. P. 1968. The photosynthesis of tomato plants in a carbon dioxide enriched atmosphere. Photosynthetica 2: 31-8.

14. Björkman, O. and Gauhl, E. 1969. Carboxydismutase activity in plants with and without β-carboxylation photosynthesis . Planta 88: 197-203.

15. Björkman, O., Gauhl, E., Hiesey, W. M., Nicholson F. and Nobs, M. A. 1969. Growth of *Mimulus, Marchantia* and *Zea* under different oxygen-carbon dioxide levels. Ann. Rep. Carnegie Inst. Wash. 1967-8: 477-9.

16. Björkman, O., Pearcy, R. W. and Nobs, M. A. 1971. Hybrids between Atriplex species with and without β-carboxylation photosynthesis. Photosynthetic characteristics. Ann. Rept. Carnegie Inst. Wash. 1969-70: 640-48.

17. Borchers, C. A., Lamm, A. B. G. and Swanson, C. A. 1979. Source-sink patterns in relation to source strength. Plant Physiol. (Supp.) 64:34.

18. Bowen, J. E. and Hunter, J. E. 1972. Sugar transport in immature internodal tissue of sugarcane. II. Mechanism of sucrose transport. Plant Physiol. 49:789-793.

19. Bowes, G., Ogren, W. L. and Hageman, R. H. 1971. Phosphoglycolate production catalyzed by ribulose diphosphate carboxylase. Biochem. Biophys. Res. Comm. 45: 716-22.

20. Boyer, P. D. 1993. The binding change mechanism for ATP synthase – some probabilities and possibilities. Biochem Biophys. Acta 1140: 215-250.

21. Briat, F. J., Liscure, A. M. and Mache, R. 1986. Transcription of the chloroplast DNA - a review. Biochimie 68, 981-90.

22. Bull, T. A. 1969. Photosynthetic efficiencies and photorespiration in Calvin cycle and C_4 dicarboxylic acid plants. Crop Sci. 9: 726-9.

23. Burry, J. A., Lorimer, G. H., Pierce, J., Seamaan, J. R., Meek, J. and Freas, S. 1987. Isolation, identification and synthesis of 2-carboxyarabinitol 1-phosphate, a diurnal regulator of ribulose bisphosphate carboxylase activity. Proc. Natl. Acad. Sci USA 84, 734-8.

24. Challa, H. 1976. An analysis of the diurnal course of growth, carbon dioxide exchange and carbohydrate reserve content of cucumbers. Publ. 20. Centre Agr. Publ. Document, Wageningen, the Netherlands.

25. Charles-Edwards, D. A. 1982. Physiological determinants of crop growth. Academic Press, New York. P. 106-112.

26. Chatterto, N., G. E. Carlson, N. J., Hungerford, W. E. and Lee, D. R. 1972. Effect of tillering and cool nights on photosynthesis and chloroplast starch in Pangola. Crop Sci. 12:206-208.

27. Chatterton, N. J. and Silvius, J. E. 1981. Photosynthate partitioning into leaf starch as affected by daily photosynthetic period duration in six species. Physiol. Plant. 49:141-144.

28. Chatterton, N. J. and Silvius, J. E. 1981. Photosynthate partitioning in soybean leaves. II. Irradiance level and daily photosynthetic period duration effects. Plant Physiol. 67:257-260.

29. Chatterton, N. J. and Silvius, J. E. 1980. Photosynthate partitioning into leaf starch as affected by daily photosynthetic period duration in six species. Physiol. Plant. 49:141-144.

30. Cheung, G. P.; Rosenblum I. Y. and Salagh, H. J. 1968. Comparative studies of enzymes related to serine metabolism in higher plants. Plant. Physiol. 43: 1813-20.

31. Chitnis, P. R. and Nelson, N. 1992. Biogenesis of photosystem I: The subunit PsaE is important for the stability of PS I complex. In: Chloroplast biogenesis (Argyroudi-Akoyunoglou, J. ed.). Plenum Press. New York, pp. 285-290.

32. Chitnis, P. R. 1997. Targeting, assembly and degradation of chloroplast proteins. In: comprechensive treatise. Cambridge University Press, Cambridge.

33. Chitnis, P. R. Chitnis, V. P., Xu, Q., Jung, Y. –S., Yu, L. and Golbeck, J. H. 1995. Mutational analysisi of photosystem I polypetides. In: (Mathis, P. ed.) Dodrecht, Vol. II, pp. 17-22.

34. Chitnis, P. R., Purvis, D. and Nelson, N. 1991. Molecular cloning and targeted mutagenesis of the gene psaF encoding subunit III of photosystem I from the Cyanobacterium synechocystis sp. PCC 6803. J. Biol. Chem., 266, 20146-20151.

35. Chitnis, P. R., Reilly, P. A. and Nelson, N. 1989. Insertional inactivation of the gene encoding subunit II of photosystem I from the Cyanobacterium synechocystis sp. PCC 6803. J. Biol. Chem., 264, 18381-18385.

36. Chitnis, P. R., Xu.Q., Chitnis, V. P. and Nechushtai, R. 1995b. Function and organization of photosystem I polypetides. Photosynth. Res., 44, 23-40.

37. Chitnis, V. P., Jung, Y.-S., Albee, L., Golbeck, J. H. and Chitnis, P.R. 1996. Mutational analysis of photosytstem I polypetides: Role of PsaD and the lysyl 106 residue in the reductase activity of photosystem I. J. Biol. Chem., 271, 11772-11780.

38. Chitnis, V. P. and Chitnis, P. R. 1993. PsaL subunit is required for the formation of photosystem I trimers in the Cyanobacterium synechocystis sp. PCC 6803. FEBS Lett., 336, 330-334.

39. Chitnis, V. P. Xu., Xu, Q., Yu, L. Golbeck, J. H., Nakamoto, H., Xie, D. L. and Chitnis, P. R. 1993. Targeted inactivation of the gene psaL encoding a subunit of photosystem I of the cyanobacterium *Synechocystis* sp. PCC 6803. J. Biol. Chem., 268, 11678-11684.

40. Chitnis, V. P., An, K. and Chitnis, P. R. 1997. The PsaD subunit of photosystem I: Mutations in the basic domain impair the assembly of PsaD subunit of photosystem I: Mutations in the basic domain impair the assembly of PsaD into the complex. Plant Physiol., 115, 1699-1705.

41. Chollet, R., Vidal, J. and O'Leary, M. H. 1996. Phosphoenolpyruvate Carboxylase: a ubiquitous, highly regulated enzyme in plants. Annu. Rev. Plant Physiol. Plant Mol. Biol. 47: 273-298.

42. Christopher, D. A., Ki , M., Mullet, J. E. 1992. A novel light regulated promoter conserved in cereal and dicot chloroplasts. Plant cell 1 : 785-789.

43. Clauss, H., Mortimer, D. C. and Gorham, P. R. 1964. Time-course study of translocation of products of photosynthesis in soybean plants. Plant Physiol. 39:269-273.

44. Claussen, W. 1983. Investigations on the relationship between the distribution of assimilates and sucrose synthase activity in *Solanum melongena*. Z. Pflanzen-physiol. 110:175-182.

45. Claussen, W. and Biller, E. 1977. The significance of sucrose and starch contents of the leaves for the regulation of net photosynthetic rates. Z. Pflanzenphysiol. 81:189-198.

46. Claussen, W. and Lenz, F. 1983. Investigations on the relationship between sucrose phosphate synthase activity and net photosynthetic rates, and sucrose and starch content of leaves of *Solanum melongena* Z. Pflanzenphysiol. 109:459-468.

47. Colombo. R., Demichelis, M. I. and Lado, P. 1978. 3-O-Methyl glucose uptake stimulation by auxin and fusicoccin in plant material and its relationship with proton extrusion. Planta 138:249-256.

48. Cook, M. G. and Evans, L. T. 1983. The roles of sink size and location in the partitioning of assimilates in wheat ears. Aust. J.Plant Physiol. 10:313-327.

49. Coombs, J. and Whittingham, C. P. 1966. The effect of high partial pressures of oxygen on photosynthesis in *Chlorella*. I. The effect on end products of photosynthesis. Phytochemistry 5: 643-51.

50. Cornic, G. and Miginiac, E. 1983. Nonstomatal inhibition of net CO_2 uptake by (±) abscisic acid in *Pharbitis nil*. Plant Physiol. 73:529-533.

51. Crane, J. C. 1964. Growth substances in fruit setting and development. Annu. Rev. Plant Physiol. 15:303-326.

52. Crompton, H. J., Lloyd-Jones, C. P. and Hill-Cotting-Ham, D. G. 1981. Translocation of labeled assimilates following photosynthesis of $^{14}CO_2$ by the field bean, *Vicia faba*. Physiol. Plant. 51:189-194.

53. Cronshaw, J. 1981. Phloem structure and function. Annu. Rev. Plant Physiol. 32:465-84.

54. Crookston, R. K., Otoole, J., Lee, R., Ozbun, J. L. and Wallace, D. H. 1974. Photosynthetic depression in beans after exposure to cold for one nigh. Crop Sci. 14:457-464.

55. Daie, J. 1984b. Characteristics of sugar transport in storage tissue of carrot. J. Amer. Soc. Hort. Sci. 109:718-722.

56. Daie, J. and Wyse, R. E. 1983. Regulation of phloem loading by plant growth regulators in *Phaseolus vulgaris*. In : Proc. *Plant Growth Regulator Soc. Amer.*

57. Daie, J., Wyse, R. E., Hein, M. and Brenner, M. L. 1984. ABA metabolism in source and sink tissues of sugarbeet. Plant Physiol. 74:810-814.

58. Daniell, H., Datta, R., Varma, S., Gray, S. and Lee, S. B. 1998. Containment of herbicide resistance through genetic engineering of the chloroplast genome. Nat. Biotechnol. 16: 345-348.

59. De Duve, C. 1969. The peroxisome: a new cytoplasmic organelle. Proc. Roy. Soc. B 173: 71-83.

60. Decker, J. P. 1955. A rapid post-illumination acceleration of respiration in green leaves. Plant Physiol. 30: 82-4.

61. Deiting, U., Zrenner, R. and Stitt, M. 1998. Similar temperature requirement for sugar accumulation and the induction of new forms of sucrose phosphate synthase and amylase in cold-stored potato tubers. Plant, Cell and Environ. 21: 127-138.

62. Delrot, S. 1981. Proton fluxes associated with sugar uptake in *Vicia faba* leaf tissue. Plant Physiol. 68:706-711.

63. Delrot, S. and Bonnemain, J. L. 1979. Proton sugar cotransport in leaf tissues of *Vicia faba*. Plant Physiol. 63:35-44.

64. Delrot, S., Despeghel, J. P. and Bonnemain, J. L. 1980. Phloem loading in *Vicia faba* leaves: effect of N-ethylmaleimide and parachloromercuribenzensulfonic acid on H^+ extrusion, K^+ and Sucrose uptake. Planta 149:144-148.

65. Dewelle, R. B., Hurley, P. J. and Pavek, J. J. 1983. Photosynthesis and stomatal conductance of potato clones (*Solanum tuberosum*). Plant Physiol. 72:172-176.

66. Dick, P. S. and Aprees, T. 1975. The pathway of sugar transport in roots of *Pisum sativum*. J. Expt. Bot. 26:305-314.

67. Dinar, M. and Stevens, M. A. 1981. The relationship between starch accumulation and soluble solids content of tomato fruits. J. Amer. Soc. Hort. Sci. 106:415-418.

68. Doehlert, D. C. and Huber, S. C. 1983. Spinach leaf sucrose phosphate synthase: activation by G-6-P and interaction with inorganic phosphate. Fed. Europ. Biol. Sci. Lett. 153:293-296.

69. Doman, D. C. and Geiger, D. R. 1979. Effect of exogenously supplied foliar potassium on phloem loading in *Beta vulgaris*. Plant Physiol. 64:528-533.

70. Donald, C. M. 1962. In search of yield. J. Austral. Inst. Agr. Sci. 28:171-178.

71. Donald, C. M. and Hamblin, J. 1976. The biological yield and harvest index of cereals as agronomic and plant breeding criteria. Adv. Agron. 28:361-405.

72. Doughty, C. C. and Scheer, W. P. A. 1975. Growth regulators increase yield and reduce length of harvest of highbush blueberries. HortScience 10:260-261.

73. Downes, R. W. and Hesketh, J. D. 1968. Enhanced photosynthesis at low oxygen concentration: differential response of temperate and tropical grasses. Planta 78: 79-84.

74. Downton, W. J. S. and Tregunna, E. B. 1968a. Carbon dioxide compensation: its relation to photosynthetic carboxylation reactions, systematics of Gramineae and leaf anatomy. Can. J. Bot. 49: 207-15.

75. Downton, W. J. S. and Tregunna, E. B. 1968. Photorespiration and glycolate metabolism:a re-examination and correlation of some previous studies. Plant. Physiol. 43:923-9.

76. Drager, R. G. Girared-Basou, J., Choquet, Y., Kindle, K. L. and Stern, D. 1998. In vitro evidence for 5′-3′ exoribonuclease degradation of an unstable chloroplast mRNA. Plant J. 13: 85-96.

77. Echeverria, E., Salvucei, M. E. Gonzalez, P., Paris, G. and Salerno, G. 1997. Physical and Kinetic evidence for an association between sucrose-phosphate synthase and sucrose-phosphate phosphatase. Plant Physiol. 115: 223-227.

78. Egle, K. and Fock, H. 1967. In: Biochemistry of Chloroplasts, Goodwin, T. W. (ed). Acad. Press, New York.

79. Ehness, R. and Roitsch, T. 1997. Coordinated induction of mRNAs for extracellular invertase and a glucose transporter in *Chenopodium rubrum* by cytokinins. Plant J. 11: 539-548.

80. Ellyard, P. W. and Gibbs, M. 1969. Inhibition of photosynthesis by oxygen in isolated spinach chloroplasts. Plant Physiol. 44: 1115-21.

81. Eschrich, W. 1980. Free space invertase, its possible role in phloem unloading. Ber. Deutsch. Bot. G. 93:363-378.

82. Eskling, M., Arvidsson, P-O. and Akerlund, H-E. 1997. The xanthophyll cycle, its regulation and components Physiol. Plant. 100: 866-816.

83. Evans, L. T. 1975. Crop physiology. Cambridge Univ. Press, New York.

84. Evert, R. F., Eschrich, W. and Heyser, W. 1978. Leaf structure in relation to solute transport and phloem loading in *Zea mays*. Plant Physiol. 65:864-870.

85. Fisher, D. B. and Outlaw, W. H. 1979. Sucrose compartmentation in the palisade parenchyma of *Vicia faba*. Plant Physiol. 64:481-483.

86. Fluhr, R. and Chua, N. H. 1986. Developmental regulation of 2 genes encoding ribulose 1,5-bisphosphate carboxylase small subunits in pea and transgenic petunia plants; phytochrome response and blue light induction. Proc. Natl. Acad. Sci. USA 83, 2358-2362

87. Fondy, B. R. and Geiger, D. R. 1983. Control of export and of partitioning among sinks by allocation of products of photosynthesis in source leaves. Proc. Ann. Plant. Bioch. Physiol. Symp:Univ. Missouri-Colombia 2:33-42.

88. Forde, B. J., Whitehead, H. C. M. and Rowley, J. A. 1975. Effect of light intensity and temperature on photosynthetic rate, leaf starch content and ultrastructure of *Paspalum dilatum*. Austral. J. Plant Physiol. 2:185-195.

89. Forney, C. F. and Breen, P. J. 1983. Dry matter partitioning in fruiting and non-fruiting strawberry plants. Hort Science 18:70.

90. Forrester, M. L., Krotokov G. and Nelson, C. D. 1966a. Effect of oxygen on photosynthesis, photorespiration and respiration in detached leaves. I. Soybean. Plant Physiol. 41: 422-7.

91. Forrester, M. L.; Krotokov G. and Nelson, C. D. 1966b. Effect of oxygen on photosynthesis, photorespiration and respiration in detached leaves. II. Corn and other monocotyledons. Plant Physiol. 41: 428-31.

92. Foyer, C., Walker, D. and Latzko, E. 1982. The regulation of cytoplasmic fructose-1, 6-bisphosphatase in relation to the control of carbon flow to sucrose. Z. Pflanzenphysiol. 107:457-465.

93. Foyer, C. H., Valadier, M. H., Migge, A. and Becker, T. W. 1998. Drought-induced effects on nitrate reductase activity and mRNA and on the coordination of nitrogen and carbon metabolism in maize leaves. Plant Physiol. 117: 283-292.

94. Francheschi, V, and Giaquinta, R. 1983. The paraveinal mesophyll of soybean leaves in relation to assimilate transfer and compartmentation. 1. Ultrastructure and histochemistry during vegetative development. Planta 157:411-421.

95. Frank, H. A. et al. 1994. Photophysics of the carotenoids associated with the xanthophylls cycle in photosynthesis. Photosyn. Res. 41: 389-395.

96. Frier. V. 1977. The relationship between photosynthesis and tuber growth in *Solanum tuberosum*. J. Expt. Bot. 28:999-1007.

97. Galliani, S., Monselise, S. P and Goren, R. 1975. Improving fruit size and breaking alternate bearing in mandarins by Ethephon and other agents. Hort-Science 10:27-28.

98. Gamble, P. E., Sexton, T. B. and Mullet, J. E. 1988. Light – dependent changes in psbD and psbC transcripts maintains psbD and psbC translation capability in mature chloroplast. EMBOJ., 7, 1289-97.

99. Gawronska, H., Dwelle, R. B., Pavek, J.J. and Row, P. 1984. The partitioning of photoassimilates by four potato clones (*Solanum tuberosum*). Crop. Sci. 24:1031-1036.

100. Geiger, D. R. 1975. Phloem loading. In: Transport in plants. I. M.H. Zimmerman and J.A. Milburn (eds.) Springer-Verlag, New York. p. 395-450.

101. Geiger, D. R. 1976. Phloem loading in source leaves. In: I.F. Wardlaw and J.B. Passioura, (eds.). Transport and transfer processes in plants. Academic Press, New York.

102. Geiger, D. R. 1979. Control of partitioning and export of carbon in leaves of higher plants. Bot Gaz. 140:241-248.

103. Geiger, D. R. and Fondy, B. R. 1980. Phloem loading and unloading: Pathways and mechanisms. What Is New in Plant Physiology 11:25-28.

104. Geiger, D. R., Giaquinta, R. T., Sovanick, S. A. and Fellow, R. J. 1973. Solute distribution in sugar beet leaves in relation to phloem loading and translocation. Plant Physiol. 52:585-589.

105. Geiger, D. R., Sovonick, S. A., Shock, T. L. and Fellow, R. J. 1974. Role of free space in translocation in sugar beet. Plant Physiol. 54:892-898.

106. Giaquinta, R. T. 1976. Evidence for phloem loading from the apoplast: chemical modification of membrane sulfhydryl groups. Plant Physiol. 57:872-875.

107. Giaquinta, R. T. 1977. Possible role of pH gradient and membrane ATPase in the loading of sucrose into the sieve tubes. Nature 267:369-370.

108. Giaquinta, R. T. 1983. Phloem loading of sucrose. Ann. Rev. Plant Physiol. 59:178-180.

109. Giaquinta, R. T. and Geiger, D. R. 1977. Mechanism of cyanide inhibition of phloem translocation. Plant Physiol. 59:178-180.

110. Giaquinta. R. T. 1980. Sucrose/proton cotransport during phloem loading and its possible control by internal sucrose concentration. In: Plant membrane transport: current conceptual issues. R.M. Spanswick, W.J. Lucas, and J. Dainty (eds.), North Holland-Elsevier, Amsterdam. pp. 273-282.

111. Giaquinta. R. T., Lin, W., Sadler, N. L. and Franceschi, V. R. 1983. Pathway of phloem unloading of sucrose in corn root. Plant Physiol. 72:362-367.

112. Gifford, R. M. and Jenkins, C. L. 1983. Prospects of applying knowledge of photosynthesis towards improving crop production. In: Photosynthesis: CO_2 assimilation and plant productivity. Vol. 2. Govindjee (ed.). Academic Press. New York. pp. 419-434.

113. Gifford. R. M. and Evans, L. T. 1981. Photosynthesis, carbon partitioning, and yield, Ann. Rev. Plant Physiol. 32:485-509.

114. Giglioli-Guivarch, N. et al. 1996. The light-dependent transduction pathway controlling the regulatory phosphorylation of C_4 phosphoenolpyruvate carboxylase in protoplasts from *Digitaria sanguinalis*. Plant Cell 8: 573-586.

115. Glasziou, K. T. and Gayler, K. R. 1972. Storage of sugars in stalks of sugar cane. Bot. Rev. 38:471-490.

116. Goldsworthy, A. 1966. Experiments on the origin of CO_2 released by tobacco leaf segments in the light. Phytochemistry 5: 1013-19.

117. Goldsworthy, A. 1969. The riddle of photorespiration. Nature (London) 224: 501-2.

118. Goldsworthy, A. 1970. Photorespiration. Bot. Rev. 36 : 321-40.

119. Goldsworthy, A. and Day, P. R. 1970a. Further evidence for the reduced role of photorespiration in low compensation point species. Nature (London) 228: 687-8.

120. Goldsworthy, A. and Day, P. R. 1970b. A simple technique for the rapid determination of plant CO_2 compensation points. Plant Physiol. 46: 850-51.

121. Griffith, S. and Wyse, R. E. 1983. Some characteristics of the phloem unloading system in stem segments of *Vicia faba*. Plant Physiol. 72:138.

122. Grof, C. P. L. Knight, D. P., McNeil, S. D., Lunn, J. E. and Campbell, J. A. 1998. A modified assay method shows leaf sucrose phosphate synthase activity is correlated with leaf sucrose content across a range of sugarcane varieties. Aust. J. Plant Physiol. 25: 499-502.

123. Gud, C., Lee, S. B. and Daniell, H. 2000. Stable expression of biodegradable protein-based polymer in tobacco chloroplasts. Plant Cell Rep. 19: 257-262.

124. Gunning, B. E. S. 1976. The role of plasmodesmata in short distance transport to and from phloem. In: B. E. S. Gunning and A.W. Robards (eds). In: Intercellular communication in plants: studies on plasmodesmata. Springer, Berlin. pp. 203-227.

125. Gunning, B. E. S. and Pate, J. S. 1974. Transfer cells. In: Dynamic aspects of plant ultrastructure. A.W. Robards (ed.). McGraw-Hill, London. pp. 441-480.

126. Habeshaw, D. 1973. Translocation and the control of photosynthesis in sugar beet. Planta 110:213-226.

127. Hall. A. J. and Milthorpe, F. L. 1978. Assimilate source-sink relationships in *Capsicum annum*. II. The effects of fruit excision on photosynthesis and leaf and stem carbohydrate. Aust. J. Plant Physiol. 5:1-13.

128. Hanson, W. D. and Grier, R. E. 1973. Rates of electron transfer and non-cyclic photophosphorylation for chloroplasts isolated from maize populations selected for juvenile productivity and leaf width. Genetics. 75:247-257.

129. Harbron, S., Foyer, C. and Walker, D.A. 1981. The purification and properties of sucrose phosphate synthase from spinach leaves: The involvement of this enzyme and fructose bisphosphatase in the regulation of sucrose biosynthesis. Arch. Biochem. Biophys. 212:237-246.

130. Hare, P. D., Cress, W. A. and Van staden, J. 1998. Dissecting the roles of osmolyte accumulation during stress. Plant Cell and Environ. 21: 535-553.

131. Hartung, W. and Dierich, B. 1983. Uptake and release of abscisic acid by runner bean root tip segments L. Naturforsch. 38:719-723.

132. Hartwell, J., et al., 1999. Phosphoenolpyruvate carboxylase kinase is a novel protein kinase regulated at the level of expression. Plant. J. 20: 333-342.

133. Hasler, K., Engelbrecht, S. and Junge, W. 1998. Three stepped rotation of subunits gamma and epsilon in single molecules of F – ATPase as revealed by polarized confocal fluorometry. FEBS leH. 426: 301-304.

134. Hatch, M. D. 1971. The C_4 pathway of photosynthesis. Evidence for an intermediate pool of carbon dioxide and the identity of the donor C_4 dicarboxylic acid. Biochem. J. 125: 425-32.

135. Hatch, M. D., Osmond, C. B. and Slatyer, R. O. 1971. Photosynthesis and Photorespiration. Wiley-Interscience, New York.

136. Hatch, M. D., Slack, C. R. and Johnson, H. S. 1967. Further studies on a new pathway of photosynthetic carbon dioxide fixation in sugarcane and its occurrence in other plant species. Biochem. J. 102: 417-2.

137. Hatch, M. D. and Slack, C. R. 1966. Photosynthesis by sugarcane leaves: a new carboxylation reaction and the pathway of sugar formation. Biochem. J. 101: 103-11.

138. Hatch, M. D. 1992. C4 photosynthesis: an unlikely process full of surprises. Plant Cell physiol. 33: 333-342.

139. Heineke, D., et al. 1992. Apoplastic expression of yeast-derived invertase in potato. Plant Physiol. 100: 301-308.

140. Heldt, H. W., Chan, C. J., Maronde, D. Herold, A. Stankovic, Z. S., Walker, D. A. Framiner, A. Kirk, M. R. and Heber, U. 1977. Role of orthophosphate and other factors in the regulation of starch formation in leaves and isolated chloroplasts. Plant Physiol. 59:1146-1155.

141. Hellebust, J. A. 1965. Excretion of some organic compounds by marine phytoplankton. Limnology Oceanography 10: 192-206.

142. Herdman, M. and Denman, S. 1977. The cyanelle; chloroplast or endosymbiotic prokaryote ? FEMS Lett, 1, 7-12.

143. Herold, A. 1980. Regulation of photosynthesis by sink activity-The missing link. New Phytol. 86:131-144.

144. Hesketh, J. D. 1963. Limitations to photosynthesis responsible for differences among species. Crop. Sci. 3: 493-6.

145. Hew, C. S. and Krotkov, G. 1968. Effect of oxygen on the rates of CO_2 evolution in light and in darkness by photosynthesizing and non-photosynthesizing leaves. Plant Physiol. 43: 464-6.

146. Hewitt, J. D. and Stevens, M. A. 1981. Growth analysis of two tomato genotypes differing in total fruit solids. J. Amer. Soc. Hort. Sci. 106:723-727.

147. Hewitt, J. D., Dinar, M. and Stevens, M. A. 1982 Sink strength of fruits of two tomato genotypes differing in total fruit solids content. J. Amer. Soc. Hort. Sci. 107:896-900.

148. Heyser, W. 1980. Phloem loading in the maize leaf. Ber. Deutsch. Bot. Ges. 93:221-28.

149. Heyser, W., Heyser, R. Eschrich, W. and Leonard, O. A. 1976. The influence of externally applied organic substances on phloem translocation in detached maize leaves. Planta 132:269-277.

150. Heyser, W., Evert, R. F., Fritz, E. and Eschrich, W. 1978. Sucrose in the free space of translocating maize leaf bundles. Plant Physiol. 62:491-494.

151. Ho. L. C. 1976. The relationship between the rates of carbon transport and photosynthesis in tomato leaves, J. Expt. Bot. 27:87-97.

152. Ho. L. C. 1979. Regulation of assimilate translocation between leaves and fruits in the tomato. Ann. Bot. 43:437-448.

153. Ho. L. C. and Shaw, A. F. 1977. Carbon economy and translocation of ^{14}C in leaflets of the seventh leaf of tomato during leaf expansion. Ann. Bot. 41:833-848.

154. Ho. L. C. and Baker, D. A. 1982. Regulation of loading and unloading in long distance transport systems. Physiol. Plant. 56:225-230.

155. Hüber, S. C. 1981. Interspecific variation in activity and regulation of leaf sucrose phosphate synthetase. Z. Pflanzenphysiol. 102:443-450.

156. Hüber, S. C. 1983. Relation between photosynthetic starch formation and dry weight partitioning between the shoot and root. Can. J. Bot. 61:2709-2716.

157. Hüber, S. C. and Moreland, D. E. 1980. *Translocation.* Efflux of sugars across the plasmalemma of mesophyll protoplasts. Plant Physiol. 65:560-562.

158. Hüber, S. C. and Moreland, D. E. 1981. Co-transport of K^+ and sugars across plasmalemma of mesophyll protoplasts. Plant Physiol. 67:163-169.

159. Hüber, S. C. and Bickett, D. M. 1984. Evidence for control of carbon partitioning by fructose-2,6-bisphosphate in spinach leaves. Plant Physiol. 74:445-447.

160. Hüber, S. C. and Israel, D. W. 1982. Biochemical basis for partitioning of photosynthetically fixed carbon between starch and sucrose in soybean leaves. Plant Physiol. 69:691-696.

161. Hubschmann, T. and Borner, T. 1998. Characterization of transcript initiation sites in ribosome-deficient barley plastids. Plant Mol. Biol. 36: 493-496.

162. Hughes, D. L., Bosland, J. and Yamaguchi, M. 1983. Movement of photosynthetes in muskmelon plants. J. Amer. Soc. Hort. Sci. 108:189-192.

163. Humphreys, T. E. 1980. Sugar-proton cotransport and phloem loading. What Is New in Plant Physiology 11:9-12.

164. Hurewitz. J. and Janes, H. W. 1983. Effect of altering the root-zone temperature on growth, translocation, carbon exchange rate, and leaf starch accumulation in tomato. Plant Physiol. 73:46-60.

165. Ingram, J., Chandler, J. N. Gallagher, L., Salamini, F. and Bartels, D. 1997. Analysis of cDNA clones encoding sucrose-phosphate synthase in relation to sugar inter-conversions associated with dehydration in the resurrection plant *Craterostigme plantaginium.* Plant Physiol. 115: 113-121.

166. Jackson, W. A. and Volk, R. J. 1970. Photorespiration. Ann. Rev. Plant Physiol. 21: 385-432.

167. Jain, H. K. 1975. Breeding for yield and other attributes in grain legumes. Ind. Agr. Res. Inst., New Dehli.

168. Jarvis, P., Chen, L-J., Li H-M., Peto, C.A. Frankhauser, C., Chory J. 1998. An *Arabidopsis* mutant defective in the plastid general protein import apparatus. Science. 282: 100-103.

169. Jenkins, B. D., Kulhanek, D. J. and Barkan, A. B. 1999. Nuclear mutations that block group II RNA splicing in maize chloroplasts reveal several intron classes with distinct requirements for splicing factors. Plant Cell. 9: 283-296.

170. Jenner, C. F. 1974. An investigation of the association between the hydrolysis of sucrose and its absorption by wheat grains. Aust. J. Plant Physiol. 1:319-329.

171. Jensen, R. G. and Bassham, J. A. 1968. Photosynthesis by isolated chloroplasts. III. Light activation of the carboxylation reaction. Biochem. Biohys. Acta. 153: 227-34.

172. Jolliot, P. and Jolliot, A. 1992. Electron transfer between photosystem II and the cytochrome bf complex: mechanistic and structural implications. Biochem. Biophys. Acta. 1102: 53-61.

173. Kaiser, W. M., Paul, J. S. and Bassham, A. J. 1979. Release of photosynthates from mesophyll cells *in vitro* and *in vivo.* Plant Physiol. 64:377-385.

174. Kato-Yamada, Y., Noje, H., Kinosita, K. and Yoshida, M. 1998. Direct observation of the rotation of the epsilon subunit in F_1-ATPase. J. Biol. Chem. 273: 19375-19377.

175. Kearney, P. C. and Tolbert, N. E. 1962. Appearance of glycolate and related products of photosynthesis outside of chloroplasts. Arch. Biochem. 98: 164-71.

176. Khan, S. M. and Maliga, P. 1999. Nat. Biotechnol. 17: 910-915.

177. Kisaki, T. and Tolbert, N. E. 1969. Glycine as a substrate for photorespiration. Plant and Cell Physiol. 11. 247-58.

178. Kisaki, T. and Tolbert, N. E. 1962. Glycolate and glyoxylate metabolism in isolated peroxisomes and chloroplasts. Plant Physiol. 44: 242-50.

179. Klann, E-M. Hall, B. and Bennett, A. B. 1996. Antisense acid investase (TIVI) gene alters soluble sugar composition and size in transgenic tomato fruits. Plant Physiol 112: 1321-1330.

180. Knoblauch, M. et al. 1999. Nat Biotechnol. 17: 906-909.

181. Kobayashi, Y., Kaiser, W. and Heber, U. 1995. Bioenergeties of carbon assimilation in intact chloroplast: Coupling of proton to electron transport at the ratio of H^+/e^- = 3 is incompatible with H^+/ATP = 3 in ATP synthesis. Plant Cell Physiol. 36: 1629-1637.

182. Kohler, R. H., Cao, J. Zipfel, W. R. Webb, W. W. and Hanson, M. 1997. Exchange of protein molecules through connections between higher plant plastids. Science. 276: 2039-2042.

183. Komor, E. 1977. Sucrose uptake by cotyledons of *Ricinus communis*: characteristics, mechanism and regulation. Planta 137:119-131.

184. Komor, E. 1982. Transport of sugar. In: Encyclopedia of Plant Physiology. Plant carbohydrates I. F.A. Loewus and W. Tanner (eds.). Springer-Verlag. Berlin. 635-676.

185. Komor, E., Thom, M. and Maretzki, A. 1982. Vacuoles from sugarcane suspension cultures. III. Proton motive potential differences. Plant Physiol. 69:1326-1330.

186. Kota, M., Daniell, H., Varma, S., Garezynski, F., Gould, F., and Moar, W. J. 1999. Overexpression of the Bacillus thuringiensis Cry2A protein in chloroplast confers resistance to plants against susceptible and Bt-resistant insects. Proc. Natl. Acad. Sci. USA. 96: 1840-1845.

187. Kouranov, A., Chen, X., Fuks, B. and Schnell, D. J. 1998. Tic 20 and Tic 22 are new components of the protein import apparatus at the chloroplast inner envelope membrane. J. Cell Biol. 143: 991-1002.

188. Kramer, D. M., Sacksteder, C. A. and Cruz, J. A. 1999. How acidic is the lumen. Photosynt. Res. 60: 151-163.

189. Krewer, G. W., Daniel, J. W., Coston, D. C., Couvillon, G. A., and Kays, S. J. 1983. Transport of ^{14}C photosynthate into young peach fruits in response to CGA-IS281 and ethylene-releasing compounds. HortScience. 18:476-478.

190. Krol, M. et al., 1995. Chloroplast a/b binding proteins pigment conversions and early light-induced proteins in a chlorophyll b-less barley mutant. Plant Physiol. 107: 873-883.

191. Kueneman. E. A., Wallace, D. H. and Ludford, P. M. 1979. Photosynthetic measurements of field grown dry beans and their relations to selection for yield. J. Amer. Soc. Hort. Sci. 104:480-482.

192. Kursanov, A. L. and Brovchenko, M. E. 1970. Sugars in the free space of leaf plates:Their origin and possible involvement in transport. Can. J. Bot. 48:1243-1250.

193. Lenz, F. 1979. Fruit effects on photosynthesis, light and dark respiration. In: R. Marcelle, C. Clijsters, and M. Van Pouke (eds.), Photosynthesis and Plant development. W. Junke, Hague, The Netherlands.

194. Lepp, N. W. and Peel, A. J. 1970. Some effects of IAA and kinetin upon the movement of sugars in the phloem of willow. Planta. 90:230-235.

195. Li, B. and Chollet, R. 1993. Resolution and identification of C_4 phosphoenolpyruvate-carboxylase protein-kinase polypeptides and their reversible light activation in maize leaves. Arch. Biochem. Biophyl. 307: 416-419.

196. Li, P – X., Bjorkman, O., Shih, C., Grossman, A. R. et al., 2000. A pigment binding protein essential for regulation of photosynthetic light harvesting. Nature 403: 391-395.

197. Lichtner, F. T. and Spanswick, R. M. 1981. Sucrose uptake by developing soybean cotyledons. Plant Physiol. 68:693-698.

198. Lis, E.K. and Antoszewski, R. 1982. Do growth substances regulate the phloem as well as xylem transport of nutrients to the strawberry receptacles? Planta. 156:492-495.

199. Loomis, R. S. and Williams, W. A. 1963. Maximum crop productivity: An estimate. Crop Sci. 3:67-72.

200. Lunn, J. E. and Furbank, R. T. 1997. Localization of sucrose-phosphate synthase and starch in leaves of C4 plants. Plants 202: 106-111.

201. Lunn, J. E., Price, G. D. and Furbank, R. T. 1999. Cloning and expression of a prokaryotic sucrose-phosphate synthase gene from the Cyanobacterium synechocystis sp. PCC 6803. Plant Mol. Biol. 40: 297-305.

202. Madore, M. and Webb, J. A. 1981. Leaf free space analysis and vein loading in *Cucurbita pepo*. Can. J. Bot. 59:2550-2557.

203. Malek, F. and Baker, D. A. 1978. Effect of fusicoccin on proton cotransport of sugars in the phloem loading of *Ricinus communis*. Plant Sci. Lett. 11:233-239.

204. Mantell, S. H., Mathews, J. A., and Mckee, R. A. 1985. Plant molecular biology. In : Mantell, S. H., Mathews, J. A., Mckee, R. A. (ed) In: Principles of plant biotechnology in introduction of genetic engineering in plants, Blackwell Scientific Publication. pp. 8-33.

205. Marini, R. P. and Barden, J. A. 1981. Seasonal correlations of specific leaf weight to net photosynthesis and dark respiration of apple leaves. Photosynth. Res. 2:251-258.

206. Marker, A. F. H. and Whittingham, C. P. 1966. The photoassimilation of glucose in *Chlorella* with reference to the role of glycolic acid. Proc. Roy. Soc. B 165: 473-85.

207. Marker, A. F. H. and Whittingham, C. P. 1967. The site of synthesis of sucrose in plant cells. J. expt. Bot. 18: 732-9.

208. Martin, G. C. and Nishijima, C. 1972. Levels of endogenous growth regulators in abscising and persisting peach fruits. J. Amer. Soc. Hort. Sci. 97-561-565.

209. Mauney, J. R., Guinn, G., Fry, K. E. and Hesketh, J. D. 1979. Correlation of photosynthetic CO_2 uptake and carbohydrate accumulation in cotton, soybean, sunflower and sorghum. Photosynthetica. 13:260-266.

210. May Field, S. P. and Taylor, W. C. 1987. Chloroplast photoxidation inhibits the expression of a set of nuclear genes. Mol. Gen. Genet. 208, 308-14.

211. Maynard, J. W. and Lucas, W. 1982. Sucrose and glucose uptake into *Beta vulgaris* leaf tissue: *A case for general apoplastic retrieval systems.* Plant Physiol. 70:1436-1443.

212. Mccashin, B. G. and Canvin, D. T. 1979. Photosynthetic and photorespiratory characteristics of mutants of *Hordeum vulgare*. Plant Physiol. 64:354-360.

213. Mcneal, F. H., Berg, M. A., Brown, P. L. and Mcguire, C. F. 1971. Productivity and quality response of five spring wheat genotypes. *Triticum aestivum* L., to nitrogen fertilizer. Agron. J. 63:908-910.

214. Mehta, M., Sarafis, V. and Critchley, C. 1999. Thylakoid membrane architecture, Aust. J. Plant Physiol. 26: 709-716.

215. Mehta, M., Sarafis, V. and Critchley, C. 1999. Thylakoid membrane architecture. Aus. J. Plant Physiol. 26: 709-716.

216. Milburn, J. A. 1974. Phloem transport in *Ricinus*: Concentration gradients between source and sink. Planta. 117:303-319.

217. Monroy, A. F., Sangwen, V. and Dhindsa, R. S. 1998. Low temperature signal transduction during cold acclimation: protein phosphatase 2A as an early target for cold-inactivation. Plant J. 13: 653-660.

218. Monselise, S. P. and Goldschmidt, E. E. 1982. Alternate bearing in fruit trees: A review, Hort. Rev. 4:128-173.

219. Monselise, S. P. and Lenz, F. 1979. Effect of fruit load on photosynthetic rates of budded apple trees. Gartenbauwissenschaft. 45:220-224.

220. Moorby, J. 1968. The influence of carbohydrate and mineral nutrient supply on the growth of potato tubers. Ann. Bot. 32:57-68.

221. Moorby, J., Troughton, J. H. and Currie, B. G. 1974. Investigations of carbon transport in plants. J. Expt. Bot. 25:937-944.

222. Morgan, D. C. 1980. Factors affecting fruit and seed development in field beans and oil seed rape. In: Aspects and prospects of plant growth regulators. B. Jeffcoat (ed.), Monograph 6. British Plant Growth Regulator Group. Wassex Press, Wantage, Oxfordshire.

223. Morris, D. A. 1982. Hormonal regulation of sink invertase activity: implications for control of assimilate partitioning. In: Plant growth substances. P.F. Wareing (ed.). Academic Press, New York. pp. 659-668.

224. Moss, D. N. and Rasmussen, H. P. 1969. Cellular localization of CO_2 fixation and translocation of metabolites. Plant Physiol. 44:1063-1068.

225. Mulligan, D. R. and Patrick, J. W. 1979. Gibberellic acid promoted transport of assimilates in stems of *Phaseolus vulgaris*. Localized vs. remote sites of action. Planta 145:233-238.

226. Nafziger, E. D. and Koller, H.R. 1976. Influence of leaf starch concentration on CO_2 assimilation in soybean. Plant Physiol. 57:560-563.

227. Nilsson, R., Brunner, J., Hoffman, N. E. and Van Wijk, K. J. 1999. Interactions of ribosome nascent chain complexes of the chloroplast encoded D1 thylakoid membrane protein with cpSRP54. EMBO J. 18: 733-743.

228. Nimmo, H. G. 2000. The regulation of phosphoenolpyruvatecarboxylase in CAM plants. Trends in plant Sci. 5: 75-80.

229. Nomura, T., Nakayama, N., Murata, T. and Akazawa, T. 1967. Biosynthesis of starch in chloroplasts. Plant Physiol. 42:327-332.

230. Ogren, W. L. and Bowes, G. 1971. Ribulose diphosphate carboxlase regulates soybean photorespiration. Nature New Biology 230: 159-60.

231. Olufajo, O. O., Daniels, R. W. and Scarisbrick, D. H. 1982. The effect of pod removal on the translocation of ^{14}C photosynthates from leaves in *Phaseolus vuigaris*, c.v. Lochness, J. Hort. Sci. 57:333-338.

232. Ongun, A. and Stocking, C. R. 1965. Effect of light on the incorporation of serine into the carbohydrates of chloroplasts and non-chloroplast fractions of tobacco leaves. Plant Physiol. 40: 819-24.

233. Owens, T. G. 1992. Processing of excitation energy by antenna pigments, In: Photosynthesis and Environment, Baker, N.R., (ed.) Kluwer Academic Publishers. The Netherlands.

234. Ozbun, J. L. 1978. Photosynthetic efficiency and crop production. HortScience 6:678-679.

235. Partington, J. R. 1957. *A short history of chemistry*, 3rd ed. London: Macmillan.

236. Patrick, J. W. 1979. An assessment of auxin-promoted transport in decapitated stems and whole shoots of *Phaseolus vulgaris*. Planta 146:107-112.

237. Patrick, J. W. 1982. Hormonal control of assimilate transport. In: P.F. Wareinged., Plant growth substances. P.F. Waringed (ed) Academic Press, New York. pp. 669-678.

238. Patrick, J. W. 1983. Photosynthate unloading from seed coats of *Phaseolus vulgaris*. General characteristics and facilitated transfer. Z. Pflanzenphysiol. 111:9-18.

239. Patrick, J. W. 1984. Photosynthate unloading from seed coats of *Phaseolus vulgaris* L. Control by tissue water relations. J. Plant Physiol. 115:297-310.

240. Patrick, J. W. and Wareing, P. F. 1978. Auxin-promoted transport of metabolites in stems of *Phaseolus vulgaris*. J. Expt. Bot. 29:359-366.

241. Patrick, J. W. and Wareing, P. F. 1980. Hormonal control of assimilate movement and distribution. In: Aspects and prospects of plant growth regulators. B. Jeffcoat (ed.), Monograph 6. British Plant Growth Regulator Group. Wassex Press, Wantage, Oxfordshire, England. pp. 65-68.

242. Patrick, J. W. and Turvey, P. M. 1981. The pathway of radial transfer of photosynthates in decapitated stems of *phaseolus vulgaris*. Ann. Bot. 47:611-621.

243. Peet, M. M., Bravo, A., Wallace, D. H. and Oxbun, J. L. 1977. Photosynthesis, stomatal resistance, and enzyme activities in relation to yield of field grown dry bean cultivars. Crop Sci. 17:287-293.

244. Pilgrim, M. L. Van Wijk K-J., Parry, D. H. Sy DAC, Hoffman, N. E. 1998. Expression of a dominant negative form of cpSRP54 inhibits chloroplast biogenesis in Arabidopsis. Plant J. 13: 177-186.

245. Poostchi, I., Rouhani, I. and Razmi, K. 1972. Influence of levels of spring irrigation and fertility on yield of winter wheat *Triticum aestivum* L. under semiarid conditions. Agron. J. 64:438-442.

246. Potter, J. R. and Breen, P. J. 1980. Maintenance of high photosynthetic rates during the accumulation of high leaf starch levels in sunflower and soybean. Plant Physiol. 66:528 531.

247. Preiss, J. 1982. Regulation of the biosynthesis and degradation of starch. Annu. Rev. Plant Physiol. 34:431-454.

248. Preiss, J. 1984. Starch, sucrose biosynthesis and partition of carbon in plants are regulated by orthophosphate and triose-phosphates. Trends Biol. Sci. 9:24-27.

249. Preiss, J. and Levi, C. 1979. Metabolism of primary products of photosynthesis. Metabolism of starch in leaves. In: Encyclopedia of plant physiology. New series. M. Gibbs and E. Latzk (eds.), Springer-Verlag, New York. pp. 282-312.

250. Preiss, J. and Boyer, C. D. 1980. Evidence for independent genetic control of the multiple forms of maize endosperm branching enzymes and starch synthesis. In: Mechanisms of saccharide polymerization and depolymerization. N.C. Brady (ed.), Academic Press, New York.

251. Rabeson, R., Tolbert, N. E. and Kearney, P. C. 1962. Formation of serine and glyceric acid by the glycolate pathway. Arch. Biochem. Biophys. 98: 154-63.

252. Randall, D. D., Nelson, C. J. and Asay, K. H. 1977. Ribulose bisphosphate carboxylase: Altered genetic expression in tall fescue. Plant Physiol. 59:38-41.

253. Richardson, K. E. and Tolbert, N. E. 1961. Phosphoglycolic acid phosphatase. J. Biol. Chem. 236: 1285-90.

254. Rocher, J. P., Prioul, J. L. Lecharrey, A, Reyss, A. and Jourssaume, M. 1989. Genetic variability in carbon fixation, sucrose-P-synthase and ADP glucose pyrophosphorylase in maize plants of differing growth rates. Plant Physiol. 89: 416-420.

255. Rodriguez, B. P. and Lambeth, V. N. 1976. Photosynthetic rate and yield component relationship in winter greenhouse tomatoes. HortScience 11:430-431.

256. Romanowska, E. and Albertsson, P-A. 1994. Isolation and characterization of the cytochrome bf complex from whole thylakoids, grana and stroma lamellae vesicles from spinach chloroplast. Plant Cell Physiol. 35: 557-568.

257. Rott, R., Lay, H., Drager, R. G., Stern, D. B. and Schuster, G. 1998. 3′-Processed mRNA is preferentially translated in *Chlamydomonas reinhardtii* chloroplasts. Mol Cell Biol. 18: 4605-4611.

258. Roy, L. M. and Barkan, A., 1998. A Sec Y homologue is required for the elaboration of the chloroplast thylakoid membrane and for normal chloroplast gene expression. J. Cell Biol. 141: 385-395.

259. Rumberg, B., Schubert, K., Strelow, F. and Tran-Ahn, T. 1990. The H^+/ATP coupling ratio at the H^+ - ATP – synthase of spinach chloroplast is four. In: Current Research in photosynthesis. Baltscheffsky, M. (ed). Kluwer Academic Publishers, Dordrecht, The Netherlands.

260. *Sacksteder, C. A., Herbert, S. and Kramer, D. M. 1998. H^+/e ratios for photosynthetic electron transfer in intact leaves in steady-state. In: Progress in Photosynthesis Research, Garab G. (ed) Kluwer Academic Publishers, Dordrecht, The Netherlands.

261. Saftner, R. A. and Wyse, R. E. 1984. Effect of plant hormones on sucrose uptake by sugarbeet root tissue discs. Plant Physiol. 74:951-954.

262. Saftner, R. A., Daie, J. and Wyse, R. E. 1983. Sucrose uptake and compartmentation in sugarbeet tap root tissue. Plant Physiol. 72:1-6.

263. Sams, C. 1980. Factors affecting the leaf and shoot morphology and photosynthetic rate of sour cherry *Prunus cerasus* L. Montmorency). Ph. D. Thesis, Michigan State Univ. East Lansing.

264. Sanwall, G. G., Greenberg, E., Hardie, J., Cameron, E.C. and Preiss, J. 1967. Regulation of starch biosynthesis in plant leaves: Activation and inhibition of ADP-glucose pyrophosphorylase. Plant Physiol. 43:417-427.

265. Sarafis, V. 1998. Chloroplasts : a structural approach. J. Plant Physiol. 152: 248-244.

266. Saraste, M. 1999. Oxidative phosphorylation at the fin de siecle. Science 283: 1488-1493.

267. Sarhan and Danyluk, J. 1998. Engineering cold-tolerant crops-throwing the master switch. Trends in Plant Science 3: 289-290.

268. Sarquis, J. I. Gonzalez, H., de Jimenez, E. S. and Dunlap, J. R. 1998. Physiological traits associated with mass selection for improved yield in a maize population. Field crops Res. 56: 239-246.

269. Sasaki, Y., Nakamura, Y., and Masuna, R. 1986. FEBS Letts. 196, 171-174.

270. Schmitt, M. R., Hitz, W. D., Lin, W. and Giaquinta, R. T. 1984. Sugar transport into protoplasts isolated from developing soybean cotyledons. II. Sucrose transport kinetics, selectivity and modeling studies. Plant Physiol. 75:941-946.

271. Schuenemann, D., Amin, P., Hartmann, E. and Hoffman, N. E. 1999. Chloroplast Sec Y is complexed to Sec E and involved in the translocation of the 33 KDa but not the 23 KDa subunit of the oxygen-evolving complex. J. Biol. Chem. 274: 12177-12182.

272. Servaites, J. C. and Geiger, D. R. 1974. Effect of light intensity and oxygen on photosynthesis and translocation in sugarbeet. Plant Physiol. 54:575-578.

273. Setter, T. L., Brun, W. A. and Brenner, M. L. 1980. Effect of obstructed translocation on leaf abscisic acid and associated stomatal closure and photosynthesis decline. Plant Physiol. 65:1111-1115.

274. Sexton, T. B., Christopher, D. A., and Mullet, J. E. 1990. Light – induced switch in barley psb D – psbC promoter utilization ; a novel mechanism regulating chloroplast gene expression. EMBO J., 9 : 4485-4494.

275. Shannon, J. C. and Dougherty, C. T. 1972. Movement of ^{14}C-labelled assimilates into kernels of *Zea mays*. II. Invertase activity of the pedicel and placentachazal tissue. Plant Physiol. 49:203-206.

276. Sharma, A. K., Singh, B. B. and Singh, S. P. 1982. Relationship among net assimilation rate, leaf area index and yield in soybean genotypes. Photosynthetica 16:115-118.

277. Shinozaki, K., Ohme, M., Tanaka, M., Wakasugi, T., Hayshida, N., Matsubayashi, T., Zaita, N., Chun Wongse, J., Obokata, J., Yamaguchi Shinozaki, H., Ohto. C. Torazawa, K., Meng, B. Y., Sugeta, M., Takaiwa, F. A., Tohdoh, N., Shemada, K., Shemada, H. and Sugiura, M. 1986. The complete neucleotide sequence of the tobacco chloroplast genome; its gene organization and expression EMBO J. 5, 2043-9.

278. Silvius, J. E., Chatterton, N. J. and Kremer, D. F. 1979. Photosynthate partitioning in soybean leaves at two irradiance levels. Plant Physiol. 64:872-875.

279. Slack, C. R.; Hatch, M. D. and Goodchild, D. J. 1969. Distribution of enzymes in mesophyll and parenchyma sheath chloroplasts in maize leaves in relation to the C_4 dicarboxylic acid pathway in photosynthesis. Biochem. J. 114: 489-98.

280. Smith, J. A. and Milburn, J. A. 1980. Phloem turgor and the regulation of sucrose loading in *Ricinus communis* L. Planta 148:42-48.

281. Sonnewald, U. et. al. 1997. Increased potato tuber size resulting from apoplastic expression of a yeast invertase. Nat. Biotechnol. 15: 794-797.

282. Sovonick, S. A., Geiger, D. R. and Fellows, R. J. 1974. Evidence for active phloem loading in the minor veins of sugarbeet. Plant Physiol. 54:886-891.

283. Stevens, M. A. 1976. Inheritance of viscosity potential in tomato. J. Amer. Soc. Hort. Sci. 101:152-155.

284. Strotman, H. and Lohse, D. 1988. Determination of the H+/ATP ratio of the H+ transport-coupled reversible chloroplast ATPase reaction by equilibrium studies. FEBS Lett. 229: 308-312.

285. Sturgis, J. N. and Rubery, P. H. 1982. The effects of indol-3-yl acetic acid and fusicoccin on the kinetic parameters of sucrose uptake by discs from expanded primary leaves of *Phaseolus vulgaris*. Plant Sci. Lett. 24:319-326.

286. Sugiura, M. 1992. The chloroplast genome. Plant Mol. Bio., 19 : 149-168.

287. Tanner, W. 1980. On the possible role of ABA on phloem unloading. Ber. Deutsch. Bot. Ges. 93:349-351.

288. Thompson, C. M. and Whittingham, C. P. 1967. Intracellular location of phosphoglycolate phosphatase and glyoxylate reductase. Biochem. Biophys. Acta 143:642-4.

289. Thompson, W. F., Everett, M., Polans, N. O., Jortegensin, R. A., Palmer, J. D. 1983. Phytochrome control of RNA levels in developing pea and mung bean leaves. Planta, 158-487-500.

290. Thorne, J. H. 1982. Characterization of the active sucrose transport system of immature soybean embryos. Plant. Physiol. 70:953-958.

291. Thorne, J. H. 1983. An *in vivo* technique for the study of phloem unloading in seed coats of developing soybean seeds. Plant Physiol. 72:268-271.

292. Thorne, J. H. and Koller, H. R. 1974. Influence of assimilate demand on photosynthesis, diffusive resistance, translocations and carbohydrate level of soybean leaves. Plant Physiol. 54:201-207.

293. Thorne, J. H. and Giaquinta, R. T. 1982. Pathways and mechanisms associated with carbohydrate translocation in plants. In: Physiology and biochemistry of storage carbohydrates in vascular plants. D. H. Lewis (ed.), Soc. Expt. Bot. Symp. Series. Cambridge Univ. Press.

294. Tolbert, N. E. and Zill, L. P. 1956. Excretion of glycolic acid by algae during photosynthesis. J. Boil. Chem. 222: 895-906.

295. Tolbert, N. E. and Yamazaki, R. K. 1969. Leaf peroxisomes and their relation to photorespiration and photosynthesis. Ann. New York Acad. Sci. 168: 325-41.

296. Tozawa, Y., Tanaka, K., Takahashi, H. and Wakasa, K. 1998. Nuclear encoding of a plastid σ-factor in rice and its tissue and light dependent expression. Nuc. Acids Res. 26: 415-419.

297. Trifa, Y., Prival, I., Gagnon, J., Baeza, L. and Lerbs-Mache, S. 1998. The nuclear RPL4 gene encodes a chloroplast protein that co-purifies with the T7-like transcription complex as well as plastid ribosomes. J. Biol. Chem. 273: 3980-3985.

298. Valpuesta, V. and Bukovac, M. 1983. Cherry fruit growth: Localization of indoleacetic acid oxidase and inhibitors in the seed. J. Amer. Soc. Hort. Sci. 108:457-459.

299. van Camp, W., Capiau, K., Van Montagu, M., Inze, D., and Slooten, L. 1996. Enhancement of oxidative stress tolerance in transgenic tobacco plants overproducing Fe-superoxide dismutase in chloroplast. Plant Physiol. 112: 1703-1714.

300. Van Niel, C. B. 1949. The comparative biochemistry of photosynthesis. In: Photosynthesis in plants, J. Franck, W.E. Loomis (ed.). Iowa: Iowa State College Press.

301. Vermass, W. F. J., Jkeuchi, M. 1991. Photosystem II. In: The photosynthetic Apparatus; Molecular Biology and Operation. L. Bogarad, I.K. Vasil, (eds.), Academic Press, San Diego, pp 25-11.

302. Walker, A. J. and Thornley, J. H. M. 1977. The tomato fruit: Import, growth, respiration, and carbon metabolism at different fruit sizes and temperatures. Ann. Bot. 41:977-985.

303. Walker, A. J. and Ho, L. C. 1977A. Carbon translocation in tomato fruit: carbon import and fruit growth. Ann. Bot. 41:813-823.

304. Walker, A. J. and Ho, L. C. 1974. Investigations of carbon transport in plants. J. Expt. Bot. 25:937-944.

305. Walker, A. J. and Ho, L. C. 1976. Young tomato fruits induced to export carbon by cooling. Nature 261:410-411.

306. Walker, A. J. and Ho, L. C. 1977b. Carbon translocation in tomato: Effect of fruit temperature on carbon metabolism and the rate of translocation. Ann. Bot. 41:825-832.

307. Walker, D. A. and Herold, A. 1977. Can the chloroplast support photosynthesis unaided? In: Photosynthetic Organelles: Structure and function.Special issue: Y. Fujita, S. Katsh, K. Shibeta, S. Miyachieds. (eds.). 1-7.

308. Walker, D. A. and Robinson, S. P. 1978. A contemporary view of photosynthetic carbon assimilation. Ber. Deutsch. Bot. Gaz. 91:513-526.

309. Walker, D. A., Leegeod, R. C. and Sivak, M. N. 1986. Ribulose bisphophate carboxylase/oxygenase, its role in photosynthesis. Phil. Trans. R. Soc. London B, 313, 305-24.

310. Wang, F. et al. (1993). Sucrose synthase, starch accumulation and tomato fruit sink strength. Plant Physiol 101: 321-327.

311. Wareing, P. F., Khalifa, M. M. and Threharne, C. J. 1968. Rate-limiting processes in photosynthesis at saturating light intensities. Nature 220:453-457.

312. Weihe, A. and Borner, T. 1999. Transcription and the architecture of promoters in chloroplasts. Trends Plant Sci. 4: 169-170.

313. Williams, A. M. and Williams, K. R. 1978. Regulation of movement of assimilates into ovules of *Pisum sativum*. Austral. J. Plant Physiol. 5:295-300.

314. Wittwer, S. H. 1974. Carbon dioxide fertilization of crop plants. In: Crop Physiology, Gupta, U. S. (ed), H. A. U., Hissar, India.

315. Wittwer, S. H. and Robb, W. 1964. Carbon dioxide enrichment of greenhouse atmospheres for food crop production. Econ. Bot. 18: 34-56.

316. Wittwer, S. H. 1974. Maximum production capacity of food crops. BioScience 24:216-224.

317. Wittwer, S. H. 1978. The next generation of agricultural research. Science 199:375-378.

318. Wittwer, S. H. 1980. The shape of things to come. In: The biology of crop productivity. P.S. Carlsoned. (ed.), Academic Press, New York. pp. 413-459.

319. Wolswinkel, P. and Ammerlaan, A. 1983. Phloem unloading in developing seeds of *Vicia faba*. The effect of several inhibitors on the release of sucrose and amino acids by the seed coat. Planta 158:205-215.

320. Wyse, R. E. 1979. Parameters controlling sucrose content and yield of sugarbeet roots. J. Amer. Soc. Sugarbeet Tech. 20:368-385.

321. Wyse, R. E. and Saftner, R. A. 1982. Reduction in sink mobilizing ability following periods of high carbon flux. Plant Physiol. 69:226-228.

322. Wyse, R. E., J. Daie, and R. A. Saftner. 1980. Hormonal control of sink activity in sugarbeet. Plant Physiol.Suppl. 65:121.

323. Zelitch, I. 1958. The role of glycolic acid oxidase in the respiration of leaves. J. Biol. Chem. 233: 1299-1303.

324. Zelitch, I. 1959. The relationship of glycolic acid to respiration and photosynthesis in tobacco leaves. J. Biol. Chem. 234: 3077-81.

325. Zelitch, I. 1966. Increased rate of net photosynthetic carbon dioxide uptake caused by the inhibition of glycolate oxidase. Plant Physiol. 41: 1623-31.

326. Zelitch, I. 1971. Photosyntheses, Photorespiration and Plant Productivity. Acad. Press, New York.

327. Zelitch, I. and Walker, D. A. 1964. The role of glycolic acid metabolism in opening of leaf stomata. Plant Physiol. 39: 856-62.

328. Zhu, Y. J. Komor, E., and Moore, P. H. 1997. Sucrose accumulation in the sugarcane stem is regulated by the difference between the activities of soluble acid invertase and sucrose phosphate synthase. Plant Physiol 115: 609-616.

329. Ziegler, H. 1974. Biochemical aspects of phloem transport. In: Transport at the cellular level. M.A. Sleigh and D. H. Jennings (eds.), Cambridge Univ. Press, London. pp. 43-62.

330. Zrenner, R. et al. (1995). Evidence for the crucial role of sucrose synthase for sink strength using transgenic potato plants. Plant J. 7: 97-107.

Subject Index